CONTEMPORARY ISSUES IN SCIENCE AND SOCIETY

Solar Energy

Solar Energy
Chemical Conversion and Storage

Edited by

Richard R. Hautala
R. Bruce King and
Charles Kutal

The Humana Press • Clifton, New Jersey

The present volume is an outgrowth of an American Chemical Society symposium held in Savannah, Georgia in November, 1978.

© 1979 The HUMANA Press Inc.
Crescent Manor
P. O. Box 2148
Clifton, NJ   07015

Printed in the United States of America

Library of Congress Catalog Card No.: 79–87568

  Hautala, R. R. et al., editors
    Solar Energy.

  Clifton, N.J. : Humana Press
  432 p.
  7904   790320

PREFACE

The intense current interest in the development of solar energy as a viable energy alternative comes as no surprise in view of the widespread awareness of impending world-wide energy shortages. After all, the magnitude of energy available from the sun is impressive, its diffuseness and intermittent nature notwithstanding. The fact that, as a source, it represents a constant and inexhaustible supply of energy is alluring. The fact that most solar application schemes are nonpolluting in nature is an attractive bonus. In spite of these impressive attributes, research and development in the area of solar energy is in its infancy, owing largely to the prior lack of any need to exploit such *diffuse* sources. Indeed efforts in this area have traditionally been within the province of solid-state physics and engineering. The problems associated with efficient light harvesting and storage, however, are not simply technological ones. Effective solutions to these problems appear to lie beyond the current forefront of the chemical sciences. Consequently input from scientists previously engaged in fundamental chemistry has begun to emerge. Thus many of the contributions in this volume represent input from research groups with a relatively short history of involvement in solar energy. On the other hand, the long-standing and perceptive commitment of Professor Melvin Calvin to research involving solar energy represents the other extreme.

This volume covers a variety of approaches to the problem of efficiently converting and storing solar energy. The scope ranges from efforts to use currently available biological systems, such as the *Euphorbia* plants described by Calvin, to the development of metal–hydride storage systems driven by the heat produced from a solar collector, as described by Gruen. Approaches to mimicking the photosynthetic system are described by Bolton and Fong. Photoinduced electron transfer and photoionization in monolayers, micelles, and other heretogeneous environments are considered by Whitten and Thomas. Contributions by Hoffman, Demas, and Johnston focus upon various aspects of photogalvanic and photovoltaic cells. The photogalvanovoltaic cell described by Tien and the tandem photoelectrolysis concept proposed by Schwerzel involve the merging of approaches. Prototype solar cells for the photoreduction of molecular nitrogen are described by Schrauzer. The use of reversible organic photoreactions to store solar energy, which can be retrieved at a later time as heat, is considered by Jones, Mariano, and Hautala. Finally, Connolly presents a brief overview of the newly established Solar Energy Research Institute (SERI), an organization charged with formulating and coordinating solar energy research in the United States.

The above contributions are based upon the invited papers presented at a symposium on the Chemical Conversion and Storage of Solar Energy that was held during the 1978 Southeastern Regional Meeting of the American Chemical Society in Savannah, Georgia on November 9th and 10th. Although research groups from within this region were represented at the symposium, the majority of participants were from elsewhere in North America. Their involvement made possible the wide scope of topics presented in this volume.

We are particularly grateful to the participants of the symposium for their stimulating contributions and involvement. Furthermore, we are pleased to acknowledge assistance by the organizing committee of the 1978 Southeastern Regional Meeting of the American Chemical Society as well as Armstrong State College (Savannah, Georgia) in securing necessary financial support for the travel expenses of the participants in this symposium as well as a gift from U.S. Industrial Chemicals Company. Finally the extensive clerical assistance of Lorri Harrison helped make possible the rapid publication of this volume.

January, 1979   Richard R. Hautala
Athens, Georgia   R. Bruce King
                                                            Charles Kutal

ACKNOWLEDGEMENTS

The Editors wish to thank the following sources for permission to reproduce the following material: In the chapter by R. R. Hautala, R. B. King, and C. Kutal (pp. 333–369), the American Chemical Society for the use of Figs. 1–6 from the article by R. R. Hautala and J. L. Little in *Interfacial Photoprocesses,* M. S. Wrighton, ed., American Chemical Society, Washington, D.C., 1979 in the chapter by W. D. Johnston, Jr. (pp. 237–260), The Electrochemical Society for the use of Figs. 2–4 from the article by W. D. Johnston, Jr. and W. M. Callahan, *J. Electrochem. Soc.* **125,** 877 (1978); in the chapter by M. Calvin (pp. 1–30), the World Meteorological Organization for the use of the figure in W. W. Kellogg, *Effects of Human Activities on Global Climate,* WMO Technical Note No. 156, Geneva, Switzerland, 1977; in the chapter by R. E. Schwerzel, E. W. Brooman, R. A. Craig, D. D. Levy, F. R. Moore, L. E. Vaaler, and V. E. Wood (pp. 83–116), the American Institute of Aeronautics and Astronautics for material from his article in *Radiation Energy Conversion in Space;* in the chapter by H. T. Tien, J. Higgins, and J. Mountz (pp. 203–236), the North-Holland Publishing Co. for the use of Fig. 10 from R. R. Saxena and R. H. Bragg; *J. Non-Cryst. Solids* **23,** 58 (1978). The publisher also acknowledges that the chapter by J. S. Connolly and J. C. Grosskreutz (pp. 395–404) was prepared under U.S. Government Contract EG–77–C–01–4042, and that the U.S. Government may reproduce it, in whole or part, for its purposes without further permission.

CONTENTS

Solar Energy

Chemical Conversion and Storage

PETROLEUM PLANTATIONS*

Melvin Calvin

Laboratory of Chemical Biodynamics

University of California, Berkeley, CA 94720

ABSTRACT

Photosynthesis is examined as an annually renewable resource for material and energy. The production of fermentation alcohol from sugar cane as a major source of materials for chemical feedstocks is examined as well as the direct photosynthetic production of hydrocarbons from known plant sources. Experiments are underway to analyze the hydrocarbons from Euphorbias and other hydrocarbon containing plants with a view toward determining their various chemical components. In addition, experimental plantings of several species of Euphorbias have begun to obtain data on which species would be most successful. Using Euphorbia lathyris, there are indications that we may expect a yield of approximately ten barrels of hydrocarbon material per acre in a seven-month growing period on semiarid land.

* The work described in this paper was sponsored, in part by the Division of Biological and Environmental Research, the Basic Energy Sciences Division and the Solar Technology Division of the Department of Energy.

1

The topic of this chapter represents an application
of previous work from this laboratory to a critical practical problem, one which is paramount to the survival of
mankind. The problem is one of energy supply, which, today,
for the most part is fossil fuel. Figure 1 shows the flow
of energy in the United States in 1976. It can be seen
that the sources (left side of the figure) consist largely
of fossilized photosynthetic product. Specifically, ninety
eight percent of our energy resources, as shown on the left
side of this figure are oil, gas or coal--all three of
which represent fossilized carbon that is first reduced by
photosynthetic organisms.

There is hardly a need to point out the fact that the
discovery and delivery of the convenient fossil fuels (oil
and gas) has become an increasing problem. In the United
States, production is now falling. In the world, discovery
is now falling. It behooves those of us who are dependent
upon fossil fuel, at least in the form of oil and gas, to
find some other way to fulfill our needs.

In the United States, there exists a very substantial
supply of fossil carbon in the form of coal, which is fossilized photosynthetic product. This is demonstrated in
Figure 2 which shows that, while we will very soon (i.e.,
in the next 20 years) be completely out of oil, the supply
of coal (represented by the lighter area of this graph)
may be expected to continue for several hundred more years.
Therefore, there is a considerable movement in the United
States, and in the rest of the world as well, to make a
greater use of coal as a source, not only of energy, but of
chemical feedstocks as well. It seems to me that there are
constraints of this use which will prevent us from continuing and expanding the use of coal as our principal
source of energy and chemical raw materials, even though it
may be physically possible to do this.

There are two principal constraints on the expanded
use of coal. The first is illustrated in Figure 3, in
which the result of liquifying coal, i.e. making oil out of
coal,is shown. This can be done, but the figure shows re-
vealing data from one 25,000 ton/per day plant (or factory)
for converting coal into oil. It is evident that, while
somewhat more than 25,000 tons of oil will be made, that oil
will contain 200,000 pounds of polycyclic aromatic hydrocarbons, because coal is so poor in hydrogen. In partic-

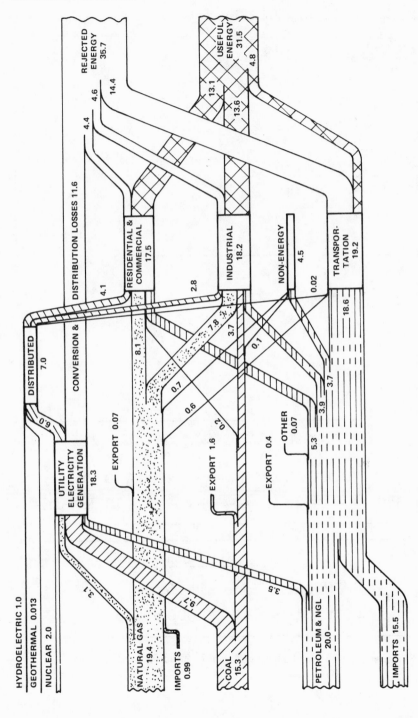

Figure 1. U. S. energy flow, 1976 (Primary resource consumption 72.1 quads).

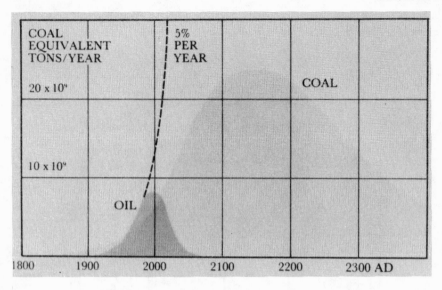

Figure 2.   Oil and Coal cycles.

(25,000 TONS / DAY OF COAL)

POLYCYCLIC   AROMATICS – – – 200,000 lbs/day

BaP – – – – – – – – – – – – – – – – – – – 10 lbs / day

BaP concentration

PETROLEUM – – – – – 1 – 5 ppm

COAL OIL – – – – – 10 – 100 ppm

Figure 3.   Potential Carcinogen Production in Coal Oil
            Plant.

ular, this oil will contain about 10 pounds of benzo-
(a)pyrene, which is known to be a potent carcinogen--one
which is produced when any organic matter is burned, but
particularly when coal is burned, simply because it is so
poor in hydrogen. Benzo(a)pyrene is, of course, a very
powerful carcinogen, and is one of the principal causes
for the rise in lung cancer among smokers. The amount of
benzo(a)pyrene (BaP) in the oil produced from coal is seen
to be of the order of 10 to 100 parts per million, whereas
the amount of BaP in natural petroleum is only 1/10 or 1/20
as much. This problem is not an insurmountable one. BaP
can be removed from the oil by purification processes so
that when the oil is burned there will be considerably less
BaP in the exhaust. There will be a cost for doing this,
but the environmental requirement for doing so will be so
great that it will be necessary. This, however, is a
soluble problem, one for which we can invent or devise a
solution which can be implemented.

The second principal constraint on the expanded use of
coal is the fact that when one burns coal (or any other
fossil carbon, for that matter, but coal in particular because it is the highest producer) carbon dioxide must be
produced. The rate at which carbon dioxide is being produced upon the surface of the earth is higher than the rate
at which it is consumed by green plants and oceans. The
inset in Figure 4 shows the rise of CO_2 from 1958 to 1976
measured at five different stations across the globe:
Point Barrow in Alaska, one at the South Pole, one on Mauna
Loa in the Hawaiian Islands, one on American Samoa, and one
over Sweden. All of these sites indicate the same result:
a rise in CO_2 from 310 parts/million to about 330 parts/
million in the last 15 years, or roughly a 5% rise in 15
years. By measuring the isotopic composition of the wood
in the rings of ancient trees, we can estimate the level of
carbon a hundred years ago. The major curve in this figure
shows what has happened in the last century, from 1860 to
about 1980: the rise has been about 15%, from 290 parts/
million to 330 parts/million. The projections here represent a number of possible consequences, depending upon the
different rates at which fossil carbon, in the form of coal,
is used. The four alternatives shown would all produce a
very large rise in the expected CO_2 level by the year 2020
or thereabouts. There is no escape from this if we continue to burn fossil carbon from underground--particularly

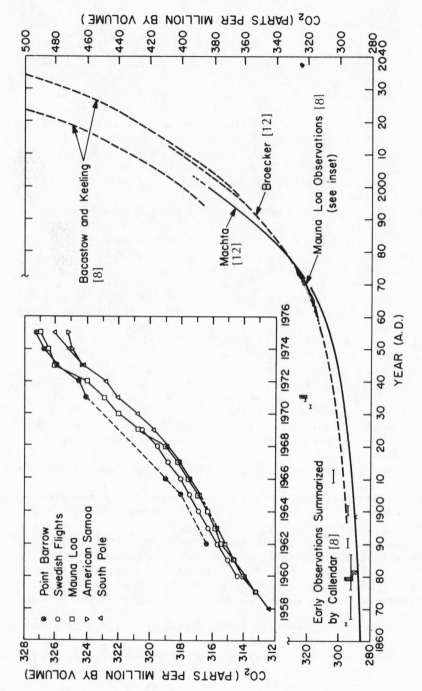

Figure 4. Carbon Dioxide in the earth's atmosphere (1860-2040).

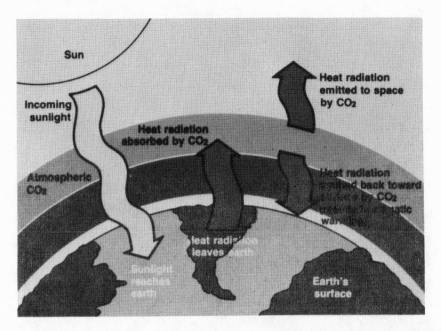

Figure 5. Greenhouse effect: CO_2 reflects heat, warms earth.

coal, since it produces more CO_2 per unit of heat or energy.

What are the possible consequences of this rise in
the CO_2 level? It is known that CO_2 is transparent to the
visible light from the sun, as shown in Figure 5. The
visible light comes right through the carbon dioxide
blanket. However, when the sunlight strikes the surface
of the earth at any point, whether it be in the green plant,
on a brown rock, or the ocean, it will eventually be converted to heat. Then an attempt will be made by the earth
to reradiate that heat, but when that happens, the CO_2
will not permit the heat or the infrared to escape from
the earth's surface--it will reflect it back. Therefore,
the surface of the earth will become warmer and, as a result of that, the entire climatic pattern of the earth's
surface can be expected to change: the rainfall patterns
will be different, and the places in which people can grow
food and live will be very different from what they are today. Thus it behooves us to find other ways to supply the
energy needs of mankind without burning the fossilized

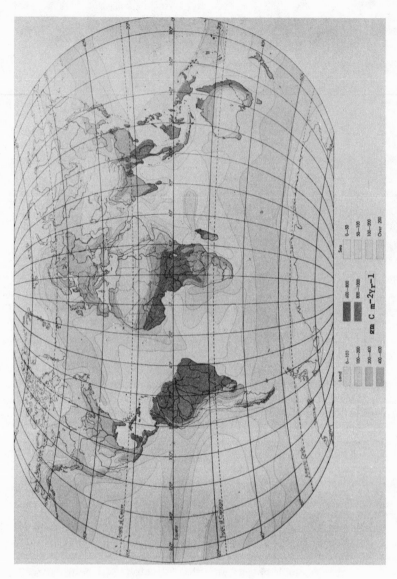

Figure 6. Natural vegetable productivity of biosphere, expressed in grams of carbon assimilated per sq. m. (After Gessner 1959 and Lieth 1965).

photosynthetic product which is now stored in the ground in
the form of coal.

The only convenient way that we have of doing this,
without producing a net increase in CO_2 level and without
producing other environmental hazards, is to use the natural vegetable productivity of the earth. The principal
way in which sunshine is converted today into chemical
energy for use as material or fuel is by the photosynthetic
process of the green plant. The places on the earth's surface in which this happens favorably are shown in Figure 6.
In those regions about one kilogram of carbon is reduced
per square meter per year. The most favorable regions lie
in the earth (the Amazon, the Congo, and Southeast Asia)
where it is humid, warm and sunny. These are the natural
regions of highest vegetable productivity. We cannot cut
the trees with impunity, but we can harvest their products.
However, there are many regions of the earth which have a
very high solar insolation but which are not very productive
today in terms of vegetable productivity for the simple
reason of lack of water (the South African desert of Kala-
hari, the Southwestern desert of the United States, a good
part of the Northeastern desert of Brazil, etc.). It is in
an attempt to use such regions that we are now engaged.

The first effort to do this has been very dramatically
demonstrated, both in Brazil and in the United States,
using a member of the grass family for this purpose. In
Brazil, the grass is sugar cane. In the United States,
sugar cane cannot be grown on the scale that is necessary,
but we can grow corn, a very close relative of sugar cane.
With both corn and sugar cane there is already a very substantial effort being made to use these products, insofar
as they do not compete for food production, as a source of
energy and materials. In both Brazil and Australia sugar
cane is grown in much of the northeastern parts of the
countries. The crop can be harvested not only for food
but, because there is more of it than is needed for that
purpose, it can be harvested also as an energy crop, and
the fermentable sugars converted into a convenient fuel
and chemical, namely ethanol. This is currently being done
on a very large scale in Brazil. Presumably enough cane
can be harvested within the next five years to produce 20
billion liters of alcohol in one year as a source of fuel
for automobiles and also as a source of chemicals for the
petrochemical industry.

Figure 7. Sugar burn, Mackay, Australia.

Figure 7 is a photograph of a sugar cane field in
Australia, showing the burning of the cane just prior to
harvest. The ripened cane has very dry leaves, and it is
not possible either for man or machine to enter into the
crop until the dry leaves are burned away. This seems to
be a tremendous waste of energy. We should invent a machine
capable of harvesting all of the cane, instead of burning
it off in this fashion.

After the cane is burned it is cut and crushed, and
the fermentable sugar can then be converted into a more
useful form--a liquid fuel or chemical, ethanol. This is
done by a simple fermentation, which, except for the alcohol-
water separation step, is a highly developed process in
Brazil. It is instructive to consider the stoichiometry of
that fermentation to illustrate what happens to the material (Figure 8). Note that in the transformation of 180
grams of sugar into 92 grams of alcohol, a solid is converted into a liquid of half the weight and the calories

$$C_6H_{12}O_6 \longrightarrow 2\,C_2H_5OH + 2\,CO_2$$

180 gm   92 gm

(673 Kcal)   (655 Kcal)

$$12.88 \text{ lbs} \longrightarrow 1 \text{ gal. (84,356 B.T.U.)}$$

Cost / gal. = raw material + ~20¢ process cost

Figure 8. Fermentation reactions.

Figure 9. Rubber tap, Brazil

are conserved. There is practically no loss of energy in this conversion, so it is a very efficient fermentation process.

This is not feasible in the United States because there is not enough cane for this purpose. However, we do

Figure 10.   Guayule, Saltillo, Mexico.

have enough corn--the corn that is being grown in Nebraska
and all through the Midwest as a source of feed for cattle.
If some fraction of that corn could be diverted to alcohol
fermentation along with the distilled dried yeast, which
is a byproduct from the fermentation for cattle feed, it
might be possible for us to use this corn not only as
cattle feed but also as an alcohol source, without any major disruption of the cattle meat industry.

While most plants store their energy as carbohydrate,
in the form of sugar or polymerized sugar (cellulose);
some plants store their energy as fully reduced carbon, in
the form of hydrocarbon. The commercial example of this,
of course, is the Hevea rubber tree, which originally grew
in Brazil as a commercial product. It produces a latex,
which consists of approximately one-third hydrocarbon emul-
sion in water. Figure 9 shows a tapping of such a tree in
Brazil. While it is true that the latex of the Hevea rubber tree is only 30% hydrocarbon, the value of that hydrocarbon is very high--because of its high molecular weight.
Because it is more valuable as an elastomer (i.e., as a

Figure 11. <u>Euphorbia</u> <u>lactea</u>, Puerto Rico.

rubber) than it would be if it were broken down into smaller
molecules to produce ethylene and polyethylene, it is not
prudent to destroy the product of <u>Hevea</u>. We must find other
sources of hydrocarbon-producing plants which make hydrocarbons of lower molecular weight.

We began such a search about four years ago, by ana-
lyzing alternative sources of hydrocarbon containing la-
texes which do not produce rubber. We wanted to use plants
that would grow in the semiarid parts of the world which
are not now productive of food crops. We knew of a plant
belonging to the Compositeae family (<u>Hevea</u> belongs to the
Euphorbiaceae family), which grows in the northern desert
of Mexico, in the southern United States, in Texas and in
Arizona (Figure 10). However, this plant, called guayule,
produces a latex very similar in molecular weight distribution to <u>Hevea</u>, and hence useful as rubber. Therefore,
we did not wish to use guayule as a possible source of
hydrocarbon for oil.

In looking further, we began exploring other members

Figure 12. Euphorbia lactea, Puerto Rico, showing flow of
            latex.

Figure 13. Euphorbia tirucalli, South Coast Field Station,
            Southern California.

of the family of Euphorbiaceae, to which <u>Hevea</u> belongs as
one genus. We encountered another genus of Euphorbiaceae
known as Euphorbia, which has about 2000 species. The
Euphorbias grow in all kinds of climates, both wet and dry,
and in all sorts of forms, small plants as well as trees.
Figures 11-17 show a few photographs of these Euphorbias
from which we have collected latex, and which we have now
analyzed.

Figure 11 shows a Euphobia (<u>Euphorbia</u> <u>lactea</u>) growing
on the South Coast of Puerto Rico, and Figure 12 shows the
flow of latex when it is tapped. Figure 13 shows a common
Euphorbia, <u>Euphorbia</u> <u>tirucalli</u>, growing on a plantation in
Southern California. It is about 20-30 cm high and about
30 cm in diameter. This is one year's growth, from the
time it is put in the ground as a single 5 cm high cutting.
Figure 14 shows a very large <u>Euphorbia</u> <u>tirucalli</u> tree, of
unknown age, growing in Brazil. Figure 15 shows a <u>Euphor-
bia</u> <u>tirucalli</u> bush, about one year old, growing in Okinawa
on a large plantation. There are indications that it is
as productive as any of the Euphorbias that we have been
growing in Southern California--perhaps even more so, be-

Figure 14. <u>Euphorbia</u> <u>tirucalli</u>, Brazil.

Figure 15. Euphorbia tirucalli, Okinawa.

cause Okinawa is very wet, whereas Southern California is
not.

    Figure 16 shows Euphorbia dendroides (a completely
different Euphorbia) growing on a hillside in Los Angeles.
This is wild; it is not planted. They grow quite large--
the whole hillside is covered with them. They are close
relatives to the ones we have planted in Southern California
for which we have measured the growth rate.

    Figure 17 is a photograph of the latter plant, called
Euphorbia lathyris. These plants are seven months old from
seed, and have grown 1½ meters in seven months. They are
also growing in Okinawa, but the climate there is too wet
for them to grow well; they grow much better in Southern
California. Figure 18 shows the data that we have obtained
on the growth rate of this plant in Southern California
over a ten month period. These plants were germinated in
the greenhouse in December, 1976 and were put in the ground
in February, 1977. The dry weight is shown on the far left
of the graph. In October, 1977 it became too cold for them

Figure 16. _Euphorbia_ _dendroides_, Southern California

Figure 17. _Euphorbia_ _lathyris_, South Coast Field Station,
            seven months old.

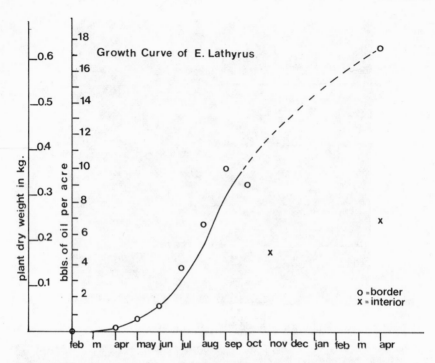

Figure 18.   Growth curve Euphorbia lathyrus, South Coast
             Field Station, fifteen months.

to continue growing, and they went into a 6-month dormant
phase. They are now going to seed. We have analyzed these
plants in the course of their growth. The second column on
the left side of Figure 18 shows the conversion of their
total oil content into barrels of oil/acre. For maximum
oil content they could have been harvested in September,
at which time they could yield 10 barrels of oil/acre (25
barrels per hectare) in about nine months. This is a substantial production.

     This oil is a mixture of hydrocarbons consisting most-
ly of terpenoid isoprenes--not only open chain isoprenes,
but cyclic isoprenes: diterpenes and triterpenes, sterols,
etc. A preliminary article describing the detailed materials present in the oil mixture has been published in Science and more information will be forthcoming.

The crude mixture molecular weight distribution of this oil can be compared to that of the hydrocarbon from Hevea. Figure 19 shows that Hevea has an average molecular weight of somewhere over a million. The molecular weight curve happens to be bimodal with modes at ∿500,000 and ∿2,000,000 whereas the other curve, for Euphorbia tirucalli (the one that grows well in Okinawa and Brazil), has a very small molecular weight--about 10,000. So, the latter is a possible candidate as a crude oil mixture of many different terpenes as well as some open chain polyisoprenes.

In this regard it is instructive to consider the last step of the biosynthetic sequence which will produce polyisoprenes. The first products of the plant are sugars, particularly phosphoglyceric acid. The pathway then proceeds to pyruvic acid, then to acetoacetic acid, and then, by successive reductions, to beta-hydroxy glutaric acid and mevalonic acid, 3,5-dihydroxy-3-methylpentanoic acid. After pyrophosphorylation, reduction and decarboxylation, isopentenyl pyrophosphate (IPP) is produced. As shown in Figure 20, part of the isopentenyl pyrophosphate, a five-carbon compound, can be isomerized to dimethylallyl pyrophosphate, DMAPP. Then these two condense to form a C_{10} pyro-

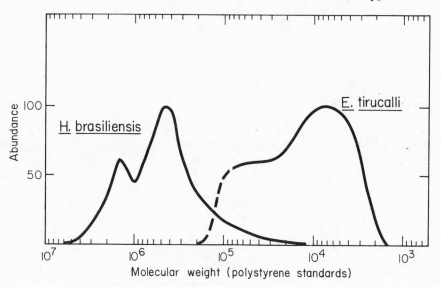

Figure 19. Molecular weight distribution of polyisoprenes isolated from Hevea brasiliensis and Euphorbia tirucalli.

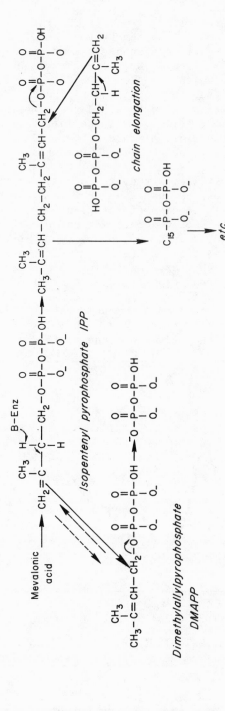

Figure 20. Isoprenoid condensation reactions.

phosphate which regenerates the incipient allylic carbonium ion of DMAPP which can then continue to grow on another isopentenyl pyrophosphate, so that the chain elongation can continue. Thus, polyisoprenes represent the terminal step in that synthesis.

As mentioned earlier we are able to grow about ten barrels of oil (mixed hydrocarbon) per acre in nine months, and this is accompanied by about 15 tons of dry lignocellulose. The cost of growing such a crop in the parts of the United States which today are non-productive of food crops (Texas, Arizona, Southern California and Southern Florida) is about $100 per acre. It should be noted that this involves wild plants that have not been improved, and in which there has been no selection. With these wild plants, we are getting ten barrels of oil per acre, which costs about $100 per acre to grow--so the cost of growing the oil in the United States is about $10 per barrel.

I have discussed with my friends in Chemical Engineering and Agricultural Engineering how much it would cost to cut the plant, to crush it, and to extract the oil. We do not yet have a commercial process for this. Using the laboratory process, which is a very ordinary kind of chemistry involving solvent extraction, and which probably could not be done on a very large scale (involving millions of barrels of oil), the cost would be about $10 a barrel to take it from the plant and to put it in the barrel. Thus, we would have oil in the United States for a total cost of $20 a barrel--$10 a barrel to grow it, and another $10 a barrel to put it in the barrel--without improvements, either in in growing or in processing.

I think we are already on the edge of economic viability with these plants, even just treating the material as crude oil, not as a chemical. In addition, this mixture of hydrocarbons will have in it many different steroids, diterpenes and diterpenols, some with very interesting biological activity which could be developed. This crude mixture of hydrocarbons and hydrocarbon alcohols will have in it many things of higher value than simply a fuel or raw material for cracking in a petrochemical factory. The total value of this product as a new raw material will become even greater than it is today. Today it is still a "wild" plant (or a whole group of "wild" plants), not yet "domesticated", not yet an agricultural staple or crop.

With genetic and agronomic improvements, I expect that
the yield will go as high as 20 or even 30 barrels per acre.
Even with only the proper selection of plants, it will rise.
When we get to the 10 or 20 acre size for our plantation,
we will be able to develop a good commercial extraction pro-
cess--one which would not use organic solvents, only water,
followed by a catalytic reformation of the whole water extract. Such modifications will undoubtedly lead to a reduction in price.

In passing, it should be noted that this very idea was
broached by the Italians in Abyssinia, back in 1938. The
following perceptive article was published in that year.

A NEW BRAND OF GASOLINE

*"One of the major problems in Italy's recently
conquered province of Ethiopia is fuel for motor
cars. The need for gasoline is urgent as almost all transportation is dependent upon motor
cars and because all motor fuels have to be im-
ported at extremely high prices. During the
subjugation of 1935, an oil company entered into
a contract with Haile Selassie for all oil rights
in that country. This gave the Italians wild
expectations of an unlimited supply of crude oil,
but these dreams have not materialized, ardent
though the search has been.*
*Alcohol, the substitute for gasoline, can be
made from palm nuts, corn and millet which are
available in sufficient quantities to produce an
estimated one million gallons per year, but petroleum is best and most desirable as a motor fuel
and so far the country has produced none.*
*By this time the reader will perhaps be wonder-
ing what all this has to do with succulents, but
the unusual association is that the Italian Fuel
Commission announces that the best assurance now
appears to be in the billions of weird cactoid
Euphorbias that grow throughout that country.
The latex of the large tree-like species is said
to yield approximately 60% of a gasoline-like
liquid which can be produced by a simple chemical
treatment at little expense, and today this "vegetable gasoline" holds out an encouraging hope to*

*the worried Italian engineers. The Italian Fuel
Commission is at least going to try it out and
are now arranging for the construction of a Euphorbia gasoline refinery at Agodat, which is the
center of the large Euphorbia forests and on the
main highway to Adis Ababa where fuel-hungry
fleets of trucks pass every day on their trips
to the sea.
The entire experiment will be watched with
interest since geologists inform us that the
worlds petroleum supply is limited and deminish-
ing rapidly, and who knows but that the answer
rests with the Euphorbiae."*

G. A. Frick*

I would like to say in conclusion that ultimately, in
the last analysis, we must be able to reproduce the light
capturing system of the plant artificially. We should not
have to grow plants in order to capture sunshine. But in
order to do that, we must understand how it is done by the
plant. If we understand exactly how the plant does it, then
we have a possibility of doing it without the plant--syn-
thetically and artificially. We can make a system for
catching the quanta from the sun and reducing organic carbon, or photolyzing water, to make a useful product. In
fact, we have already taken one step in that direction,
which I would like to show you. Figures 21 and 22 show a
model of a photosynthetic membrane in a green plant. It is
a bilipid membrane, with two surfaces for the two quantum
act: Photosystem I and Photosystem II (two different sensitizers). We cannot use chlorophyll, because it is not
stable enough; so we have made many artificial surfactant
dyestuffs. We have made the membrane in the form of a ves-
icle, and have put the dyestuffs on the inside and the outside surface. We have put a donor system inside and the
acceptor outside, and have shown that we can photochemically
transfer an electron from one side of that membrane to the
other. What we cannot yet do is to generate molecular oxygen. That would require a manganese complex, and we do not
yet know what that is; but we can make the hydrogen, using
methylviologen and platinum. So the manganese complex is
not yet known, and that is the big problem which remains to
be solved before we can make a synthetic system work. I am

*G. A. Frick, Cactus Succulent Journal, Vol. 10,
         p. 60 (1938).

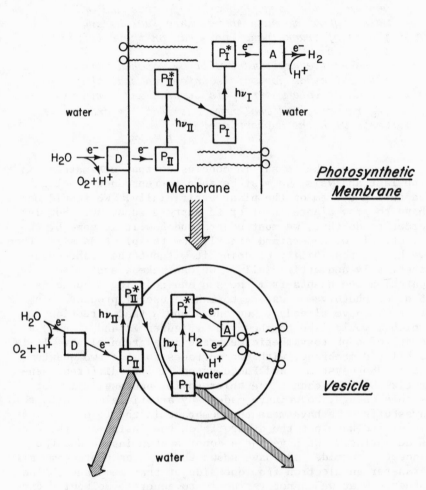

Figure 21. Photoelectron transfer scheme.

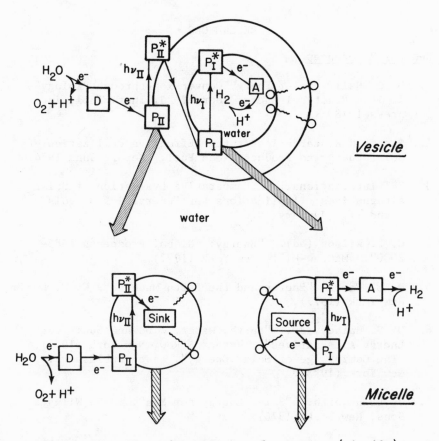

Figure 22.   Photoelectron transfer scheme (micelle).

sure that somebody will find out what it is and how to make it within the next decade or so. Then perhaps we will be free of the need of arable land to capture the sunshine for energy and materials.

REFERENCES

General References

1. D. O. Hall. Solar Energy Conversion through Biology—— Is It a Practical Energy Source? J. Inst. Fuel, in press, 1978.

2. Stanford Research Institute. Effective Utilization of Solar Energy to Produce Clean Fuel. Report June 1974.

3. SRI International. A Comparative Evaluation of Solar Alternatives: Implications for Federal R & D, Vols. I and II. January 1978.

4. C. L. Wilson (ed). "Energy: Global Prospects 1985– 2000". McGraw-Hill, New York (1977).

5. D. E. Carr. "Energy and the Earth Machine". W. W. Norton, New York (1977).

6. D. S. Halacy, Jr. "Earth, Water, Wind and Sun: Our Energy Alternatives". Harper & Row, New York (1977). "The Coming Age of Solar Energy". Avon Publishers, New York (1973).

7. J. Brinkworth. "Solar Energy for Man". John Wiley & Sons, New York (1976).

8. R. Buvet et al. (eds). "Living Systems as Energy Con- verters". Elsevier Publishing Co., New York (1977).

9. A. Mitsui, S. Miyadhi, A. San Pietro and S. Tamura (eds). "Biological Solar Energy Conversion", Academic Press, New York (1977).

10. J. R. Bolton (ed). "Solar Power and Fuels". Academic Press, New York (1977).

Specific references in a given area of discussion in the paper are given below:

Effect of Increased CO_2 Concentration in the Atmosphere

1. M. Stuiver. Atmospheric Carbon Dioxide and Carbon Reservoir Changes. Science, 199, 253 (1978).

2. G. M. Woodwell, R. H. Whittaker, W. A. Reiners, G. E. Likens, C. C. Delwiche and D. B. Botkin. The Biota and the World Carbon Budget. Science, 199, 141 (1978).

3. J. H. Mercer. West Antarctic Ice Sheet and CO_2 Greenhouse Effect: A Threat of Disaster. Nature 271, 321 (1978).

4. U. Siegenthaler and H. Oeschger. Predicting Future Atmospheric Carbon Dioxide Levels. Science, 199, 388 (1978).

5. J. S. Olson, et al. Changes in the Global Carbon Cycle and the Biosphere. ORNL/EIS-109, Sept. 1978.

6. B. R. Strain (ed). Report of the Workshop on Anticipated Plant Responses to Global Carbon Dioxide Enrichment. Duke University, Durham, North Carolina, June 1978.

7. Irene Smith (ed). Carbon Dioxide and the "Greenhouse Effect"--An Unresolved Problem. Report No. ICTIS ER-01, April 1978. (IEA Technical Information Service, London)

8. G. M. Woodwell. The Carbon Dioxide Question. Scientific American, 238(1) 34 (1978).

Rubber

1. Agricultural Research Service, U. S. Dept. of Agriculture. Plants Collected and Tested by Thomas A. Edison as Possible Sources of Domestic Rubber. Report ARS-34-74, July 1967.

2. H. M. Hall and F. L. Long. "Rubber Content of North American Plants". Carnegie Institution of Washington, Washington, D. C. (1921).

3. B. M. Vanderbile. Rubber from Goldenrod. In "Thomas
    Edison, Chemist" (Chapter 9). American Chemical Society,
    Washington, D. C. (1971).

4. International Rubber Conference, Kuala Lumpur, Malaysia,
    Oct. 1975 proceedings. Five volumes. Rubber Research
    Institute, Kuala Lumpur, Malaysia (1976).

Guayule

1. National Academy of Sciences. Guayule: An Alternative
    Source for Natural Rubber. Washington, D. C. (1976).

2. W. G. McGinnies and E. F. Haase (eds). International
    Conference on Utilization of Guayule. University of
    Arizona, Tucson (1975).

3. E. Campos-Lopez (ed). Guayule (proc. int. conference,
    Saltillo, Mexico, Aug. 1976). Centro de Investigacion
    en Quimica Aplicada, Saltillo, Mexico (1978).

4. E. Campos-Lopez (ed). Larrea (proc. int. conference,
    Saltillo, Mexico, Aug. 1978).

5. H. Yokoyama, E. P. Hayman, W. J. Hsu, S. M. Poling and
    A. J. Bauman, Chemical Bioinduction of Rubber in the
    Guayule Plant. Science, $\underline{197}$, 1076 (1977).

Sugar Cane

1. E. Inojosa (ed). Açucar e Alcool: Um Grande Projeto
    Economico do Brasil. APEC/COPERFLU, Rio de Janeiro
    (1976).

2. F. H. M. Kelly. A Feasibility Study of the Production
    of Ethanol from Sugar Cane. Private communication, 1977.

Gasohol

1. U.S. Senate Republican Conference. Alcohol: The Re-
    newable Fuel from Our Nation's Resources. Washington,
    D. C. (1977).

2. W. A. Scheller. Gasohol: Food and Fuel for the Future.
    University of Nebraska, Omaha (1977).

3. E. V. Anderson. Gasohol: Energy Mountain or Molehill?
    Chem. Eng. News, July 31, 1978.

Petroleum Plantations

1. M. Calvin. Photosynthesis as a Resource for Energy and
    Materials. Photochem. Photobiol. 23, 425 (1976).

2. M. Calvin. The Sunny Side of the Future. Chem. Tech.
    7, 352 (1977).

3. M. Calvin. Energy and Materials via Photosynthesis.
    In "Living Systems As Energy Converters". R. Buvet et
    al. (eds). Elsevier Publishing Co., New York (1977),
    pp. 231-259.

4. M. Calvin. Hydrocarbons Via Photosynthesis. Energy
    Res. 1, 299 (1977).

5. P. E. Nielsen, H. Nishimura, J. W. Otvos and M. Calvin.
    Plant Crops as a Source of Fuel and Hydrocarbon-like
    Materials. Science, 198, 942 (1977).

6. M. Calvin. Chemistry, Population, Resources. Pure &
    Appl. Chem., 50, 407 (1978).

7. M. Calvin. Green Factories. Chem. Eng. News, March 20,
    1978. pp. 30-36.

8. G. J. Calvin and M. Calvin. Green Factories for Liquid
    Fuel. Lawrence Berkeley Laboratory Report LBL 7584,
    April 1978.

9. E. Lipinsky. Fuels from Biomass: Integration with Food
    and Materials Systems. Science, 199, 644 (1978).

10. R. A. Buchanan, I. M. Cull, F. H. Otey and C. R. Russell.
    Hydrocarbon and Rubber Producing Crops: Evaluation of
    U.S. Plant Species, Econ. Botany, 32, 131 (1978).

11. R. A. Buchanan, I. M. Cull, F. H. Otey and C. R. Russell.
    Hydrocarbon and Rubber Producing Crops. Evaluation of
    100 U.S. Plants Species. Econ. Botany, 32, 146 (1978).

12. R. A. Buchanan, F. H. Otey, C. R. Russell and I. M. Cull.
    Whole Plant Oils, Potential New Industrial Raw Materials.
    J. Amer. Oil Chem. Soc., 55, 657 (1978).

Artificial Photosynthesis

1. G. Porter and M. D. Archer. In vitro Photosynthesis.
    Interdisciplin. Sci. Rev., 1, 119 (1976).

2. G. Porter. In vitro Models for Photosynthesis (The
    Bakerian Lecture). Proc. Roy. Soc. 362A, 281 (1978).

3. T. Matsuo (ed). "Biomimetic Chemistry" (symposium,
    Kyushu University, Fukuoka, Japan, Sept. 1977). In
    press.

4. J. R. Bolton. Solar Fuels. Science, 202, 705 (1978).

5. M. Calvin. Simulating Photosynthetic Quantum Conver-
    sion. Accts. Chem. Res., 11, 369 (1978).

6. W. Ford, J. W. Otvos and M. Calvin. Photosensitized
    Electron Transport across Phospholipid Vesicle Walls.
    Nature, 274, 507 (1978).

7. M. Calvin. Synthetic Chloroplasts. Energy Res., in
    press.

SOLAR ENERGY CONVERSION IN PHOTOSYNTHESIS - FEATURES RELEVANT TO ARTIFICIAL SYSTEMS FOR THE PHOTOCHEMICAL CONVERSION OF SOLAR ENERGY[*]

James R. Bolton

Photochemistry Unit, Department of Chemistry

University of Western Ontario, London, Canada N6A 5B7

ABSTRACT

Photosynthesis is the only known process which can photochemically convert and store solar energy with a reasonable efficiency. Thus any attempts to develop artificial systems will probably be most successful when modelled on the natural process. In this paper, I attempt to examine the mechanism and efficiency of photosynthesis as a solar energy converter and extract those features which are most relevant to the design of artificial systems.

After a brief review of the thermodynamic and kinetic limitations on any photochemical conversion and storage process, the overall gross efficiency of photosynthesis is analyzed and found to be $9.2 \pm 0.8\%$. This is reduced to $5.6 \pm 1.2\%$ when dark- and photo-respiration are considered. A wavelength threshold relation is derived for a general photochemical solar energy process and it is shown that if photosynthesis were to proceed with one photosystem, then the threshold wavelength would be about 610 nm. Since photosynthesis is known to have a threshold wavelength of $\sim$ 700 nm, it is seen that two photosystems are necessary not only to carry out the reaction but also to maximize the fit to the solar spectrum and hence to maximize the efficiency.

*Publication No. 196 of the Photochemistry Unit, University of Western Ontario, London, Canada, N6A 5B7.

The last part of the paper will analyze the mechanism of photosynthesis, particularly the mechanism of the primary photochemical electron transfer process in the reaction center protein. Although highly speculative, I attempt to extract the important features of the structure and mechanism which lead to a very high quantum yield and efficiency.

INTRODUCTION

Photosynthesis is the most efficient system known for the conversion and storage of solar energy as chemical energy and outranks by far any artificial system yet devised. Thus, it is most important to understand the mechanism and efficiency of photosynthesis if we hope to be able to design efficient artificial systems to store solar energy.

In this paper, I examine photosynthesis from the standpoint of an energy converter and try to show how nature has coped with certain basic limitations on the conversion of light to chemical energy in order to achieve a process of optimal efficiency.

GENERAL LIMITATIONS ON ENERGY STORAGE FOR ANY PHOTOCHEMICAL REACTION

Many investigators (1-5) have considered the general requirements for useful solar photochemical energy storage reactions. In summary, they are

1. The photochemical reaction must be endergonic.
2. The process must be cyclic.
3. Side reactions leading to the irreversible degradation of the photochemical reactants must be totally absent.
4. The reaction should be capable of operating over a wide bandwidth of the visible and ultraviolet portions of the solar spectrum with a threshold wavelength well into the red or near infrared.
5. The quantum yield for the photochemical reaction should be near unity.
6. The back reaction must be extremely slow under ambient conditions to permit long-term storage, but should proceed rapidly under special controlled catalytic conditions or elevated temperatures so

as to release the stored energy when needed.

7. The product(s) of the photochemical reaction should be easy to store and transport.

8. The reagents and any container material should be cheap and non-toxic and the reaction should operate under aerobic conditions.

At present the only photochemical energy storage systems which meets nearly all of the above requirements is photosynthesis. However, before we delve into the details of the photosynthesis reaction, let us consider some general thermodynamic and kinetic limitations on the conversion of light energy to chemical energy.

Consider the endergonic photochemical reaction

$$R \rightarrow P$$

illustrated in Fig. 1 where R is the absorber and P is the primary product of the photochemical step. The reaction may involve more than one reactant but the overall process will be considered to be first order, as is the case in photosynthesis.

There are three fundamental loss processes which must be incurred in the photochemical step(s) (6).

1. Photons having a wavelength λ such that $\lambda > \lambda_g$, where λ_g is the wavelength corresponding to the band-gap energy E_g, do not have sufficient energy to raise R to R* and thus cannot contribute to the photochemistry.

2. Photons which are absorbed with $\lambda < \lambda_g$ may raise R to R* but the very efficient process of internal conversion will reduce R* to its lowest vibrational level within a few picoseconds and hence the excess photon energy above E_g will be lost as heat.

3. The conversion of R* to P must be an exergonic reaction so that an activation energy E_r* for the back reaction will be established (see Fig. 1). Otherwise, P would have no stability to allow for storage or subsequent reactions leading to chemical storage.

The quantitative evaluation of the first two loss terms has been carried out by Ross and Hsiao (7) for air

mass (AM) 0 (outside the earth's atmosphere) solar radiation
and for AM 1.2 (a bright sunny day near noon on the earth's
surface) by Bolton (5). The fraction of solar power η_E
available at the band gap energy for a perfect absorber (all
light absorbed for $\lambda \leq \lambda_g$) is shown as curve E in Fig. 2.
It is a common misconception among photochemists that η_E
represents the fraction of solar power which can be conver-

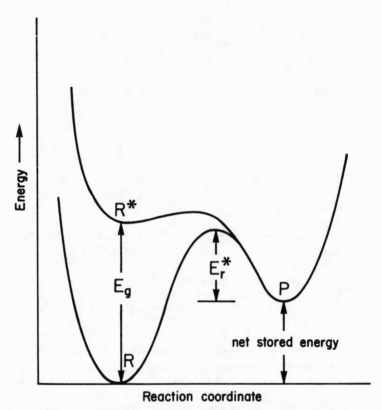

Fig. 1. Energy profile for a general endergonic photochemical reaction R → P. E_g is the minimum energy gap
between the lowest vibrational states of the
excited state R* and the ground state of R. E_r^*
is the activation energy for the back reaction
P → R.

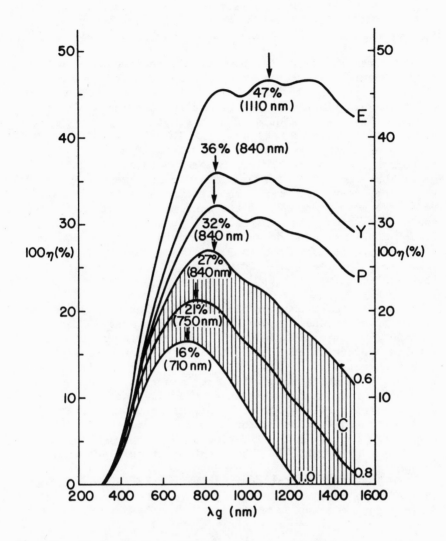

Fig. 2. Plots of the efficiencies η_E, η_Y, η_P and η_C as a function of the wavelength λ_g corresponding to the energy gap E_g (taken from reference 5). η_C has been calculated for 0.6 eV, 0.8 eV and 1.0 eV energy loss respectively as indicated in the figure.

ted to chemical energy. This view is erroneous because there are additional losses necessary to take account of the Second Law of Thermodynamics.

Several investigators have considered the problem of thermodynamic limits on the conversion of light to work (5,7-10) with some giving special attention to the reaction of photosynthesis (11-14). The maximum thermodynamic efficiencies have been calculated as a function of λ_g by Ross and Hsiao (7) for AM0 and by Bolton (5) for AM1.2. The latter results are shown as curves Y and P in Fig. 2. η_Y (curve Y) corresponds to the efficiency one would obtain under conditions where the quantum yield for storage $\phi_{st} \to 0$, i.e. where the energy conversion rate is infinitesimally slow. η_P (curve P) corresponds to the condition of optimal energy conversion rate. For all practical purposes Curve P in Fig. 2 should be considered as the maximum thermodynamic efficiency. It maximizes at 32% for λ_g = 840 nm. η_P is 28% at 700 nm, the threshold wavelength for green plant photosynthesis.

η_P just takes account of conversion losses but does not allow for additional losses required for storage. Here a kinetic limitation is imposed as seen by reference to Fig. 1. The conversion of R* to P must be exergonic so as to create an activation energy E_r^* for the back reaction. If the photochemical reaction is first order (i.e., either unimolecular or bimolecular where the two reactants are confined to react only with each other as in a solid state or on a membrane) then as Bolton showed (5) there is a direct relationship between the lifetime of P and the energy loss E_r^* as shown in Table 1. The creation of this activation barrier must be done at the expense of the excitation energy and the loss is constant and independent of λ_g. Thus, we can define a chemical efficiency η_C (which, of course, must be less than η_P) as

$$\eta_C = \eta_E \eta_{chem} \qquad (1)$$

where

$$\eta_{chem} = \frac{\Delta G}{E_g} \qquad (2)$$

ΔG being the net free energy storage in the photochemical step (see Fig. 1).

Table 1

Lifetime τ (s) of the product P	Required activation energy E_r^* (eV)[a]
1.0×10^{-12}	0.05
1.0×10^{-9}	0.23
1.0×10^{-6}	0.40
1.0×10^{-3}	0.58
1.0	0.76
1.0×10^{3}	0.94
1.0×10^{6}	1.12

[a]E_r^* is calculated from unimolecular rate theory

$$E_r^* = - \frac{kT}{e} \ln\left[\frac{h}{kT\tau}\right] \qquad \text{(in eV)}$$

where k is Boltzmann's constant, h is Planck's constant, τ is the lifetime of P and e is the electronic charge.

η_C is plotted as curves C in Fig. 2 assuming that $E_r^* = 0.6$, 0.8 and 1.0 eV respectively. For $E_r^* = 0.8$ eV, η_C has a maximum value of 21% at $\lambda_g = 750$ nm.

If the photochemical step is second order, as for the bimolecular reaction of two components in solution, then E_r^* can be quite small because of the short lifetime of the collision complex. However, in this case the back reaction will be virtually diffusion controlled and no net storage will occur. Anything which will slow down the back reaction must either involve exergonic secondary reactions or conversion of the process to first order (e.g., by confining the reactants to a surface or a membrane). Hence, it is unlikely that one can avoid a loss of perhaps 0.6 - 1.0 eV in any photochemical endergonic step.

An Estimate of Chemical Storage Efficiency

Curves P and C in Fig. 2 represent ideal limits of conversion efficiencies; however, there are other loss factors to be considered as indicated in the following general expression for the energy storage efficiency

$$\eta_{storage} = \eta_{abs}\eta_{C}\eta_{\phi}\eta_{coll} \qquad (3)$$

where η_{abs} is the fraction of incident photons with $\lambda \leq \lambda_{g}$ which are absorbed; η_{C} is defined above, η_{ϕ} is the quantum yield for the photochemical step(s) and η_{coll} is the fraction of product produced which can be collected and stored.

η_{abs} is unlikely to be greater than ~ 0.75 for most absorbers and η_{ϕ} and η_{coll} are unlikely to be greater than 0.9 each. Thus, if we take all of the factors together with $\eta_{C} = 0.21$ from Fig. 2, then we find that the net yield of product in a photochemical energy storage reaction is unlikely to be greater than 12 - 13%.

SOLAR ENERGY STORAGE EFFICIENCY IN PHOTOSYNTHESIS

Now let us examine the efficiency of the reaction of photosynthesis which can be written as

$$CO_2(g, 0.00032 \text{ atm}) + H_2O(\ell) \rightarrow \frac{1}{6} C_6H_{12}O_6(s)$$

$$+ O_2(g, 0.21 \text{ atm}) \qquad (4)$$

for which $\Delta H = 467$ kJ mol^{-1} and $\Delta G = 496$ kJ mol^{-1} as calculated from standard enthalpies and free energies of formation corrected to the pressures shown (15).

The efficiency will be calculated using the equation

$$\eta = \frac{\int_{\lambda_{min}}^{\lambda_{max}} \dfrac{N_s(\lambda)\alpha(\lambda)\phi(\lambda)\Delta H}{N_0} d\lambda}{\int_{0}^{\infty} E(\lambda) d\lambda} \qquad (5)$$

where $N_s(\lambda)$ = incident solar photon flux (photons m^{-2}s^{-1}nm^{-1}) in the wavelength band from λ to $\lambda + d\lambda$.

$\alpha(\lambda)$ = absorption coefficient of a leaf from λ to $\lambda + d\lambda$.

$\phi(\lambda)$ = quantum yield for production of O_2 or consumption of CO_2 in the wavelength band from λ to $\lambda + d\lambda$.

ΔH = enthalpy change (kJ mol^{-1}) in Eq. (1).

$E(\lambda)$ = incident solar irradiance (Wm^{-2} nm^{-1}) in the wavelength band from λ to $\lambda + d\lambda$.

N_O = Avogadro's number

and the denominator is the total incident solar irradiance (Wm^{-2}).

$N_s(\lambda)$ is calculated from $E(\lambda)$ by

$$N_s(\lambda) = E(\lambda) \cdot \frac{\lambda}{hc} \qquad (6)$$

The analysis of efficiency was carried out at 10 nm intervals using Eqs. (5) and (6) and the following data and conditions:

(a) $E(\lambda)$ values were obtained from the air mass (AM) 1.2 (T/S) solar distribution calculated by Boer (16). This distribution corresponds to a bright sunny summer day near noon.

(b) $\phi(\lambda)$ values were taken from the measurements of CO_2 absorption (corrected for respiration) versus wavelength by Balegh and Biddulph (17) on red kidney bean leaves (Phaseolus vulgaris L.). As their $\phi(\lambda)$ was defined in terms of the incident rather than absorbed light intensity $\alpha(\lambda)$ is included in their measurement of $\phi(\lambda)$. For the region 680 - 720 nm oxygen quantum yields (corrected for absorption) for Chlorella pyrenoidoisa (18) in the presence of supplementary short wavelength light were used to simulate the effect of white light for the long wavelength region.

(c) λ_{min} was taken as 360 nm and λ_{max} as 720 nm. The threshold or band-gap wavelength λ_g for photosynthesis is 700 nm; however, some yield beyond 700 nm is possible due to thermal activation.

(d) Only the forward reaction was considered - i.e., no allowance was made for respiration.

With the above assumptions, the overall gross efficiency of photosynthesis was calculated to be $9.2 \pm 0.8\%$. The largest uncertainties are in the values of $\phi(\bar{\lambda})$ and $E(\lambda)$. It is possible to get values from 9.0% to 9.6% depending on the solar distribution used. If ΔG is used instead of ΔH the efficiency becomes 9.8%. These values compare well with values obtained by others (19-21).

The value of 9.2% is a realistic estimate of the gross efficiency of photosynthesis to be expected in a healthy leaf or alga suspension. However, it is interesting to calculate two more "ideal" efficiencies. First, if we assume that $\alpha(\lambda) = 1.0$ over the photosynthetically active region (i.e., all photons in this region are absorbed), the efficiency would be 11.2%. Second, if we, in addition to assuming $\alpha(\lambda) = 1.0$, assume that $\phi(\lambda)$ has its "ideal" value of 0.125 (8 photons per O_2 molecule evolved) in the photosynthetically active region, the efficiency rises to 13.3%. When this "ideal" efficiency is compared with the maximum thermodynamic efficiency of 28% (at 700 nm) and the maximum efficiency under kinetic control of 21% (see Fig. 2) it is seen that photosynthesis approaches the "perfect" systems quite closely.

Of course, the gross efficiency calculated above can never be achieved in a real organism because of the fact that respiration is always occurring. Since the respiration reaction is the back reaction of photosynthesis, the effect of respiration is to reduce the efficiency of photosynthesis to a net value somewhat below the gross value of 9.2%. For a healthy leaf in bright sunlight, there are two types of respiration occurring - dark respiration and photorespiration. The rate of dark respiration varies from 0.5 - 4.0 mg CO_2 dm^{-2} hr at 25° C (22). This will have a relatively small effect on the efficiency in bright sunlight and will perhaps subtract $\sim 0.4\%$ from the gross efficiency. On the other hand, photorespiration usually occurs at rates considerably greater than dark respiration and at a rate which is about 30% to 40% of the gross rate of photosynthesis (19). Thus, the combination of dark respiration and photorespiration results in a net efficiency of about $5.6 \pm 1.2\%$.

WHY DOES GREEN PLANT PHOTOSYNTHESIS REQUIRE TWO PHOTOSYSTEMS?

The reaction of photosynthesis, Eq. (4), is endergonic and all of the input energy must come from sunlight. There are two fundamental limitations which apply:

1. A maximum of one electron is transferred per photon absorbed in any photochemical electron transfer reaction.
2. Not all of the excitation energy can be converted to chemical energy of the products.

To answer the question of why two photosystems, let us first examine the consequences of the photosynthesis reaction being driven by a single photochemical system. In such a system, if every photon absorbed resulted in one electron being transferred, the maximum quantum yield for the evolution of oxygen would be 1/4 since four electrons must be transferred per oxygen molecule evolved. At 700 nm four Einsteins of photons have an energy of 684 kJ mol^{-1}. Thus, the efficiency of conversion of excitation energy (at 700 nm) to chemical energy would be 496/684 = 0.73. This compares with the thermodynamic limit (η_P/η_E from Fig. 2) of 0.75 at 700 nm. Thus photosynthesis would have to operate at almost the thermodynamic limit of efficiency to accommodate one photosystem.

Photosynthesis is an energy storage process and so will almost certainly be kinetically controlled. As we have seen, a constant loss must be incurred to build in an activation barrier for the back reaction. Indeed, in the primary photochemical reactions of photosynthesis to produce products with at least millisecond stability, it appears that from the difference in redox potentials of donor and acceptor a loss of ~ 0.8 eV is incurred (23). This would drop the maximum efficiency of conversion of excitation energy at 700 nm to 0.56 (η_C/η_E in Fig. 2). Clearly, the operation of the photosynthesis reaction with one photosystem is virtually impossible.

It is useful to define a threshold wavelength λ_{max}, above which the photosynthesis reaction would be impossible

$$\lambda_{max} = \frac{nN_o \, hc\eta_{chem}}{\Delta G} \qquad (7)$$

where n is the number of photons required per oxygen molecule evolved, N_o is Avogadro's number, h is Planck's constant in kJs, ΔG is the Gibbs free energy change in the reaction and η_{chem} (see Eq. (2)) is the fraction of the photon energy at λ_{max} which is converted to chemical free energy of the products. If we assume that 0.8 eV must be lost from each photon, then

$$\eta_{chem} = \frac{\Delta G}{\Delta G + 0.8 \, neN_o} \qquad (8)$$

If n = 4 (one photosystem) the λ_{max} = 607 nm. However, if n = 8 (two photosystems), then λ_{max} = 872 nm.
This leaves plenty of scope for secondary reactions in
photosynthesis where λ_{max} = 700 nm. Of course, it is now
well established that photosynthesis does employ two photosystems and that experimental minimum values of n are 8 -
10 (19,24-26).

THE CONVERSION OF LIGHT ENERGY TO CHEMICAL
ENERGY IN THE PRIMARY PHOTOCHEMICAL PROCESS

The primary photochemical reactions in photosynthesis
involve electron transfer from a chlorophyll (or bacteriochlorophyll) species to some acceptor. This is the key step
in the conversion of light energy to chemical energy and
yet we still do not know all the details of how it happens.
Nevertheless, enough is known to allow at least a speculative picture of this remarkable process.

In photosynthetic bacteria, it has been possible to
isolate a single protein complex, called the reaction center
protein, which is still able to carry out the primary photochemical reaction (27-30). It seems likely that similar
reaction center proteins exist in the two photosystems of
green plant photosynthesis, but as yet, no clearly defined
photoactive reaction center from either photosystem has
been isolated. As the bacterial reaction center is the
best studied, it will be useful to examine this system in
some detail with the tacit assumption that similar structures exist for green plants.

The bacterial reaction center protein has been thoroughly analyzed (31) and it is possible to place the components as is shown in Fig. 3. In the primary photochemical
process an electron is removed from P870 (a dimer of two
bacteriochlorophyll (BChl) molecules) to form P870$^+$. The
evidence is now quite strong that the ultimate electron
acceptor is a ubiquinone-iron (UQ/Fe) complex (32). Analysis of the back-electron transfer rate at low temperatures
indicates that electron tunnelling is occurring and from a
model calculation a barrier width of $\sim$ 40 Å is obtained
(33,34).

Recent picosecond spectroscopy (35-37) has shown that
there is at least one intermediate in the forward electron-

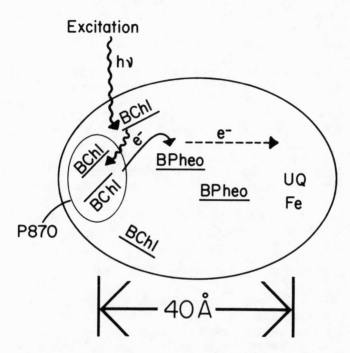

Fig. 3. A model for the bacterial reaction center protein
from Rhodopseudomonas sphaeroides. The structure
and positioning of components in this diagram are
highly speculative.

transfer process which is thought to be the anion radical of
one of the bacteriopheophytin (BPheo) molecules (38). The
lifetime of this intermediate is ∿ 200 ps (36,37). This
lifetime is consistent with the energy loss (∿ 0.4 eV) in
forming the P870$^+$ BPheo$^-$ state. A further ∿ 0.4 eV is lost
by the time the electron reaches the UQ/Fe complex. Its
lifetime is ∿ 100 ms at room temperature, again consistent
(see Table 1) with the overall energy loss of ∿ 0.8 eV.

The BPheo molecule clearly plays a crucial role in the electron-transfer process in two ways: (1) by allowing the electron transfer to occur in small energy drops so as to enhance the forward transfer rate, and (2) by allowing the electron to travel some distance away so as to inhibit the wasteful back-electron transfer process. There may be other shorter-lived intermediates or possibly the porphyrin rings and perhaps some aromatic amino acids may be arranged in the protein to provide an overlapping system of low-lying vacant π molecular orbitals. This would, in effect, create a narrow conduction band and allow electron transfer to occur through the 40 Å of protein to the UQ/Fe acceptor.

This discussion of the reaction center protein is necessarily brief; however, it should illustrate that the efficiency with which solar energy storage is achieved in photosynthesis is primarily due to the structure and functioning of the reaction center protein and perhaps the nature of the surrounding membrane as well.

ARTIFICIAL SYSTEMS FOR PHOTOCHEMICAL SOLAR ENERGY STORAGE

One of the important requirements listed in the second section is that the reactants be cheap and readily available. Naturally, the constituents of the atmosphere and liquid water fill this requirement admirably. Bolton (5) has considered most of the endergonic fuel generation reactions which involve N_2, CO_2 and H_2O as reactants. These are listed in Table 2. Note that the potential stored per electron transferred (column 5) is between 1.06 V and 1.48 V for all of these reactions. We have already noted that the limitations on energy conversion forced photosynthesis to use two photosystems. Thus, we must conclude that if we wish to have an efficient photochemical energy-storage reaction which will utilize the abundant photons in the visible region of the solar spectrum, it is likely that more success will be achieved if attention is given to photochemical processes employing two coupled reactions so that two photons are used for every electron transferred in the ultimate reaction (6).

THE WATER DECOMPOSITION REACTION

In the past few years there has been considerable

Table 2.　Some Endergonic Fuel Generation Reactions Starting with N_2, CO_2 and H_2O.

Reaction	$\Delta H°$ (kJ mol⁻¹)[a]	$\Delta G°$ (kJ mol⁻¹)[a]	n[b]	$\Delta E°$ (V)	λ_{max} (nm)[c] one photosystem	λ_{max} (nm)[c] two photosystems
$H_2O(\ell) \rightarrow H_2(g) + \frac{1}{2}O_2(g)$	286	237	2	1.23	611	877
$CO_2(g) \rightarrow CO(g) + \frac{1}{2}O_2(g)$	283	257	2	1.33	581	845
$CO_2(g) + H_2O(\ell) \rightarrow HCOOH(\ell) + \frac{1}{2}O_2(g)$	270	286	2	1.48	543	804
$CO_2(g) + H_2O(\ell) \rightarrow HCHO(g) + O_2(g)$	563	522	4	1.35	576	840
$CO_2(g) + 2H_2O(\ell) \rightarrow CH_3OH(\ell) + \frac{3}{2}O_2(g)$	727	703	6	1.21	616	881
$CO_2(g) + 2H_2O(\ell) \rightarrow CH_4(g) + 2O_2(g)$	890	818	8	1.06	667	932
$N_2(g) + 3H_2O(\ell) \rightarrow 2NH_3(g) + \frac{3}{2}O_2(g)$	765	678	6	1.17	629	895
$CO_2(g) + H_2O(\ell) \rightarrow \frac{1}{6}C_6H_{12}O_6(s) + O_2(g)$	467	480	4	1.24	607	872

a.　All thermodynamic data have been obtained from "Selected Values of Chemical Thermodynamic Properties" Part 1 National Bureau of Standards Circular 500 U.S. Government Printing Office, Washington, D.C. (1961) except data for $C_6H_{12}O_6(s)$ which were obtained from D.R. Stull, E.F. Westrum and G.C. Sinke, "The Chemical Thermodynamics of Organic Compounds", John Wiley and Sons, New York, 1969, p. 680.

b.　n is the number of electrons which should be transferred in an electrochemical reaction for the reaction as written.

c.　λ_{max} is calculated from Eq. (7) assuming that n_{chem} is given by Eq. (8) and n doubled for two photosystems.

interest in the photochemically sensitized decomposition of
water (the first reaction listed in Table 2). This interest
is quite understandable since hydrogen is an almost ideal
fuel and the starting material is certainly cheap. Almost
all of the processes studied have employed a single photochemical process. However, from Table 2 we see that this
reaction has almost exactly the same energy requirements
per electron transferred as does photosynthesis and indeed
the threshold wavelengths are almost the same. These
threshold wavelengths are based only on the primary electron transfer reaction. When one considers subsequent exergonic secondary reactions, in a reaction which is almost certainly very complex, then the true threshold wavelength for
a single photochemical process to split water is likely to
be less than ~ 425 nm (5). Thus, we are forced to conclude
that research should be directed toward the development of
processes involving two photochemical systems operating in
series.

Such a two-photosystem scheme has been developed by the
author (5,6) and is shown in Fig. 4. The right-hand compartment contains an aqueous solution of a dye D which is a
strong donor (reducing agent) in the excited state D*. D*
transfers an electron to a charge storage catalyst M. The
cycle is repeated again with another photon so that M ultimately accepts two electrons. In the M^{2-} state the catalyst
is capable of reducing two H^+ ions to form a molecule of H_2
without the formation of hydrogen atoms.

The left-hand compartment contains an aqueous solution
of a dye A which is a strong acceptor (oxidizing agent) in
the excited state A*. A* accepts electrons from the charge
storage catalyst N, one at a time, until four electrons
have been removed. In the state N^{4+} the catalyst is capable
of oxidizing two water molecules to produce one molecule of
oxygen.

For every four electrons cycled in the system, four H^+
ions are produced on the left and consumed on the right.
Thus, there must be a membrane connecting the two compart-
ments which is permeable to protons. It must also allow for
transfer of electrons from A^- to D^+ so as to return the dyes
to their photoactive states. In the thylakoid membrane of
photosynthesis, a pool of plastoquinone serves the function
of transferring electrons and protons across the membrane.
Perhaps a similar ploy could be used in an artificial system

using a quinone impregnated membrane.

An essential feature of the scheme in Fig. 4 is the need for charge storage catalysts. Hydrogenase, which catalyzes the reduction of H^+ ions to H_2 is an example of an electron charge storage catalyst; in photosynthesis a manganese containing enzyme is thought to perform the function of storing four positive charges to enable oxygen to be produced directly from water.

As yet, no one has assembled a complete working system according to the scheme in Fig. 4. However, some groups (39,40) have reported success with the reduction half reaction to generate hydrogen, but as yet there has been little progress in developing an oxidation half reaction to generate oxygen.

The limitations and considerations that have been developed here for the photochemical decomposition of water by solar energy apply also to the photoelectrolysis of water which has attracted much attention (4,41,42) since the original experiments by Fujishima and Honda (43). The analog to a single photochemical system is a cell in which only one electrode is illuminated. Indeed, the only cells which

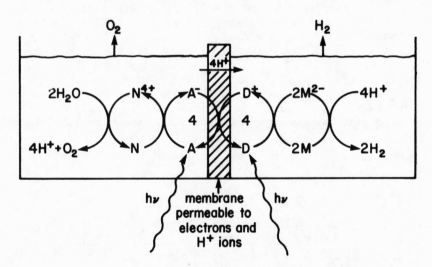

Fig. 4. A scheme for the sensitized photolysis of water using two photochemical reactions coupled in series. See text for an explanation of symbols.

work, unassisted by an external potential, are those which require ultraviolet illumination.

The electrochemical analog to two photochemical systems is a cell in which both electrodes participate in the photochemistry. The requirement for two photosensitive electrodes was pointed out by Manassen et al. (44). Nozik (45) has proposed such a cell and has had some limited success.

Recently, Fong et al. (46) have described a photogalvanic water splitting system using one electrode coated with a chlorophyll a dihydrate with a maximum absorption at $\sim$ 740 nm. This system requires further investigation as a λ_{max} of 740 nm is well beyond the limit for a single photochemical system. It is possible that the photochemical process is biphotonic involving a long-lived excited state intermediate. However, at normal sunlight intensities, that intermediate would have to have a lifetime of > 0.1 s to permit an efficient biphotonic process.

CONCLUSIONS

When considered as a solar energy storage process, photosynthesis performs remarkably well with a gross efficiency of $\sim$ 9% and a net efficiency of $\sim$ 6%. This compares with an estimate of $\sim$ 13% for a reasonable maximum efficiency for any photochemical solar energy storage process. The reaction of photosynthesis has an energy requirement which has forced the green plant to devlop two photosystems operating in series. Most of the artificial reactions have similar energy requirements and thus more success is likely to be achieved with processes involving two photochemical systems.

REFERENCES

1. S. Levine, H. Halter and F. Mannis, Solar Energy 2, 11 (1958).
2. J.G. Calvert, "Photochemical Processes for Utilization of Solar Energy", in Introduction to the Utilization of Solar Energy, ed. by A.M. Zarem and D.D. Erway, McGraw Hill, New York, 1964, pp. 190-210.
3. V. Balzani, L. Moggi, M.F. Manfrin, F. Bolletta and M. Gleria, Science 189, 852 (1975).
4. G. Porter and M.D. Archer, Interdiscip. Sci. Rev. 1, 119 (1976).

5. J.R. Bolton, Science (in press).
6. J.R. Bolton, Solar Energy 20, 181 (1978).
7. R.T. Ross and T.-L. Hsiao, J. Appl. Phys. 48, 4783 (1977).
8. R.T. Ross, J. Chem. Phys. 45, 1 (1966); ibid 46, 4590 (1967).
9. R.G. Mortimer and R.M. Mazo, J. Chem. Phys. 35, 1013 (1961).
10. M. Almgren, Photochem. Photobiol. 27, 603 (1978).
11. L.N.M. Duysens, Brookhaven Symp. Biol. No. 11 (1958), p. 18.
12. R.T. Ross and M. Calvin, Biophys. J. 7, 595 (1967).
13. R.S. Knox, Biophys. J. 9, 1351 (1969).
14. R.S. Knox, in Primary Processes of Photosynthesis, ed. by J. Barber, Elsevier/North Holland Biomedical Press, Amsterdam, 1977, Chap. 2, pp. 55-97.
15. All thermodynamic data have been obtained from "Selected Values of Chemical Thermodynamic Properties", Part 1, National Bureau of Standards Circular 500, U.S. Government Printing Office, Washington D.C., 1961, except for data for $C_6H_{12}O_6$(s) which were obtained from D.R. Stull, E.F. Westrum and G.C. Sinke, "The Chemical Thermodynamics of Organic Compounds", John Wiley and Sons, New York, 1969, p. 680.
16. K.W. Böer, Solar Energy 19, 525 (1977).
17. S.E. Balegh and O. Biddulph, Plant Physiol. 46, 1 (1970).
18. R. Emerson, R. Chalmers and C. Cederstrand, Proc. Natl. Acad. Sci. U.S.A. 43, 133 (1957).
19. D.O. Hall, in Solar Power and Fuels, ed. by J.R. Bolton, Academic Press, N.Y., 1977, pp. 27-52.
20. J.A. Bassham, in "Clean Fuels from Biomass Sewage, Urban Refuse and Agricultural Wastes", Inst. Gas Technology, 3424 South State St., Chicago, Ill., 1976; T.R. Schneider, Energy Conversion 13, 77 (1973).
21. J.R. Bolton, in Photosynthesis 78, Proceedings of the Fourth International Congress on Photosynthesis, D.O. Hall, J. Coombs and T.W. Goodwin, eds., The Biochemical Society, London, 1978, pp. 621-634.
22. I. Zelitch, Photosynthesis, Photorespiration and Plant Productivity, Academic Press, New York, 1971, p. 131.
23. J.R. Bolton, in Topics in Photosynthesis, Vol. 2, "Primary Processes of Photosynthesis" ed. by J. Barber, Elsevier, New York, 1977, pp. 188-201.
24. R. Emerson and C.M. Lewis, Am. J. Bot. 30, 126 (1943).
25. R. Govindjee, E. Rabinowitch and Govindjee, Biochim. Biophys. Acta 162, 539 (1968).

26. Govindjee and J.T. Warden, J. Am. Chem. Soc. 99, 8088 (1977).
27. D.W. Reed and R.K. Clayton, Biochem. Biophys. Res. Commun. 30, 471 (1968).
28. R.K. Clayton and R.T. Wang, Methods Enzymol. 23, 696 (1971).
29. G. Feher, Photochem. Photobiol. 14, 373 (1971).
30. L. Dutton, in Light Induced Charge Separation in Biology and Chemistry, eds. H. Gerischer and J.J. Katz, Berlin: Dahlem Konferenzen, 1978, in press.
31. K. Sauer, in Bioenergetics of Photosynthesis, ed. by Govindjee, Academic Press, New York, 1975, pp. 116-181.
32. J.R. Bolton, in The Photosynthetic Bacteria, R.K. Clayton and W.R. Sistrom, eds. Plenum Press, New York (in press).
33. J.D. McElroy, D.C. Mauzerall and G. Feher, Biochim. Biophys. Acta 333, 261 (1974).
34. B.J. Hales, Biophys. J. 16, 471 (1976).
35. K.J. Kaufman, P.L. Dutton, T.L. Netzel, J.S. Leigh and P.M. Rentzepis, Science 188, 1301 (1975).
36. M.G. Rockley, W.W. Windsor, R.J. Cogdell and W.W. Parson, Proc. Natl. Acad. Sci. U.S.A. 72, 2251 (1975).
37. T.L. Netzel, P.M. Rentzepis, D.M. Tiede, R.C. Prince and P.L. Dutton, Biochim. Biophys. Acta 460, 467 (1977).
38. J. Fajer, D.C. Brune, M.S. Davis, A. Forman and L.D. Spaulding, Proc. Natl. Acad. Sci. U.S.A. 72, 4956 (1975).
39. M.S. Chan, C.A. Evans, S. Markiewicz, R. Sparks and J.R. Bolton, in "Book of Abstracts", First International Conference on the Photochemical Conversion and Storage of Solar Energy, London, Canada, August 1976, p. E7.
40. J.M. Lehn and J.P. Sauvage, Nouveau J. Chim. 1, 449 (1977).
41. M.D. Archer, J. Appl. Electrochem 5, 17 (1975).
42. H. Gerischer, in Solar Power and Fuels, J.R. Bolton, ed. Academic Press, New York, 1977, pp. 77-117.
43. A. Fujishima and K. Honda, Bull. Chem. Soc. Jap. 44, 1148 (1971); Nature 238, 37 (1972).
44. J. Manassen, D. Cahen, G. Hodes and A. Sofer, Nature 263, 97 (1976).
45. A.J. Nozik, Appl. Phys. Lett. 29, 150 (1976).
46. F.K. Fong, J.S. Polles, L. Galloway and D.R. Fruge, J. Amer. Chem. Soc. 99, 5802 (1977).

THE CHLOROPHYLL A WATER SPLITTING LIGHT REACTION

L. Galloway, D. R. Fruge, L. M. Fetterman,
and F. K. Fong

Chemistry Department, Purdue University
West Lafayette, Indiana 47907

I. INTRODUCTION

The vast difference in temperature between the sun
and the earth is equalized through the medium of light,
which is converted by green plants into the driving force
for photosynthesis, providing sustenance for life in the
animal world. According to currently accepted interpretation, photosynthesis is given by the reaction

$$CO_2 + 2H_2O \xrightarrow[\text{Chl a}]{h\nu} CH_2O + H_2O + O_2 \tag{1}$$

in which the cleavage of the water molecule by the chlorophyll is accompanied by the reduction of the carbon dioxide.
In a recent survey, Levanon and Norris suggested[1] that
understanding of reaction 1 may ultimately lead to the
elimination of carbon dioxide resulting in 'simpler' processes such as the water decomposition reaction

$$H_2O \xrightarrow[\text{Chl a}]{h\nu} H_2 + \frac{1}{2}O_2 \tag{2}$$

This suggestion summarizes the goal of many laboratories to
gain insight into the basic principles and mechanisms of
photosynthesis.[1] One purpose of modeling complex in vivo
processes is to provide an overview that correlates detailed
developments in the field. For a chemist a second, perhaps
more important, reason for the interest in a molecular view
of photosynthesis is to delineate features of Chl a

photochemistry that render a facsimile of the in vivo water
splitting reaction, in the hope that the structural and
chemical relationships of model compounds other than Chl a
capable of photocatalyzing reaction 2 may be unraveled.

Few scientific endeavors have a background as colorful
as the search for the origins of the Chl a water splitting
reaction, the beginnings of which coincided with the emergence of modern chemistry through the discovery by Priestley
of oxygen evolution from plants in 1773. In 1796 Ingenhousz[2]
concluded, based on the observations of CO_2 assimulation
and O_2 liberation in photosynthesis, that the green parts
of plants absorb from "carbonic acid in the sunshine, the
carbon, throwing out at that time the oxygen alone, and
keeping the carbon itself as nourishment." This conclusion
influenced photosynthesis research for nearly two centuries.
As recently as 1956 Warburg believed that a bound carbon
dioxide molecule transferred an oxygen atom to the carbonyl
oxygen of the chlorophyll to form a peroxide and then
molecular oxygen.[3] That water is the source of molecular
oxygen was suggested by several authors around 1930.[4-7]
Indirect experimental support for this suggestion was provided by workers[8,9] who showed that no correlation could
be established between the uptake of CO_2 and the appearance
of O_2. Direct evidence was provided by Ruben et al[10] in
1941 by the mass spectrometric determination of O_2 evolution from H_2O. Much of the work on the in vivo oxygen
evolution process was focused on the requirement [11-13] of
Mn and Cl^-, bedimming the question regarding the intrinsic
ability of the chlorophyll to catalyze the water splitting
reaction. Remarkably little[1,14] was known of the molecular
mechanisms for reaction 1. The lack of understanding was
deemphasized by the once widely-held belief in the two-
light-reaction hypothesis,[15-18] "the central dogma of photosynthesis,"[16] in which molecular events were denoted by a
"zig-zag" display of arrows connecting two Chl a light
reactions via a dark bridge of charge carriers.[15]

We have based our understanding of the photosynthetic
primary light reactions on the photochemical properties of
hydrated Chl a aggregates.[19] In earlier work we attributed
Chl a photogalvanic effects to water splitting reactions
that result from illumination of the chlorophyll a dihydrate
aggregate $(Chl\ a \cdot 2H_2O)_n$.[19] The photooxidation of
$(Chl\ a \cdot 2H_2O)_{n \geq 2}$ by water was subsequently observed in ESR

experiments.[20] These observations were corroborated by the finding that the midpoint reduction potential of $(Chl\ a\cdot2H_2O)_n$ is 0.92 V, exceeding that, 0.81 V, required for the water oxidation (O_2/H_2O) half-reaction at pH 7.[21] The demonstration of water splitting by $(Chl\ a\cdot2H_2O)_{n\geq2}$ is of current interest. It has been suggested[19] that the water splitting reaction in vivo results directly from the primary light reaction of $(Chl\ a\cdot2H_2O)_2$. The water splitting question is topical in view of the current search for a direct process for harvesting solar energy to produce gaseous hydrogen for fuel. Considerable attention has been focused on n-type semiconducting photoanodes such as TiO_2[22] and $SrTiO_3$.[23,24] However, these materials operate in the near-ultraviolet wavelength region where the solar radiant energy density is low. In contrast, the action spectrum of the photoreactivity of $(Chl\ a\cdot2H_2O)_n$ with water spans the visible and far-red wavelength regions.[19]

In this paper we review the finding that water-saturated chlorophyll photocatalyzes the decomposition of water according to reaction 2.[25] We describe the finding that the absence of the Mg atom in pheophytin a, in which the Mg atom in Chl a has been replaced by two protons, obliterates the capability of Chl a to catalyze water photolysis. In contrast with the water splitting reaction of $(Chl\ a\cdot2H_2O)_n$ no significant amounts of molecular hydrogen were detected in the photoreaction of polycrystalline Pheo a with water, which indicates that the pheophytin is reduced by the water oxidation process. The Chl a water splitting reaction is assisted by the presence of Pt, which presumably catalyzes the H_2 and O_2 evolution processes through the lowering of their overvoltages. The properties of $Pt-H_2O$ reactions are therefore of interest in the delineation of properties exclusively attributable to the water photolysis reaction of $(Chl\ a\cdot2H_2O)_n$. We report the observation of hydrogen evolution when Pt and water, in the absence of the chlorophyll, were heated at relatively low temperatures in the range 130 - 210°C. In sharp contrast with the water photolysis results we obtained on illuminating platinized Chl a, no significant amounts of oxygen were detected in the thermochemical $Pt-H_2O$ reaction. No molecular hydrogen and oxygen were detected when Pt in water was illuminated in the visible wavelength region.

II. THE Pt/Chl a WATER SPLITTING LIGHT REACTION

The splitting of water by the chlorophyll in the photo-galvanic cell $Pt/H_2O(0.1M\ KCl)/(Chl\ a\cdot 2H_2O)_n/Pt$ may be given in terms of the half-cell reactions:

Chl a-free anode: $2H_2O \longrightarrow 4H^+ + O_2 + 4e^-$ (3)

Chl a-Pt photocathode: $2H_2O + 2e \longrightarrow H_2 + 2OH^-$ (4)

A Pt foil was platinized by passing a 30-mA current for 10 min through a $7 \times 10^{-2}M$ chloroplatinic acid solution containing $6 \times 10^{-4}M$ lead acetate. Pure $(Chl\ a\cdot 2H_2O)_n$, a polycrystalline aggregate, was prepared in the manner described by Brace et al.[25]   A layer of the polycrystalline chlorophyll, containing 1.5×10^{17} Chl a molecules, was deposited on the platinized electrode surface in the usual manner.[19]   The Chl a plated electrode was then platinized again in the same chloroplatinic acid solution by passing the 30-mA current for 15 s. The resulting sample electrode was baked in distilled water at 60°C under atmospheric pressure for several hours in order to be rid of any adventitious gaseous occlusion during the platinization procedure.

The action spectrum of the photogalvanic response of the platinized Chl a electrode at pH 7, measured in a cell[19] employing as the second half-cell a platinized electrode not covered with Chl a, shows a 740-nm maximum indicative of the fact that $(Chl\ a\cdot 2H_2O)_n$[25] is primarily responsible for the observed photogalvanic effects. Prolonged illumination using the entire output from a 1000-W tungsten-halogen lamp led to the observation of gaseous evolution from the Pt-Chl a electrode.

A mass spectrometric study of the simultaneous generation of H_2 and O_2 from $63.6:36.4\ H_2^{18}O-H_2^{16}O$ in a pH 11 sodium hydroxide-phosphate buffer, in which no KCl was added, is reproduced in Figure 1. The relative line intensities at masses $2(H_2^+)$, $34(^{16}O^{18}O^+)$, and $36(^{18}O_2^+)$, absent in the blank are in excellent agreement between the photolytic and electrolytic runs. The experiment was performed under a positive pressure of Ar in order to generate the $^{36}Ar^+$ (0.337% natural abundance) line, which serves as a useful internal standard in the mass 36 region. The intensity ratio of the lines at masses $28(^{14}N_2^+)$ and $32(^{16}O_2^+)$ in the blank agrees with that of an air sample. Based on this ratio we correct

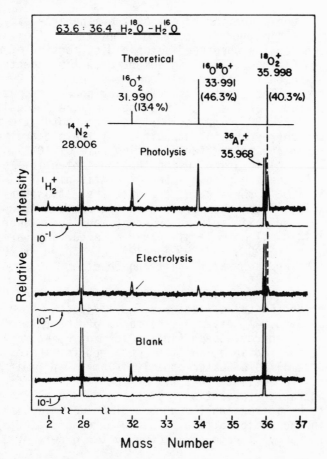

Figure 1. Mass spectrometric study demonstrating the simultaneous generation of molecular hydrogen and oxygen in water photolysis. The light reaction was carried out for 20 min. The control experiment corresponds to the electrolysis of 5×10^{-7} mol of H_2O. The arrows at mass 32 in the photolytic spectrum indicate the $^{16}O_2^+$ signals due to water photolysis after correction for air contamination. After a similar correction in the electrolytic spectrum the isotopic distributions of O_2 obtained in photolysis and electrolysis are seen to be in agreement with the theoretical pattern. The scale about the indicated mass numbers has been expanded by a factor of 3 to facilitate the representation of the line intensities and spectral resolution. The mass spectra in this figure were obtained with the ion source focused at the $^{40}Ar^+$ line.

for the air contamination effect on the line intensity at mass 32 ($^{16}O_2^+$) of the photolysis sample spectrum. The resulting pattern for the observed isotopic distribution of molecular oxygen is in excellent agreement with the expected pattern. The photolytic sample was measured after 20 min of illumination. The electrolysis was carried out by passing a 20-mA current for 5 s. Calibrating the line intensities at masses 34($^{16}O^{18}O^+$) and 36($^{18}O_2^+$) of the photolytic sample against the corresponding intensities of the electrolytic sample, we arrive at a water splitting rate of 10^{-5} mol/h. From the same calibration, using the line intensity at mass 28 (N_2^+) and the known mass intensity ratio of 28(N_2^+):32($^{16}O_2^+$) for air, we estimate that atmospheric contamination in the photolytic sample amounts to 3×10^{-7} mol of $^{16}O_2$, approximately an order of magnitude lower than the 2.5×10^{-6} mol of molecular oxygen, occurring variously in the forms of $^{16}O_2$, $^{16}O^{18}O$, and $^{18}O_2$, produced in 20 min of the light-induced water splitting reaction.

The incident irradiance of photons ("white light") focused on the Pt-Chl a electrode was 1.74W cm^{-2}. An energy efficiency of 0.04%, given by the ratio of the rate of chemical free energy stored (in the form of H_2 and O_2) and the incident irradiance, was thus obtained for the observed rate ($\sim 10^{-5}$ mol/h) of water photolysis employing 10^{17} Chl a molecules. We estimate that <1% of the light was effectively engaged in the photoreaction. It seems reasonable to suppose that, by increasing the number of Chl a molecules engaged in the photoreaction, it should be possible to enhance the energy storage efficiency by two orders of magnitude.

In view of the observed photoreactivity of (Chl a$\cdot 2H_2O)_n$ with water in the absence of Pt,[20] we believe that the experimental effects described in the present work are attributable, in part at least, to the intrinsic photochemical properties of the chlorophyll. In water splitting work on inorganic semiconductors[22-24] the indispensable role[22] of the platinum black was interpreted in terms of a synergistic mechanism that relates properties of the semiconductor with those of the platinum. However, we have evidence that the platinum itself may be partially responsible for our observations in the UV wavelength region. In the following section we focus our attention on the effects of visible light and heat on the chemical reactivity of Pt with water in the absence of Chl a.

III. THE EFFECTS OF RED AND INFRARED RADIATION ON THE WATER
      PHOTOLYSIS REACTION OF $Pt/(Chl\ a \cdot 2H_2O)_n$

        In this section we examine the relative contributions
of the effects of visible and infrared radiation to the
water splitting results shown in Figure 1. It is known that
Pt is oxidized to yield hydroxides and oxides in the presence of air and water.[26-32] In earlier studies workers investigated the corrosion of Pt in the presence of O_2 and H_2O
at ordinary temperatures, and reported that no H_2 was
evolved.[29] However, it was unclear whether hydrogen would
be evolved from the thermochemical reaction of Pt with H_2O,
in the absence of O_2, at temperatures comparatively lower
than those required for the thermal cleavage of water.[32-34]

        Gas chromatographic analyses of three experiments, in
which the thermal reaction between Pt and deoxygenated water
was carried out for 30, 45, and 60 min at 206°C are compared in Figure 2 with a corresponding analysis of H_2 produced by electrolysis. From this comparison, we estimate
that approximately $4 \times 10^{-8} h^{-1}$ moles of H_2 was liberated
by the $Pt-H_2O$ reaction. No H_2 was detected after the Pt
black-water sample was irradiated by the Argon ion laser for
1 h, which indicates that Pt is photochemically inactive in
the visible wavelength region. The temperature of the
$Pt-H_2O$ sample was approximately 35°C under continuous illumination by the Ar ion laser radiation. Pyrolytic analyses
(using the Hersch method) of experiments in which Pt was
illuminated with the tungsten halogen lamp or heated at 130°C
for 20 min in pure water and in pH 9 and 11 $Na_2B_4O_7/NaOH$
buffer solutions are shown in Figure 3. The results show
that the broad-band visible-IR light source, capable of
heating the $Pt-H_2O$ sample to 150°C, has effectively produced
the same outcome as the heat. From a comparison of these
results to corresponding results obtained for H_2 and O_2
generated by electrolysis, it became evident that little O_2
was produced in the $Pt-H_2O$ reaction. From a calibration
of the hydrogen evolution in the sample reaction against
electrolysis we observed that H_2 was produced at a rate of
2×10^{-6} mole h^{-1}, about two orders of magnitude greater than
the production rate measured in the GC study under comparable
experimental conditions.[36]

        The above observations on the $Pt-H_2O$ reaction were
corroborated by a quantitative mass spectrometric analysis
of this reaction using a mixture of

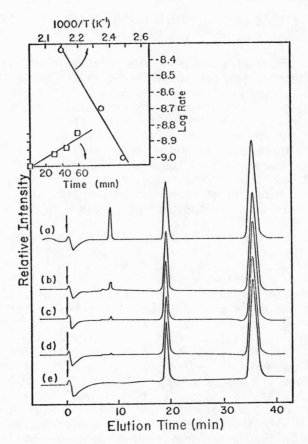

Figure 2. Gas chromatographic determination of hydrogen evolution in the Pt-H_2O reaction: (a) generation of 8×10^{-8} mole H_2 by passing a 4-mA current for 4 s; (b,c,d) H_2 generated by heating doubly-distilled water at 205°C in the presence of a platinized Pt foil ($0.8 \times 1.2 \times 0.02$ cm^3) for 60, 45, and 30 min, respectively; (e) gas chromatogram of sample after 1 h irradiation with Ar ion laser. The arrows indicate the time of sample introduction into the gas chromatograph. The position of the hydrogen peak in a provides definitive evidence for the presence of H_2 from the Pt-H_2O reaction. The electrolysis experiment also permits sample calibration. Air leakage was responsible for the oxygen and nitrogen peaks at 19 and 35 min, respectively. The temperature and time dependences of the hydrogen evolution rate are shown in the inset.

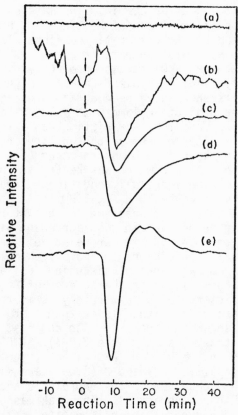

Figure 3. Pyrolytic analyses of hydrogen generated from the Pt-H$_2$O reaction: (a) doubly-distilled water in the absence of Pt; (b) doubly-distilled water heated at 130°C for 30 min in the presence of Pt; (c) pH 11 borax/NaOH buffer illuminated with a 1000-W tungsten-halogen lamp for 30 min in the presence of Pt; (d) pH 9.5 borax/NaOH buffer illuminated as in c in the presence of Pt; (e) H$_2$ and O$_2$ generated by passing a 10-mA current for 35 s. The arrows indicate the time of sample introduction into the Hersch analyzer. The sample after the light or heat treatments were cooled for 10 min prior to analysis. The narrow hydrogen band in e is due to the onset of the oxygen peak which is registered in a direction opposite to the H$_2$ signal. The lack of any detectable effects due to O$_2$ in samples b-d indicates that only H$_2$ was evolved from the Pt-H$_2$O reaction.

$D_2{}^{16}O$ and $H_2{}^{18}O$ (H:D ratio = 1:1; $^{16}O:^{18}O$ ratio = 68.7:31.3)
after heating at 130°C for 30 min. The mass spectra of
sample reactions after 30 min of heating at 130°C are given
in Figures 4a and b. On calibration against electrolysis
we obtained the H_2 evolution rate of 2.7 x 10^{-6} mol h^{-1}.[36]
The results of the Pt-H_2O reaction are compared with those
of water photolysis by platinized Chl a in Figure 4, in
which the results shown have been obtained using a 1000-W
tungsten-halogen projection lamp with a 10"-water filter to
eliminate heating from the IR radiation and a Corning
CS 2-63 cut-off filter to remove photons at wavelengths
λ< 580 nm. Also shown is a blank spectrum (Figure 4d) obtained after the isotopic water was heated in the absence
of Pt under otherwise identical conditions. The relative
line intensities at masses 2($H_2{}^+$), 3(HD^+), 4($D_2{}^+$),
34($^{16}O^{18}O^+$), and 36($^{18}O_2{}^+$) observed in the water photolysis
experiments (Figure 4c) are in good agreement with the
electrolysis experiments. Similar relative line intensities
for the various isotopic hydrogen and oxygen molecular ions
were observed from water photolysis using (Chl a·$2H_2O$)
films on microscope slides in the absence of Pt, although
the quantum efficiency of water photolysis in this case is
nearly two orders of magnitude lower than that corresponding
to the results shown in Figure 4c. In contrast, for comparable line intensities at masses 2($H_2{}^+$), 3(HD^+), and
4($D_2{}^+$) obtained in the Pt-H_2O experiments, the intensity of
lines at masses 34($^{16}O^{18}O^+$) and 36($^{18}O_2{}^+$) are significantly
weaker than those observed in the photolysis and electrolysis
runs (compare Figures 4b and c). The weak mass 34($^{16}O^{18}O^+$)
line observed in Figure 4a and b is greater than that expected from the 0.2% natural abundance of ^{18}O, however, and
may be a result of isotopic redistribution of ^{18}O in the
water and O_2 present as contamination from the atmosphere.
The possibility that this line may have resulted from water
decomposition appears rather unlikely from thermodynamic
considerations.[37,38]

We rule out H_2O_2 as a product in the Pt-H_2O reaction
by mass spectrometric determinations and by the colorimetric
triiodide method.[39] Among possible explanations for the
experimental effects described above is the thermodynamically
characterized reaction[40]

$$2H_2O + Pt \xrightarrow{\Delta} Pt(OH)_2 + H_2 \quad \Delta G^\circ = 49.7 \text{ kcal/mole} \quad (5)$$

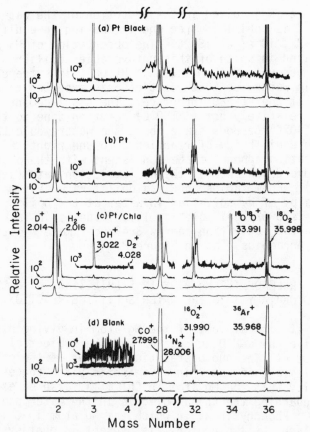

Figure 4. Comparison of the mass spectrometric determination of products from the Pt-H$_2$O reaction with that of products from the photoelectrolysis of water by platinized Chl a: (a) doubly distilled water heated for 30 min in the presence of Pt black; (b) pH 11 borax/NaOH buffer heated for 30 min in the presence of shiny Pt; (c) platinized Chl a illuminated for 30 min using red light from a 1000-W tungsten-halogen lamp; (d) doubly-distilled water heated for 30 min in the absence of Pt. The ratio of the mass 34(^{18}O^{16}O$^+$) line to the mass 36(^{18}O$_2^+$) line in c is 37.0, in agreement with the ratio obtained by electrolysis. The mass 2(D$^+$) line is attributable, in part at least, to water fragmentation.[42] The mass 1 (H$^+$) line was not observed due to instrumental limitations. The spectra in the low- and high-mass regions were measured with the spectrometer focused at masses 2 and 40, respectively.

in which Pt is oxidized in the reduction of the water.
Reaction 2 is uphill in free energy, having an equilibrium
constant K ~ 10^{-27} at 150°C. The observation of H_2 may be
rationalized in spite of this unfavorable equilibrium constant on account of the fact that $Pt(OH)_2$ is extremely insoluble in water, so that reaction 5 is expected to occur
entirely at the Pt-H_2O interface. The displacement of H_2
from the relatively hot ($\gtrsim$ 150°C) reaction zone to the
cooler (~40°C) gaseous space above the water would thus
shift the equilibrium of reaction 2 to the right. This
interpretation appears to be consistent with the observed
temperature dependence of the H_2 evolution rate (see inset,
Figure 2). From tables of H_2 solubility[41] we estimate that
the enthalpy of H_2 solvation in water at the boiling point
of water is 3.62 kcal mole^{-1}. This value is in fair agreement with the activation energy 3.48 kcal mole^{-1} obtained
from the Arrhenius plot in Figure 2.

IV.   ROLE OF THE Mg ATOM IN STABILIZING Chl a AS
      PHOTOCATALYST IN THE WATER SPLITTING REACTION.

The structural role of the Mg atom in giving rise to
photoactive Chl a-H_2O aggregates was the subject of a recent
discussion.[25] The photocatalytic function of the chlorophyll
in the cleavage of water in vivo and in vitro suggests the
possibility that the Mg atom in Chl a may also be essential
for safeguarding the Chl a molecule in the water photolysis
reaction. Pheophytin a, in which the Mg atom in Chl a has
been replaced by two protons, is present in photosynthetic
organisms.[43,44] Pheophytin a was prepared from purified
Chl a extracted from spinach in the manner given by Brace
et al.[25] An ethereal solution of Chl a was treated with
1 ml of concentrated hydrochloric acid for about 3 minutes
during which the chlorophyll was converted to Pheo a. The
preparation was then washed with distilled water to remove
the acid. After evaporation to dryness the pheophytin was
suspended in n-pentane and cooled to 0°C. Pheophytin a
films were deposited on flame-cleaned (shiny) Pt foils by
solvent evaporation. Polycrystalline Pheo a films obtained
from water-saturated n-pentane suspensions are structurally
similar to (Chl a·$2H_2O$)$_n$.[47] Pheo a deposits prepared from
ethereal solutions, on the other hand, are of an amorphous
nature (see Figure 5 for absorption spectra of various preparations of Pheo a). The sample films of Pheo a and
(Chl a·$2H_2O$)$_n$ employed in this work contained about 10^{16}

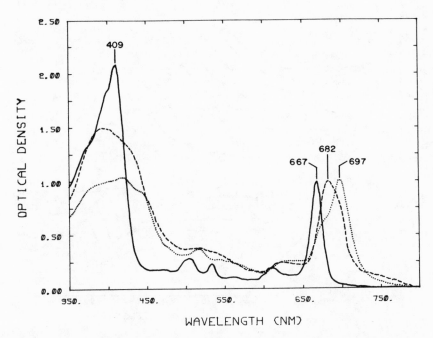

Figure 5. Normalized absorption spectra of three different forms of Pheo a:—5 x 10^{-5} M solution of Pheo a in ether; - - - - amorphous Pheo a film on microscope slide prepared from solvent evaporation of an etheral solution;. . . . polycrystalline Pheo a film on microscope slide prepared from solvent evaporation of an n-pentane suspension.

molecules cm^{-2}.

On photoactivating the Pt/Pheo a electrode in the cell, Pt/Pheo a/H_2O(0.1M KCl)/Pt, a photoanodic current was obtained. Unlike the $(Chl\ a \cdot 2H_2O)_n$-H_2O light reaction, in which photoexcitation of $(Chl\ a \cdot 2H_2O)_n$ results, as a primary step, in the reduction of H_2O and oxidation of the chlorophyll,[48] Pheo a apparently acts as the primary electron acceptor in the Pheo a-H_2O light reaction

$$Pheo\ a + \frac{1}{2}\ H_2O \xrightarrow{h\nu} Pheo\ a^- + H^+ + \frac{1}{4}\ O_2 \qquad (6)$$

The observed photoanodic current shown in Figure 1 presumably has resulted from a partial restoration of Pheo a^- to

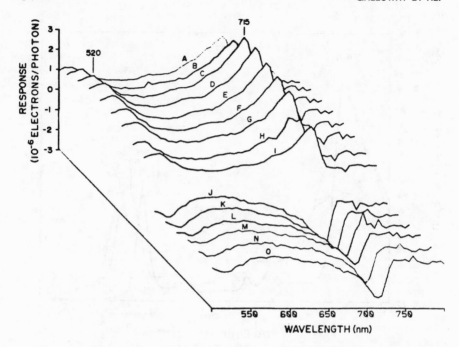

Figure 6.   Photoanodic response of polycrystalline Pheo a
on shiny Pt and the effect of O_2: A - I, 0.3, 2.4, 5.3, 8.2,
11.5, 13.7, 16.2, 19.1, and 23.5 h after the photoelectrochemical cell was rigorously deoxygenated. The increase in
intensity of the 520-nm band relative to the 715-nm band (see
also Figure 3) suggests possible photoreductive degradation
of the pheophytin. J - 0, 0.4, 2.8, 6.1, 9.3, 12.6, and
15.2h after the cell in I was saturated with O_2. The photocathodic current indicates the photooxidation of the pheophytin by the oxygen.

Pheo a by the water reduction process occurring in the Pheo
a-free, dark anodic cell:

$$H_2O + Pheo\ a^- \longrightarrow OH^- + Pheo\ a + \frac{1}{2} H_2 \qquad (7)$$

in which the transfer of electrons from Pheo a^- to H_2O
occurs through the external circuit of the photogalvanic
cell.   The action spectra of the light reaction between polycrystalline Pheo a and H_2O are shown in Figure 6, in which
the sequence of spectra A - I in the 500 - 800 nm wavelength

region span a 24-h period of observation. The maximum quantum efficiency at 715 nm is about 0.0001%, nearly two orders of magnitude lower than the corresponding maximum efficiency observed at 740 nm for the $(Chl\ a \cdot 2H_2O)_n$ $-H_2O$ light reaction under identical conditions.[49] The photoreduction of polycrystalline Pheo a by H_2O is inhibited by the presence of O_2. On saturation of the Pt/Pheo a half cell with O_2 we observed a dramatic sign reversal in the photoresponse (see Figure 6, J - 0), resulting in a photocathodic current that suggests the primary light reaction

$$Pheo\ a + O_2 \xrightarrow{\ h\nu\ } Pheo\ a^+ + O_2^- \qquad\qquad (8)$$

followed by the electrochemical cell reaction

$$\frac{1}{2} H_2O + Pheo\ a^+ \longrightarrow H^+ + Pheo\ a + \frac{1}{4} O_2 \qquad\qquad (9)$$

in which the water oxidation process occurs in the dark H_2O/Pt half cell. The above-described results indicate that Pheo a is capable of water photolysis at a quantum efficiency about two orders of magnitude lower than that of $(Chl\ a \cdot 2H_2O)_n$.

In order to ascertain details of the Pheo a-H_2O light reaction, we subjected the gaseous products of illuminated platinized Pheo a samples to mass spectrometric analysis. The samples, immersed in isotopically enriched water (H:D = 1.0; ^{16}O:^{18}O = 1.82) contained in a sample cell that was connected on-line to a Consolidated Electrodynamics Corporation 21-110-B mass spectrometer, were illuminated for 30 minutes with a 1000-W tungsten-halogen lamp equipped with a 10"-water filter to eliminate heating from IR radiation and a Corning CS 2-63 cut-off filter to remove photons at wavelengths $\lambda < 580$ nm. The $^{36}Ar^+$ line, originating from purging of the sample cell with Ar gas for 30 min under controlled conditions prior to the mass spectrometric determination, was used as an internal reference. The results of the Pheo a-H_2O light reaction are compared in Figure 7 with those obtained from $(Chl\ a \cdot 2H_2O)_n$ water photolysis in the presence or absence of the pheophytin. From a calibration of the H_2 and O_2 evolution rates observed in sample light reactions against conventional electrolysis, we obtained the rate 3×10^{-6} mole h^{-1} for water photolysis by by Pt/$(Chl\ a \cdot 2H_2O)_n$ in Figure 7A, in agreement with the determination in Figure 4b, in which only visible light in the red wavelength region was employed. Similarly we mea-

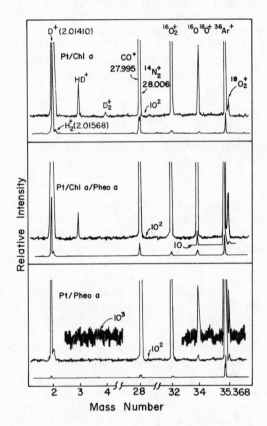

Figure 7. Mass spectrometric analyses of photochemical
decomposition of water by platinized polycrystalline Chl a
and Pheo a: A. Photolytic products of platinized
$(Chl\ a \cdot 2H_2O)_n$. The observed isotopic mixtures of molecular
hydrogen and oxygen ions are indistinguishable from those
observed in conventional electrolysis. B. Photolytic products of platinized polycrystalline Pheo a layered over a
film of $(Chl\ a \cdot 2H_2O)_n$. C. Photolytic products of platinized
polycrystalline Pheo a.

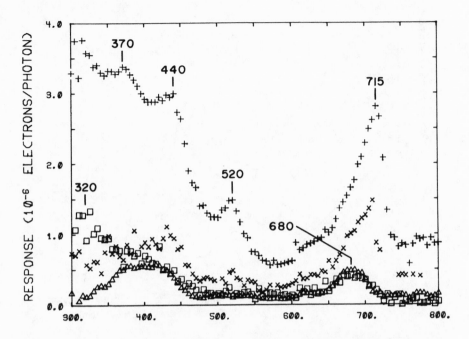

Figure 8. Spectral distribution of the photoanodic response
on shiny Pt: ××××, freshly-prepared film of polycrystalline
Pheo a; ++++. polycrystalline Pheo a after 5.3h of illumination; △△△△, freshly-prepared film of amorphous Pheo a;
◻◻◻◻, amorphous Pheo a after 4.7 h of illumination.

sured the O_2 evolution rate of 2.1 x 10^{-7} mole h^{-1} in the
Pt/Pheo a-water light reaction (Figure 7C), in which no
significant amounts of molecular hydrogen ions, H_2^+, HD^+ and
D_2^+, were detectable. The results of water photolysis using
a platinized sample of polycrystalline Pheo a layered over
an underlying (Chl a·$2H_2O)_n$ film, are shown in Figure 7B,
from which we observe a two-fold enhancement in O_2 evolution
and a corresponding diminution in hydrogen evolution, in
contrast with the water photolysis results obtained from
Pt/(Chl a·$2H_2O)_n$ in the absence of Pheo a (compare Figure
7A and B). The results thus described suggest that Pheo a,
in the presence of the primary products of water photolysis,
becomes reduced to H_2Pheo a,[50] i.e.,

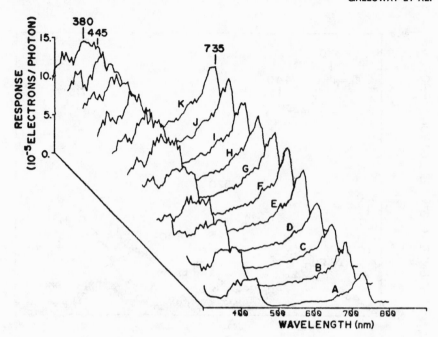

Figure 9.   Photocathodic response of polycrystalline-Pheo a-covered $(Chl\ a\cdot2H_2O)_n$ on shiny Pt: A - K, 0.2, 2.5, 5.3, 7.4, 10.4, 12.5, 14.6, 16.7, 19.3, 22.1, and 24 h after cell assembly.

$$H_2O + Pheo\ a \xrightarrow[\ (Chl\ a\cdot2H_2O)_n\ ]{h\nu} H_2Pheo\ a + \frac{1}{2}\ O_2 \qquad (10)$$

The conclusions summarized in 10 are further corroborated by the comparison of the time dependence of the photo-anodic action spectra of amorphous and polycrystalline Pheo a on shiny Pt foils (Figure 8) with that of the photocathodic response of a shiny $Pt/(Chl\ a\cdot2H_2O)_n/$ polycrystalline-Pheo a electrode (Figure 9). The spectral distribution of the poly-crystalline Pheo a-H_2O photoreaction (Figure 8) is seen to undergo substantive changes at wavelengths $\lambda < 520$ nm after the Pt/Pheo a was exposed to light for 5.3 h. The photo-cathodic response of $Pt/(Chl\ a\cdot2H_2O)_n/$ Pheo a (Figure 9) is indistinguishable in magnitude and spectral distribution from that of a comparable electrode consisting of pure $(Chl\ a\cdot2H_2O)_n$ on shiny Pt.[49] The photoresponse of Pheo a, being two orders of magnitude lower than that of

(Chl a·2H$_2$O) , apparently has no order-of-magnitude effect
on the photogalvanic action spectrum of the chlorophyll.
It is of interest to note that the photocatalytic role of
(Chl a·2H$_2$O)$_n$, in contrast with the reductive degradation
of Pheo a, is underscored by the constancy of the spectral
distributions (in Figure 9) over a 24-h period of sample
illumination.

We have thus delineated the role of the Mg atom in
safeguarding the photocatalytic function of Chl a in reaction 2. From an x-ray photoelectron spectroscopic determination it was ascertained[25] that, unlike the four approximately equivalent N atoms in Chl a, the two protonated N
atoms in Pheo a are distinctly different from the remaining
two N atoms whose exposed lone-pair electrons are evidently
vulnerable to the H atom abstraction process in 10.   This
interpretation is consistent with previous conclusions
drawn from investigations into the photoreduction of Pheo a
to H$_2$Pheo a in protic solvents.[50]

V. TWO-PHOTON MECHANISM AND QUANTUM EFFICIENCY

The illumination of the Chl a/Pt electrode resulted in
a flow of electrons from the Chl a free electrode (anode)
to the Chl a/Pt electrode (cathode). The initial (t = 0)
readings of the photogalvanic response were recorded. The
photogalvanic response indicates a linear flux dependence
under weak light conditions and a sublinear dependence at
higher light intensities. To establish the power dependence
of the (Chl a·2H$_2$O)$_n$ light reaction in the strong flux
limit, we employed the total output from a 1000-W tungsten-
halogen lamp. The incident flux was varied by placing a
series of five neutral density filters in the path of the
exciting light at several lamp power supply voltage settings.
The flux at each voltage setting was calibrated by a Spectra
Physics power meter. The resulting experimental points,
represented by log i_p vs. log I_0, where i_p and I_0 are the
photogalvanic current and incident flux, respectively, and
fitted to the empirical equation

$$i_p = [(1.58 \times 10^{-29})I_0/(1 + 10^{15}I_0^{-1})]^{\frac{1}{2}} \qquad (11)$$

are shown in Figure 10. It is apparent that the
(Chl a·2H$_2$O)-water photoreaction rate varies as the square
root of I_0 in the strong light limit.

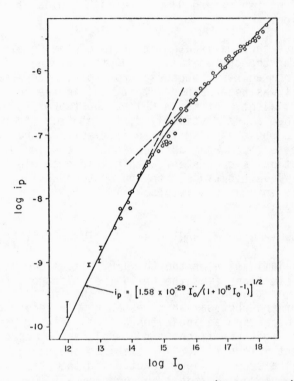

Figure 10. The flux dependence of the $(Chl\ a \cdot 2H_2O)_n - H_2O$ photogalvanic response. The low-flux data points are given in error bars (99% confidence intervals) corresponding to $\sigma = \pm 2.5$ and noise excursions about the mean. The doubling of the exponential flux dependence from semilinearity in the strong light limit to linearity in the weak light limit is attributed to the two-photon upconversion activation in eq 14. The fluxes were estimated by a power integration of the lamp function.

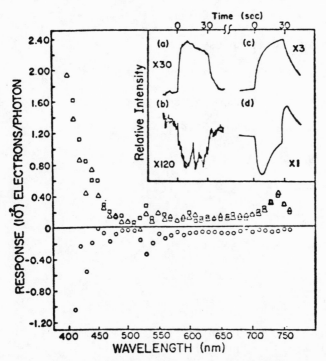

Figure 11.   The action spectrum of a degassed cell (Δ) is compared with that of the same cell after O_2 saturation. The O_2-saturated cell produced a photoanodic spectrum (o) which shows little activity in the red.   On deoxygenation with Ar for 30 min, the initial action spectrum is restored (□). Photoresponse signals are shown in the insets with the light turned on and off at 0 and 30 s, respectively.   Insets a and b give the 740-nm photoresponses in the absence and presence of O_2, respectively.   Inset c shows the white-light photoresponse of the cell 20 h after the cell in inset b was deoxygenated.   After a 4-h period of white-light illumination of the sample in inset c the photocathodic signal of inset c reverses in sign as shown in inset d.

With proper consideration of the two-photon activation mechanism, the present observations may be interpreted in terms of a steady-state scheme in which a generalized chlorophyll light reaction is given in a sequence of six consecutive steps with corresponding rate, W, given as follows:

initial excitation of photoactive aggregate {Chl a}

$$\{Chl\ a\} + h\nu \longrightarrow \{Chl\ a\}^* \quad W_a = k_a I_o \tag{12}$$

radiative or radiationless decay

$$\{Chl\ a\}^* \longrightarrow \{Chl\ a\}(+h\nu') \quad W_b = k_b[\{Chl\ a\}^*] \tag{13}$$

upconversion to a tautomeric charge transfer (CT) state[51]

$$\{Chl\ a\}^* + h\nu \longrightarrow CT \quad W_c = k_c I_o[\{Chl\ a\}^*] \tag{14}$$

nonradiative decay of the tautomeric state

$$CT \longrightarrow \{Chl\ a\} \quad W_d = k_d[CT] \tag{15}$$

reaction with the primary electron acceptor A

$$CT + A \longrightarrow \{Chl\ a\}^+ + A^- \quad W_e = k_e[CT][A] \tag{16}$$

back-reaction or regeneration

$$\{Chl\ a\}^+ + A^- \longrightarrow \{Chl\ a\} + A \quad W_f = k_f[\{Chl\ a\}^+]^2 \tag{17}$$

The steady-state solution of steps 12 - 17 results in the expression for the photogalvanic current

$$i_p \propto [\{Chl\ a\}^+] = \left[\frac{k_c k_e[A]}{k_f(k_d + k_e[A])} \times \frac{k_a I_o}{(k_c + k_b I_o^{-1})} \right]^{1/2} \tag{18}$$

which reproduces the empirical fit (eq 11) for the photo-response of $(Chl\ a \cdot 2H_2O)_n$. We observe that the presence of the term $k_b I_o^{-1}$ in eq[n] 18, resulting from the photochemi-upconversion (step 14), is responsible for the doubling in the exponential flux dependence from semilinearity in the strong light limit to linearity in the weak light limit. In the absence of step 14 the corresponding steady-state solution yields a semilinear dependence throughout the entire range of fluxes, a condition that has long been established for one-photon light reactions involving monomeric Chl a.[53]

The power of the light from the 1000-W tungsten-halogen lamp incident on the platinized $(Chl\ a \cdot 2H_2O)_n$ sample was determined to be 1.7 W, yielding a photocurrent of about 1.5 μAmp corresponding to that shown in inset c, Figure 11. Approximately one half of the incident power occurs beyond

700 nm, out of the reach of Chl a photochemistry. The observed photocurrent, 1.5 μAmp, amounts to a conversion efficiency of about 10^{-6}, assuming that the average energy of the photochemically active photons is 2V. The peak monochromatic quantum efficiency at 740 nm is 4×10^{-3} (see Figure 11). In the "green gap," where the Chl a absorption is weak, the average monochromatic conversion efficiency is about 8×10^{-4}. The average quantum efficiency using the monochromatic light is thus three orders of magnitude greater than that obtained using white light. At fluxes corresponding to the 1.7 W incident power, the photoresponse of the Pt-Chl a cell observes a semilinear flux dependence (see Figure 10). The monochromatic incident fluxes employed in the experiment represented in Figure 11 were $\lesssim 10^{14}$ photons $s^{-1}cm^{-2}$ at which the photoresponse varies linearly with flux. It is evident from Figure 10 that a linear extension of the low-flux photoresponse data at the maximum available power of the source would extrapolate to a quantum efficiency about 10^3 times higher than that observed. The 740-nm monochromatic efficiency 4×10^{-3} corresponds to the photocurrent originating from the O_2 evolution process, reaction 3, in the Chl a-free half cell. That reaction 3 occurs also in the half-cell containing the platinized $(Chl\ a \cdot 2H_2O)_n$ sample is evidenced by the data shown in inset d, Figure 11. The lower limit for the quantum efficiency of water photolysis by platinized $(Chl\ a \cdot 2H_2O)_n$ is thus at least twice that given in Figure 11, leading to the estimated quantum efficiency of 1% for reaction 2 at 740 nm.

VI. CONCLUDING REMARKS

The two-photon interpretation of the flux dependence in Figure 10 is consistent[21] with the two quanta/electron requirement observed for the water splitting reaction in vivo. The in vitro characterization[19,20] of two hydrated Chl a dimers, viz., $(Chl\ a \cdot H_2O)_2$ and $(Chl\ a \cdot 2H_2O)_2$, having physical and photochemical properties indistinguishable from the P700 and the water splitting Chl a complexes in vivo has led to a molecular description of the photosynthetic primary light reaction.[19] The schematic concept of two light reactions connected by a dark bridge of charge carriers as an obligatory pathway for the Chl a water splitting in vivo[14-18] is not sustained by unambiguous experiment. It is contrary to the present finding of water

photolysis by a single Chl a light reaction. Arguments based
on indirect observations on in vivo systems using Biochemical inhibitors such as 3-(3', 4'-Dichlorophenyl) l,
1-dimethylurea (DCMU)[18] or by detergent separations of
photosynthetic reaction centers[14] are unreliable on account
of the alteration of the indigenous photosynthetic apparatus
and of the sensitivity of Chl a light reactions to impurities, including atmospheric oxygen. For example, the removal
of antenna chlorophylls in reaction-center preparations is
expected to minimize the role of the two-photon upconversion
mechanism.[21] In view of the results in Figure 11 we suppose
that the $(Chl\ a \cdot 2H_2O)_n$ water photolysis reaction described
in this paper might have been unveiled years ago had atmospheric oxygen been rigorously excluded from the Chl a preparations of earlier investigators. Reflecting on the research on oxygen evolution by green plants Priestley com-
mented[54] in 1777: "This is a new wide field of experiments
and speculations, and a premature attachment to hypothesis
is the greatest obstruction we are likely to meet with in
our progress through it. . . .Speculation is a cheap com-
modity. New and important facts are most wanted, and therefore of most value." The two-light-reaction series scheme,
introduced nearly two decades ago by investigators as a
working hypothesis,[17],[18] in time became ensconced as the
"central dogma."[15],[16] It has also in recent years become a
dominant influence among workers seeking viable schemes to
the solar conversion question.[14b],[55] The observations described in this paper do not appear to corroborate the tenets
of these schemes.

In energy storage work, we have been concerned with the
photocatalytic decomposition of water according to reaction
2. In plant photosynthesis, however, solar conversion en-
compasses a broader scope. In addition to the cache of
energy in reaction 1, a very considerable portion of the
light energy captured by green plants goes to satisfy the
instantaneous energy demands of the living organism. In
this latter capacity the light reaction of P700, a Chl a
dimer[20] having properties indistinguishable from those of
$(Chl\ a \cdot 2H_2O)_2$, is envisaged to be engaged in cyclic photophosphorylation.[15],[19] In the water splitting reaction
eight moles of light quanta, carrying a free energy content
in excess of twice the free energy (56.7 kcal/mole) required by reaction 2, are consumed in photolyzing two moles
of H_2O. The excess energy may conceivably be employed in

photophosphorylation involving electron flow along the
chain of charge carriers thought earlier[14-18] to be the
dark bridge connecting the water splitting and the P700
light reactions. The details of secondary electron flow in
photophosphorylation lie outside the purview of Chl a light
reactions. The current understanding of photosynthesis is
reasonably summarized by a balanced synthesis of the established physiological facts and the observations on the
photochemical behavior of Chl a-H_2O aggregates.[56]

REFERENCES AND NOTES

(1) H. Levanon and J. R. Norris, Chem. Rev., 78, 185 (1978).

(2) J. Ingenhousz, "Essay on the food of plants and renovation of soils," London (1796).

(3) O. Warburg and G. Krippahl, Z. Naturforsch, 11B, 179 (1956).

(4) R. Wurmser, "Oxidations et Reductions," Presses Univ. France, Paris (1930).

(5) K. Shibata, "Carbon and Nitrogen Assimilation" translated by H. Gest and R. K. Togasaki, Japan Science Press, Tokyo in 1975 (1931).

(6) C. B. Van Niel, Arch. Mikrobiol., 3, 1 (1931).

(7) C. B. Van Niel, Cold Spring Harbor Symp. Quant. Biol. 3, 138 (1935).

(8) R. Hill and R. Scarisbrick, Nature (London), 146, 61 (1940).

(9) R. Emerson and C. M. Lewis, Am. J. Botany, 28, 789 (1941).

(10) S. Ruben, M. Randall, M. Kamen, and J. Hyde, J. Am. Chem. Soc., 63, 877 (1941).

(11) Govindjee, "Bioenergetics of Photosynthesis," Academic Press, New York, N.Y., 1975.

(12) G. M. Cheniae, Ann. Rev. Plant Physiol., 21, 467 (1970).

(13) B. A. Diner and P. Joliot, in "Photosynthesis I," Ed. A. Trebst and M. Avron, Springer-Verlag, Heidelberg, Germany, 1977, p. 187.

(14) (a) It was stated in a 1977 textbook on photosynthesis that the P700 reaction center of photosynthesis may turn out to be other than chlorophyll, see, W. A. Cramer, Photochem. Photobiol., 27, 108

(1978). (b) For a pictorial representation of the currently accepted model of photosynthetic bacterial reaction centers, see Figure 6, K. Sauer, Acc. Chem. Res., 11, 257 (1978).

(15) D. I. Arnon, in ref. 12, p 44.

(16) J. A. Raven, Nature, (London) 227, 1170 (1970).

(17) R. Hill and F. Bendall, Nature (London), 186, 136 (1960).

(18) L. N. M. Duysens, J. Amesz, and B. M. Kamp, Nature, 190, 510 (1961).

(19) F. K. Fong, J. S. Polles, L. Galloway, and D. R. Fruge, J. Am. Chem. Soc., 99, 5802 (1977).

(20) F. K. Fong, A. J. Hoff, and F. A. Brinkman, J. Am. Chem. Soc. 100, 619 (1978).

(21) L. Galloway, J. Roettger, D. R. Fruge, and F. K. Fong, J. Am. Chem. Soc., 100, 4635 (1978).

(22) A. Fujishima and K. Honda, Nature, 238, 37 (1972).

(23) M. S. Wrighton, P. T. Wolczansdi, and A. B. Ellis, J. Solid State Chem., 22, 17 (1977).

(24) J. G. Mavroides, J. A. Katalas, and D. F. Kolesar, Appl. Phys. Lett., 28, 241 (1976).

(25) J. Brace, F. K. Fong, D. H. Karweik, V. J. Koester, A. Shepard, and N. Winograd, J. Am. Chem. Soc., 100, 5203 (1978).

(26) F. R. Hartley, "The Chemistry of Platinum and Palladium," John Wiley and Sons, New York, N.Y., 1973.

(27) J. Thomsen, "Thermochemistry," Longmans, Green and Co., London, 1908.

(28) N. V. Sidgwick, "The Chemical Elements and Their Compounds, Vol. 2, Oxford, 1950.

(29) P. Hersch, Nature, 180, 1407 (1957).

(30) K. J. Ketter and D. Berndt, Z. Electrochem., 62, 378 (1957).

(31) J. VanMuylder, N. de Zoubov, and M. Pourbaix, Rappt. Tech., No. 63 (1958).

(32) Shigeo Shibata, Nippon Kagaku Zasshi, 80, 453 (1959).

(33) R. H. Wentorf, Jr. and R. E. Hanneman, Science, 185, 311 (1974).

(34) G. De Beni, Trans. Am. Nucl. Soc., 20, 717 (1975).

(35) T. N. Veziroglu, ed., "Hydrogen Energy," Part A, Plenum Publishing Corp., New York, N.Y., 1975.

(36) The discrepancy between the rate determination from the GC analysis and the determination in mass spectrometric and pyrolytic studies could have originated from variations in the quality of the platinum surface and the extent to which air had been excluded from the sample. More probably, however, the difficulty may be attributable to systematic errors arising from the sample transfer procedure in the GC analysis. The mass spectrometric and Hersch determinations, unlike the GC work, were on-line experiments.

(37) Patents issued to French workers (Chem. Abstract, 86,:157711f, 191950t (1977)) described the thermochemical cleavage of water in the temperature range 100-1200°C using a finely divided Pt group metal, preferably Pd, as catalyst. The H_2 and O_2 were reportedly generated in a chamber, the walls of which were selectively permeable to H_2. According to the present observations it would appear that Pt, being consumed in the Pt-H_2O reaction, would not be a suitable catalyst for reaction under conditions specified by these French workers.

(38) The effect of heat on Pt immersed in water was examined by comparing the rate of the catalytic

decomposition of H_2O_2 by a Pt black sample heated in
water for 20 h at 180°C with that by an identically
prepared Pt black sample stored in water at room
temperature for 20 h. Both electrodes were maintained under an Ar atmosphere during the heat
treatments. The rate of H_2O_2 decomposition
catalyzed by each sample was subsequently determined
by the colorimetric triiodide method.[39]   The
unheated Pt sample decomposed H_2O_2 at a rate 7.24 x
10^{-4} mole $cm^{-2}s^{-1}$, approximately twice that 4.20 x
10^{-4} mole $cm^{-2}s^{-1}$, using the heat treated platinum.
Details of these results will be published elsewhere.

(39)   A. O. Allen, C. J. Hochandel, J. A. Ghormley, and
       T. W. Davis, J. Phys. Chem., 56, 575 (1954).

(40)   R. C. Weast, ed., "Handbook of Chemistry and Physics,"
       54th edition, CRC Press, Cleveland, Ohio, 1973, D-121.

(41)   J. A. Dean, ed., "Lange's Handbook of Chemistry,"
       Eleventh edition, McGraw-Hill, New York, N. Y.,
       1973, 10-6.

(42)   See, however, F. K. Fong, L. M. Fetterman, L. Galloway,
       A. B. Coddington, and D. R. Fruge, J. Am. Chem. Soc.,
       in press.

(43)   N. Periasamy, H. Linschitz, G. L. Closs, and S. G.
       Boxer, Proc. Nat. Acad. Sci. USA, 75, 2563 (1978).

(44)   M. R. Wasielewski, M. H. Studier, and J. J. Katz,
       Proc. Natl. Acad. Sci. USA, 73, 4282 (1976).

(45)   A. A. Krasnovskii and M. G. Shaposhnikova, Soviet
       Plant Physiology, 17, 357 (1970).

(46)   M. F. Bacon and M. Holden, Phytochem, 6, 193 (1967).

(47)   C. Kratz and D. Duntz, J. Mol. Biol., 113, 431 (1977).

(48)   This photocathodic response of
       Pt/H_2O(KCl)/(Chl a·$2H_2O$)/Pt results from the reduction
       of $(Chl\ a·2H_2O)_n^+$ in the photocathodic half cell
       by the water oxidation reaction in the dark, Chl a-free,
       anodic half cell.

(49)   L. M. Fetterman, L. Galloway, N. Winograd, and
        F. K. Fong, J. Am. Chem. Soc., 99, 653 (1977).

(50)   G. R. Seely and A. Folkmanis, J. Am. Chem. Soc.,
        86, 2763 (1964).

(51)   Here the biphotonic upconversion mechanism[52] is
        simply denoted by the uptake of a second photon.

(52)   F. K. Fong, "Theory of Molecular Relaxation," Wiley-
        Interscience, New York, N.Y., 1975, Chapter 9.

(53)   J. J. McBrady and R. Livingston, J. Phys. Colloid
        Chem., 52, 662 (1948).

(54)   J. Priestley, "Experiments and Observations on
        Different Kinds of Air," Vol. III, J. Johnson, London
        (1777).

(55)   For typical discussions, see, (a) J. R. Bolton, J.
        Solid State Chem., 22, 4 (1977). A realistic
        assessment is given by (b) R. K. Clayton, Brookhaven
        Symposia in Biology, 22, 1 (1976).   The
        hypothetical devices, which Clayton called a "vision-
        ary variation of agriculture," are based on the series
        scheme of photosynthesis and on other concepts de-
        rived from work on bacterial reaction centers.   Sug-
        gested ingredients include "a dried film of reaction
        centers. . .fully capable of performing its photo-
        chemistry efficiently," a semitransparent film of
        some suitable metal, hydrogenase for the evolution
        of molecular hydrogen, and "a kind of dialysis
        machine in which oxygen and other components separate,
        but soluble reactants can diffuse across the mem-
        branes." Those who would consider attempting to
        make a solar cell from purified chlorophyll and other
        components were advised to "consider adding pheo-
        phytin to the mixture, in view of our present pic-
        ture of the early steps in the photochemistry of
        bacterial reaction centers." A major difficulty is
        "in keeping the oxygen away from the hydrogenase, and
        one can imagine trying to use other catalysts and not
        hydrogenase - perhaps platinum or palladium, but these
        are poisoned by oxygen, and they can promote immediate
        combustion (of the hydrogen)." The technological
        problems of a large scale application of the device

were evaluated. Professor Clayton concluded the discussion with the remark: "This model has not been
built or tested. . . .I'd put my money on silicon."

(56)   This paper has been based on the preliminary reports:
(a) F. K. Fong and L. Galloway, J. Am. Chem. Soc.,
100, 3594 (1978); (b) L. Galloway, D. R. Fruge,
G. M. Haley, A. B. Coddington, and F. K. Fong, J.
Am. Chem. Soc., in press; (c) F. K. Fong, L. M.
Fetterman, D. R. Fruge, and A. B. Coddington, J. Am.
Chem. Soc., in press.

THE "TANDEM PHOTOELECTROLYSIS PLANT" CONCEPT: A STRATEGY

FOR FUEL PRODUCTION VIA BIOMASS CONVERSION WASTES

R. E. Schwerzel, E. W. Brooman, R. A. Craig,
D. D. Levy, F. R. Moore, L. E. Vaaler, V. E. Wood

BATTELLE Columbus Laboratories
505 King Avenue, Columbus, Ohio 43201

ABSTRACT

One often-overlooked difficulty associated with the
solar photoelectrolysis of water is the potential explosion
hazard posed by the simultaneous evolution of hydrogen and
oxygen. This problem can be addressed, in principle, by
the addition of suitably-chosen redox electrolytes to the
photoelectrolysis cell. We have found, as have A. J. Bard
and his coworkers, that the evolution of oxygen can be
suppressed completely in certain aqueous photoelectrolysis
cells, with no adverse effect on either the photocurrent or
the rate of cathodic hydrogen evolution, by simple carboxylate ions such as acetate or butyrate. In basic solutions,
water-soluble products derived from the carboxylate ions are
formed at the photoanode, but in acidic solutions, hydrocarbons such as ethane or hexane are formed (along with CO_2)
via a "photo-Kolbe" reaction. Such cells can therefore
produce a valuable fuel mixture, consisting of hydrogen
(from the cathode) and hydrocarbons (from the photoanode).
Because both acetic acid and butyric acid are abundantly
available in the aqueous effluent from the enzymatic
digestion of biomass material, these observations lead to
the concept of a "tandem photoelectrolysis plant", in which
a photoelectrolysis device would derive its input water
supply from a biomass conversion device. While the problems
of electrode stability and light absorption still remain to
be solved, the approach described here completely eliminates

the necessity of separating the anodic and cathodic gases
produced in photoelectrolysis. This, in turn, can pave the
way for the design of simple and less expensive photoelectrolysis plants than would otherwise be possible.

INTRODUCTION

In 1912, the great Italian photochemist, Giacomo
Ciamician, published a remarkable paper entitled "The
Photochemistry of the Future"[1] in which he considered the
wealth of benefits which might be gained by the photochemical
utilization of solar energy for the production of useful
chemical materials. In discussing the role of plant crops
(or biomass, as we would say now) as solar energy transducers,
he suggested that: "The harvest, dried by the sun, ought to
be converted, in the most economical way, entirely into
gaseous fuel, taking care during this operation to fix
the ammonia (by the Mond process for instance) which should
be returned to the soil as nitrogen fertilizer together
with all the mineral substances contained in the ashes".
This elusive goal of efficient, economical fuel production
from renewable biomass resources has stimulated research
efforts around the world since Ciamician's time. While
much progress has been made, the problems involved are far
from solved, and the production of gaseous fuels from biomass is
still too expensive to be economically feasible on a large scale

This paper describes a new approach, based on the
photoelectrolysis of water, by which the conversion of
biomass materials to gaseous fuels might be accomplished.
We begin with a review of the photoelectrolysis technique,
and then discuss the results on which the "tandem
photoelectrolysis plant" concept is based.

THE PHOTOELECTROLYSIS OF WATER

Photoelectrolysis is a recently-discovered technique
which utilizes the photoelectric properties of certain
semiconducting materials for the conversion of optical
energy into either electrical energy or the chemical
potential energy of high-energy products such as hydrogen.
The photoelectrolysis of water is thus a method for
converting optical energy into the chemical energy of
hydrogen and oxygen.

It has been only some nine years since Fujishima and
Honda published, in 1969, the first of a series of papers[2-6]
which were to stimulate the emergence of this vigorous
new area of scientific research. These workers found that
a crystalline wafer of n-type titanium dioxide (TiO_2),
when connected to a platinum counter-electrode and exposed
to ultraviolet light while immersed in an aqueous
electrochemical cell, was capable of producing a photovoltage sufficient to cause the dissociation of water into
hydrogen and oxygen (so long as a sufficient bias potential
or pH gradient was maintained across the cell). A
schematic illustration of their experimental apparatus,
which has by now been widely adopted and modified, is
shown in Figure 1.

The key implication of these observations is that
improved semiconducting electrode materials might be capable
of catalyzing the dissociation of water (and the formation
of hydrogen gas) solely under influence of solar radiation.
If this could be achieved with sufficiently high efficiency,
it would open the way to the large-scale use of hydrogen,
both as a fuel and as a chemical feedstock. Hydrogen

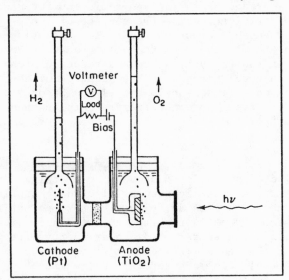

FIGURE 1.   SCHEMATIC ILLUSTRATION OF A CLASSICAL
            PHOTOELECTROLYSIS CELL

produced in this way would be an exceedingly attractive
fuel, in particular, as it would be attainable in virtually
unlimited supply from water (which would be regenerated
when the hydrogen was burned) and solar energy. This
possibility, together with the inherent scientific
challenge which the study of photoelectrolysis offers, has
been largely responsible for the rapid expansion of research
in this field during the past several years.

The "classical" photoelectrolysis cell configuration
shown in Figure 1 has been used with a variety of electrode
combinations. These include (a) an n-type semiconductor
(such as titanium dioxide) in combination with a metal
counter-electrode;[2-6] (b) a p-type semiconductor (such as
gallium phosphide) in combination with a metal counter-
electrode;[7] and (c) an n-type semiconductor in combination
with a p-type semiconductor.[7,8] Regardless of the details
of design, the fundamental process is the same.

Upon the absorption of light by the photoactive
semiconductor electrode(s), a photocurrent flows through
the external circuit connecting the two electrodes such
that the surface of the n-type semiconductor becomes
electron-deficient, while the surface of the p-type semiconductor becomes electron-rich. The resulting reactive
sites on the electrode surfaces then serve to oxidize water
(to O_2) at the (n-type) anode and reduce protons (to H_2)
and the (p-type) cathode. This sequence of events is
depicted schematically below in Figure 2 for the "Schottky
barrier" (n-type anode/electrolyte/metal cathode) type of
photoelectrolysis cell first developed by Fujishima and
Honda.

Despite its promise, however, photoelectrolysis has
been hampered since its discovery by a vexing dichotomy in
the properties of semiconductor electrode materials. This
problem may be stated as follows: Those potentially
useful semiconductors reported to date which are highly
colored, and therefore absorb solar energy effectively,
tend to be unstable with respect to corrosion or dissolution
under photoelectrolysis condictions, while those materials
which have been found to be intrinsically stable also
tend to be transparent, or nearly so, to most of the solar
spectrum. Their efficiency of utilization of solar energy
is therefore disappointingly low. There is an urgent need,
then, for materials which are both stable and highly colored,

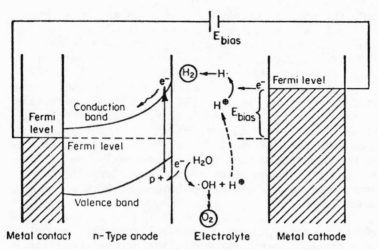

FIGURE 2.   SCHEMATIC ENERGY-LEVEL DIAGRAM FOR A
            SCHOTTKY-TYPE PHOTOELECTROLYSIS CELL

and which will promote the photoelectrolysis of water under
solar illumination with high efficiency. Without such
materials, it is doubtful whether photoelectrolysis can
ever become a commercially important source of hydrogen
or other fuels.

There is another, more pragmatic, problem with
photoelectrolysis as well. This is the requirement that the
oxygen and hydrogen produced in the cell must be physically
separated to avoid the explosion hazard which a stoichiometric mixture of these gases would pose. While this is a
trivial problem in a laboratory-scale device, it would
greatly increase the complexity and cost of a commercial
plant. Thus, there is an additional need for the development of electrolytes which can suppress the formation of
oxygen (and, hopefully, produce a useful product in its
stead). We shall return to this point shortly.

Recent Developments

The electrode materials problem described above has
stimulated an enormous increase in the research effort
directed toward photoelectrolysis during the past few years.

Indeed, the photoelectrolysis field has not only grown
dramatically but has also begun to evolve into two distinct
branches. While these necessarily have much in common,
they differ in thrust and in the ultimate end-use of the
solar energy captured by the semiconductor electrodes.
This evolutionary trend is summarized by the qualitative
chronology shown in Figure 3. In the years immediately
following Fujishima and Honda's first publications, the
field developed (with the discovery of improved materials
and new cell designs) largely as a result of inputs from
research done independently, and previously, on the
properties of semiconductor electrodes, dye-sensitized
photogalvanic cells, and photoelectrochemical processes
in general.[9,10]

More recently, however, several research groups began
to explore the inherent capabilities of the photoelectrolysis
technique for the production of electrical power;[11,12] this
approach utilizes specially chosen redox electrolytes to
suppress both the evolution of hydrogen and oxygen and the
degradation of the semiconductors,[13] and concentrates on
the electrical power delivered to the external circuit.
This type of cell is discussed in detail by several
papers in this volume.

Still more recently, the concept of a "photochemical
diode"[14] for the production of hydrogen and oxygen has
evolved as an extension of the development of p-n heterotype
photoelectrolysis cells.[7,8] Such cells, in which one
electrode is an n-type semiconductor and the other is a
p-type semiconductor, are capable of causing the dissociation of water upon irradiation in the absence of any
externally-applied pH gradient or bias potential.[7] In a
"photochemical diode", the p and n-type electrodes are in
direct ohmic contact with each other, rather than being
connected through a discrete external circuit as in the
classical cell design shown in Figure 1. These "diodes"
can, in principle, be fabricated in very small sizes. One
can thus begin to think in terms of a "photocatalytic
powder",[15,16] which could be slurried with water, perhaps
with the addition of appropriate redox electrolytes to
suppress the evolution of oxygen, to produce hydrogen
directly upon exposure to sunlight. Indeed, it has been
shown recently that "Schottky-analog" powders can be made
easily by the partial platinization of n-type semiconductor
powders such as TiO_2, and that these powders will

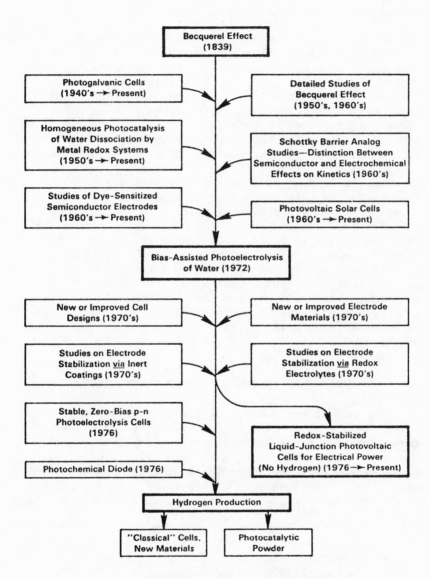

FIGURE 3.   CHRONOLOGICAL EVOLUTION
OF PHOTOELECTROLYSIS

photocatalyze a variety of electrochemical reactions.[17,18] We shall return to this point shortly.

A considerable amount of research is still being carried out on new classes of electrode materials[19] and new cell designs.[20] This work is also important in that it provides a better understanding of the limitations which optical considerations and the inherent properties of materials impose on the efficiency of photoelectrolysis and the related power-producing techniques. Further, new advances in electrode materials can rapidly advance the state of the art for both power production and the production of fuels.

Thus, it can be seen that photoelectrochemistry is a vigorous and rapidly-developing field of research, which is attempting to solve several problems of prime social and scientific significance. Our own work in photoelectrolysis has sought to address the two major problems which confront the successful production of fuels by photoelectrolysis, namely the instability (and/or poor absorbance of sunlight) exhibited by many semiconductor electrodes, and the formation of stoichiometric mixtures of hydrogen and oxygen in normal photoelectrolysis. The latter area is of primary concern here, and we now discuss our results in this area.

OXYGEN SUPPRESSION VIA REDOX ELECTROLYTES

The role of redox electrolytes in the suppression of oxygen evolution can best be understood by a consideration of the detailed electrochemical processes which occur during photoelectrolysis. In the photoelectrochemical decomposition of water to generate hydrogen and oxygen, the water itself comprises a redox system of primary interest. The conventional oxygen evolution reaction is written as:

$$2H_2O + 4p^+ \rightleftharpoons O_2 + 4H^+ \tag{1}$$

and the reversible potential, E_O, for the reaction is 1.23 V vs NHE (normal hydrogen electrode potential, 0.0 V). The hydrogen reaction is described by:

$$4H^+ + 4e^- \rightleftharpoons 2H_2 \tag{2}$$

and by definition, E_o = 0.00 V vs NHE). For water, the redox reactions may thus be written as:

$$H_2O/O_2 \quad ..... \quad E_o = 1.23 \text{ V} \qquad (3)$$

$$H^+/H_2 \quad ..... \quad E_o = 0.00 \text{ V} \qquad (4)$$

and these potentials are pH dependent, being shifted 59 mV per pH unit according to the Nernst Equation.[21]

These are not the only redox potentials which must be taken into account, however. The redox potentials of the ions which comprise the semiconductor electrodes are important, since the presence of electrode constituents which are too easily oxidized or reduced can lead to the corrosion or dissolution of the electrode. Another important consideration is that in the case of certain cell configurations – particularly those which would utilize a fine slurry of "photochemical diode" particles – the simultaneous evolution of stoichiometric amounts of hydrogen and oxygen could present a serious explosion hazard, as noted above. It may thus be not only desirable, but essential, to identify suitable redox species which would suppress both the evolution of oxygen and the electrochemical degradation of the electrodes.

At the n-type photoanode, holes can only be transferred to the electrolyte if a donor energy level exists in the electrolyte which is equal to or above the relative energy level of the valence band at the interface, as shown in Figure 4. Conversely, for p-type electrodes (photocathodes), electrons can only be transferred to the electrolyte if an acceptor energy level exists which is equal to or below the relative energy level of the conduction band, as shown in Figure 5. The energy levels in the electrolyte corresponding to the donor and acceptor redox system states may be determined with a knowledge of the standard redox potentials by taking vacuum as a zero energy reference point, then using the following approximation:

$$E_{FR} = E_o - 4.5 \text{ eV} \qquad (5)$$

where E_{FR} represents the Fermi energy of the redox system, and E_o, the redox potential vs NHE, as expressed in volts.

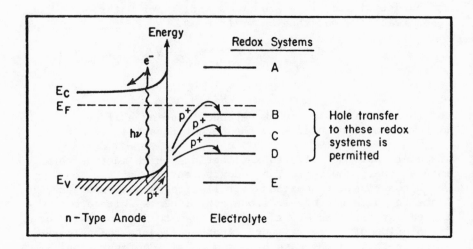

FIGURE 4. ENERGETICS OF HOLE TRANSFER FROM A PHOTOANODE
TO REDOX SYSTEMS IN THE ELECTROLYTE

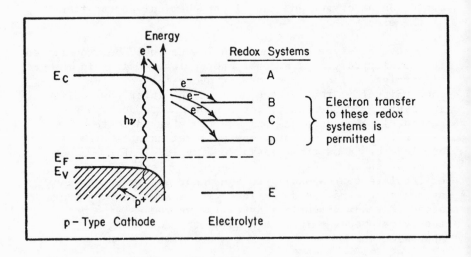

FIGURE 5. ENERGETICS OF ELECTRON TRANSFER FROM A
PHOTOCATHODE TO REDOX SYSTEMS IN THE
ELECTROLYTE

For n-type semiconducting materials, the Fermi energy, E_F, may be approximated by the flat band energy, E_{FB}, and the maximum degree of band bending, V_B^{max}, thus:

$$E_F \approx E_{FB} + V_B^{max} \approx E_c \qquad (6)$$

where E_{FB} is the energy level (or equivalent potential) at which $V_B = 0$. E_{FB} hence defines the original Fermi energy in the semiconducting material at the electrolyte interface <u>before</u> a photoelectrochemical cell is formed by completing the external circuit (e.g., before the Schottky junction analogs are established). Band bending can apply to both the conduction band energy (E_c) and the valence band energy (E_V). The difference between E_c and E_V is simply the bandgap energy, E_g:

$$E_g = E_c - E_v \qquad (7)$$

Returning to the situation for the photoelectrolysis of water, Figure 6 depicts schematically the ideal situation (with zero external bias) before the Schottky junctions are formed at the respective interfaces.[22] Similar diagrams may be drawn for other redox systems added to the electrolyte when either the H_2O/O_2 reaction needs to be suppressed, or when surface stabilization is the objective, as shown in Figure 7. For the latter objective it is important to match the redox system with the electronic characteristics of the semiconducting material in question, so as to prevent dissolution of the photoanode and promote oxygen evolution or some other acceptable anodic reaction. With photocathodes, supplanting the hydrogen evolution reaction obviously is to be avoided, but it would be desirable to prevent the precipitation of metallic films on the surface. The type of data presented in Figure 7 thus makes it a simple matter to anticipate <u>a priori</u> which redox systems may be effective with a given electrode/electrolyte combination. This is not the whole picture, of course, since kinetic factors at the electrode surface may dominate the reaction despite apparently favorable energetics. Nonetheless, one can obtain a very informative grasp of a given cell configuration in this way.

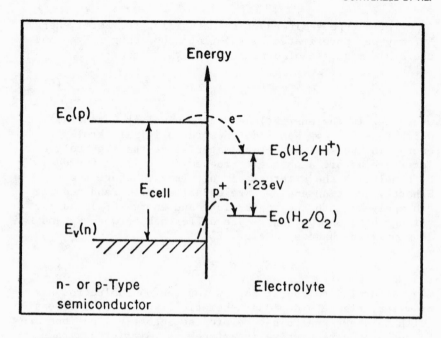

FIGURE 6.   OPTIMAL DISTRIBUTION OF ENERGY LEVELS FOR
THE PHOTOELECTROLYSIS OF WATER

The Photoelectrolytic Kolbe Reaction

A consideration of prime importance which has been
widely overlooked is that, be definition, an operational
photoelectrolysis plant will be an <u>open system,</u> through
which water must flow as hydrogen and oxygen are produced.
Therefore, the use of exotic redox reagents which must be
added continuously to the water input may well be economically and/or environmentally prohibitive. For this reason,
we have sought to identify redox reagents which would be
effective at preventing either oxygen evolution or electrode
deterioration, and which would already be present in
readily available water supplies.

As a first step, we have begun an investigation of the
effects of adding acetic acid (or acetate ion) to the
electrolyte.[18,23,24] This system was selected because its
electrochemistry is well known[25] and because it is
abundantly available, along with other low molecular

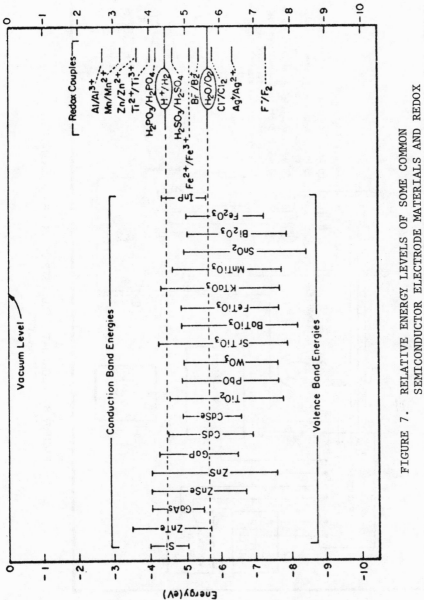

FIGURE 7. RELATIVE ENERGY LEVELS OF SOME COMMON
SEMICONDUCTOR ELECTRODE MATERIALS AND REDOX
SYSTEMS IN ACID SOLUTION

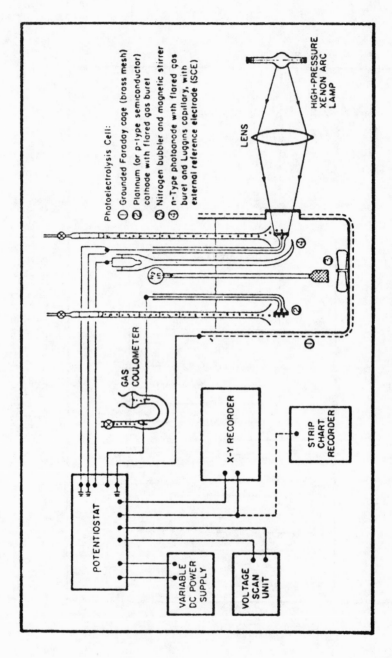

FIGURE 8.   SCHEMATIC ILLUSTRATION OF EXPERIMENTAL
APPARATUS

weight organic acids, from the microbiological hydrolysis of renewable biomass resources.[26]

At high current densities the anodic oxidation (Kolbe reaction) of acetate ion can dominate the oxygen evolution reaction,[25] leading to the formation of ethane as follows:

$$CH_3CO_2H \; \overset{\leftarrow}{\rightarrow} \; CH_3CO_2^- + H^+ \; \xrightarrow[-e]{\overset{+}{p}} \; \overset{+}{H} + CH_3CO_2 \cdot \quad (8)$$

$$CH_3CO_2 \cdot \; \xrightarrow{\text{fast}} \; CH_3 \cdot + CO_2 \quad (9)$$

$$2CH_3 \cdot \; \xrightarrow{\text{fast}} \; C_2H_6 \quad (10)$$

This process has the potential, then, of not only preventing the formation of oxygen but of also enhancing the fuel value of the hydrogen produced by virtue of the ethane byproduct. One might therefore envision that a photoelectrolysis plant could be combined with an enzymatic digester plant, so as to produce a mixture of hydrogen and light hydrocarbons from the renewable biomass products, if it were possible to carry out the Kolbe reaction in a photoelectrochemical cell.

Our experiments to date in this area have been most encouraging. In these studies, we have utilized single crystal slabs of reduced (n-type) strontium titanate (n-SrTiO$_3$) as the photoanode, and a platinized platinum foil electrode as the cathode. The experimental apparatus, which consists of a Wenking Model PCA-7211 Potentiostat in combination with an Aardvark Instruments Model SCAN-1 voltage scanning unit and a homemade variable DC power supply in conjunction with both X-Y and strip-chart recorders, is shown schematically in Figure 8. The photo-electrolysis cell itself is equipped with an optically-flat Pyrex window, a nitrogen bubbler and stirrer for deaerating the solutions prior to photoelectrolysis, a Luggins capillary connected to an external standard calomel electrode (SCE), and flared gas burets over the semiconductor and counter electrodes. A gas coulometer is connected in series with the cell. The focused, Pyrex-filtered output of either a PEK 75-watt high-pressure xenon arc lamp or an Eimac 150-watt high-pressure xenon arc lamp was used to illuminate the photoanode. These light sources closely approximate the spectral distribution of direct sunlight at sea level. This apparatus permits the measurement of current-voltage curves, the potential of either electrode

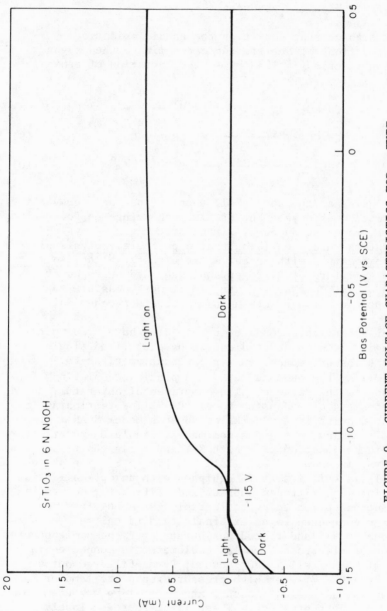

FIGURE 9.   CURRENT–VOLTAGE CHARACTERISTICS FOR n–TYPE
STRONTIUM TITANATE IN 6N SODIUM HYDROXIDE

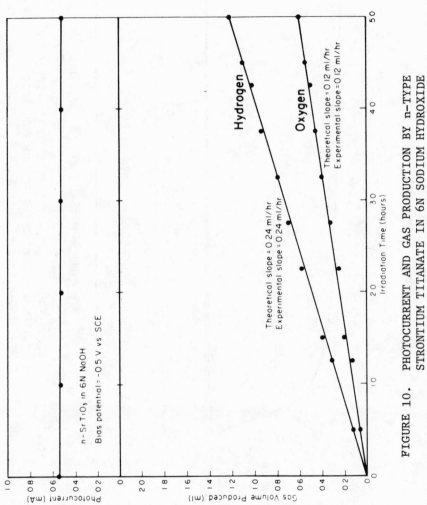

FIGURE 10.   PHOTOCURRENT AND GAS PRODUCTION BY n-TYPE
STRONTIUM TITANATE IN 6N SODIUM HYDROXIDE

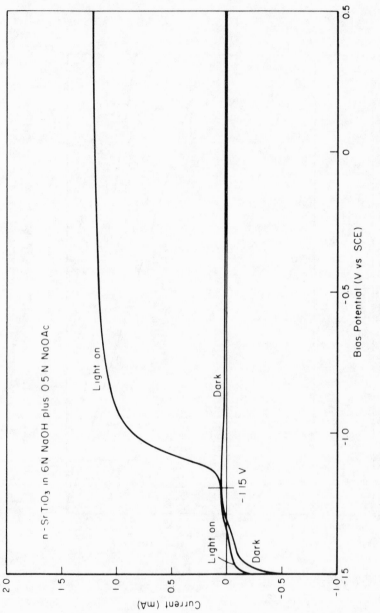

FIGURE 11.   CURRENT-VOLTAGE   CHARACTERISTICS FOR n–TYPE
STRONTIUM TITANATE IN 6N SODIUM HYDROXIDE
CONTAINING 0.5N SODIUM ACETATE

with respect to a standard calomel electrode (SCE), the volume of gas evolved at either electrode, the total current passed through the cell, and the photocurrent developed in the cell as a function of time, bias potential, or the wavelength and intensity of illumination.

Representative results for a 5 x 8 mm crystal of $SrTiO_3$ in 6N sodium hydroxide solution are shown in Figures 9 and 10. It can be seen that the photocurrent remains relatively consistent over a period of several hours, and that hydrogen and oxygen are evolved in precisely the stoichiometric ratio. The absolute volume of gas produced is equal to that calculated from the total current passed through the cell, within the experimental error.

In the presence of 0.5N sodium acetate in 6N sodium hydroxide, however, the system behaves quite differently, as Figures 11 and 12 illustrate. While the current-voltage characteristics of the cell appear to be unchanged, the evolution of oxygen is virtually eliminated. The efficiency of hydrogen production is unaffected, and still proceeds at essentially the theoretical rate, based on the total photocurrent passed through the cell. (The apparent increase in photocurrent in the presence of sodium acetate in this experiment is due entirely to an increase in the intensity of illumination). These results demonstrate that we have successfully prevented O_2 formation in a photo-electrolysis cell without simultaneously inhibiting the formation of hydrogen. It is interesting that the Kolbe electrolysis product, ethane, is not formed under these highly basic conditions. Presumably, the anode reactions lead instead to the well-known formation of water-soluble side products such as ethanol and oxalic acid, by means of the Hofer-Moest reaction,[25] although our analyses of these solutions is not yet complete.

It is clear, however, that this result demonstrates the sensitivity of the anode reactions to acetate, or at least to simple carboxylate ions similar to acetate. For example, the addition of 0.5N sodium iodide to 6N sodium hydroxide electrolyte has no effect on the photoelectrolysis process whatsoever; as Figure 13 illustrates, hydrogen and oxygen are formed in precisely stoichiometric amounts, at the theoretical efficiency. This is particularly surprising, as the iodide ion should be easily oxidized under these conditions. It may well be the case that the kinetic

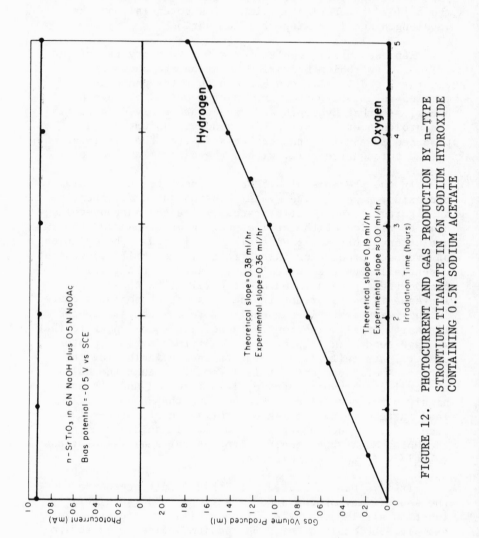

FIGURE 12. PHOTOCURRENT AND GAS PRODUCTION BY n-TYPE
STRONTIUM TITANATE IN 6N SODIUM HYDROXIDE
CONTAINING 0.5N SODIUM ACETATE

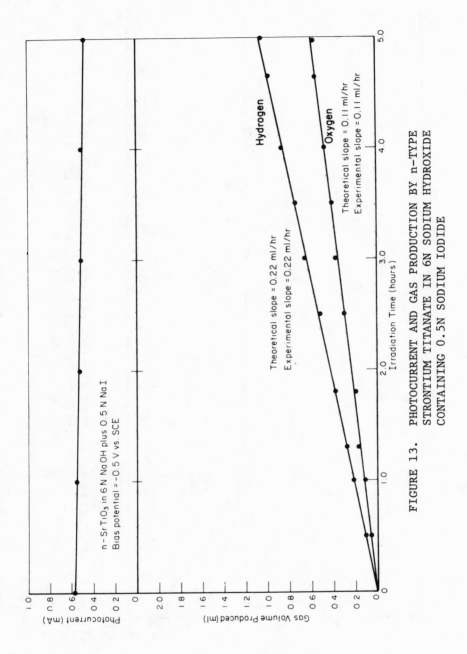

FIGURE 13. PHOTOCURRENT AND GAS PRODUCTION BY n-TYPE
STRONTIUM TITANATE IN 6N SODIUM HYDROXIDE
CONTAINING 0.5N SODIUM IODIDE

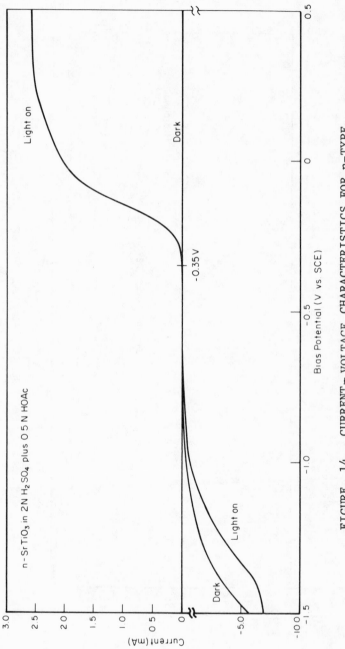

FIGURE 14. CURRENT–VOLTAGE CHARACTERISTICS FOR n–TYPE
STRONTIUM TITANATE IN 2N SULFURIC ACID
CONTAINING 0.5N ACETIC ACID

competition between the several possible electrode reactions
is the dominant factor in these systems, with the relative
rates of reaction for these three redox systems being
Acetate >> Water >> Iodide. We note in passing that
the photocatalytic oxidation of cyanide and sulfite ions
has been reported to occur when alkaline aqueous solutions
of these ions are irradiated in the presence of a slurry
of powdered n-type semiconductor, using such materials as
TiO_2, and ZnO_2, and Fe_2O_3.[27]

A strikingly different, and highly significant,
result is obtained when the photoelectrolysis experiment
is performed with dilute acetic acid in 2N sulfuric acid.
This is illustrated in Figures 14 and 15. While the current-
voltage curve shows no exceptional features beyond the
normal pH-induced shift of the onset of photocurrent, the
rate of gas production from the photoanode is virtually
equivalent to that from the cathode. This means that the
course of the anode reaction under these conditions has
been diverted from oxygen formation to the photocatalytic
oxidation of acetic acid; i.e., the Kolbe reaction. Like
the formation of hydrogen, the formation of ethane from
acetic acid is a 2-electron process, while the production
of oxygen from water requires the transfer of 4 electrons.
Thus, a shift from the 2:1 (cathode:anode) stoichiometry
observed for the photoelectrolysis of water to the 1:1
stoichiometry observed here is diagnostic for the
occurrence of the Kolbe reaction.

Indeed, we find (as have Kraeutler and Bard, in
similar experiments carried out with n-TiO_2 photoanodes[18])
that the predominant anodic products under these conditions
are ethane and carbon dioxide, with only residual amounts
of oxygen being formed. This clean diversion of the anodic
reaction toward the Kolbe reaction is particularly striking
in view of the low concentration (ca. 4% by volume) of
acetic acid in these experiments. A possible rationale may
lie in the fact that the simple Kolbe reaction is
exothermic:[18,28]

$$2CH_3CO_2H \rightarrow CH_3CH_3 + 2CO_2 + H_2 \qquad (11)$$
$$\Delta G° = -18.0 \text{ kJ/mol } (-4.3 \text{ kcal/mol})$$

Thus, the reaction is thermodynamically favored over the
formation of oxygen, which is substantially endothermic.[28]

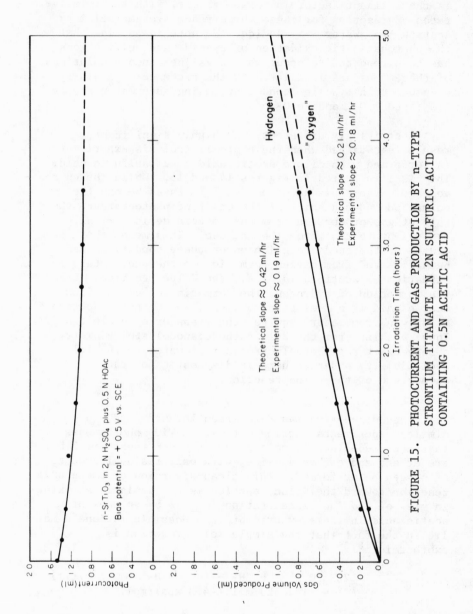

FIGURE 15.   PHOTOCURRENT AND GAS PRODUCTION BY n–TYPE
STRONTIUM TITANATE IN 2N SULFURIC ACID
CONTAINING 0.5N ACETIC ACID

Nonetheless, the irreversible potential required to electrolyze acetic acid is more <u>positive</u> than that required to oxidize water.[25] It may be, therefore, that specific surface kinetic effects (such as preferential adsorption of acetate at the electrode surface) may play a dominant role.

A particularly striking result in this context has been obtained recently by Krauetler and Bard.[18] They have found that the irradiation of acetic acid in the presence of a "Schottky-type" photocatalytic powder consisting of partially-platinized TiO_2 (anatase) particles results in the efficient formation of <u>methane,</u> rather than the ethane which was observed in the experiments described above. These photocatalytic powders are prepared simply by the irradiation of the anatase powder in a solution of hexachloroplatinic acid[18], and consist of particles which have domains of metallic platinum (typically about 1-5% platinum by weight[18]) distributed on the surface of the TiO_2. Thus, each particle of powder is capable of acting like a miniature photoelectrolysis cell, the platinized regions serving as cathodes and the unmodified areas serving as photoanodes. Under these conditions, it appears that the diffusional combination of adsorbed methyl radicals and hydrogen atoms produced in close proximity on the particle surface is the dominant process:

$$CH_3CO_2H \xrightarrow{-CO_2} CH_3 \cdot (ads) + H \cdot (ads) \rightarrow CH_4 \qquad (12)$$

Both the photocatalytic powder systems and conventional electrode systems are effective in promoting the "photo-Kolbe" reaction with a variety of simple organic acids, including propionic acid,[18] butyric acid,[18,23,24] n-valeric acid[18], and pivalic acid.[18] Thus, this appears to be a general method for the clean, efficient photocatalytic conversion of simple organic acids to the corresponding hydrocarbons.

THE "TANDEM PHOTOELECTROLYSIS PLANT" CONCEPT

In considering the implications of these observations for solar energy conversion, the most striking fact is the ease with which the photoelectrolytic formation of oxygen can be diverted to the formation of hydrocarbons by the addition of simple organic acids to the aqueous electrolyte. It

matters little whether a "photocatalytic powder" or a
series of discrete electrodes are used; the resulting
products are safer and more easily collected and handled than
the stoichiometric mixture of hydrogen and oxygen which is
formed in the normal photoelectrolysis of water.

To understand the signficance of this point, consider
the schematic concepts illustrated below in Figure 16 for the
design of solar photoelectrolysis collectors.

These represent an evolution of design fron conventional
p-n heterotype cells[7,8] (with different semiconductor
materials for the photocathode and photoanode), to the
"photochemical diode",[14] to large solar collectors consisting
of either strip arrays of alternating n-type and p-type
semiconductors or photocatalytic powders comprised
of particles containing either both n-type and p-type
regions or n-type and metallized regions. The p-n type
of strip-array device has been depicted here rather than
the Schottky-analog type because p-n cells (and photochemical
diodes) have been shown to be capable of photoelectrolysis
with no applied bias potential.[7,14]

In either case, it is clear that the design of cells
which make maximal use of the available sunlight will
more than likely be cells in which the physical separation
of the gases evolved from the anode and cathode would be a
major burden. As was pointed out earlier in this paper, this
is a necessity in a device which produces hydrogen and oxygen
in stoichiometric amounts. It is completely obviated in a
device which produces, instead, a mixture of hydrogen,
hydrocarbons, and carbon dioxide.

Because carbon dioxide can be removed easily from the
product mixture (by adsorption or by freezing), one can
envision the use of such devices, in principle, for the
large-scale production of a rich gaseous fuel consisting
of mixed hydrocarbons and (perhaps) hydrogen, with the
byproduct carbon dioxide being sold separately. But
where to get the required supply of acids? It mav well be
economically prohibitive to purchase even crude materials
for addition to the electrolyte. However, it is known[26]
that acetic acid, butyric acid, and other organic acids
can be made in high yield by the enzymatic fermentation of
simple sugars. Similarly, the production of simple sugars
can be achieved by the microbiological hydrolysis of

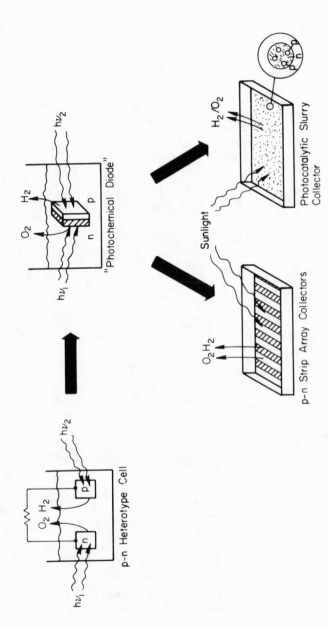

FIGURE 16.   PHOTOELECTROLYSIS COLLECTOR CONCEPTS

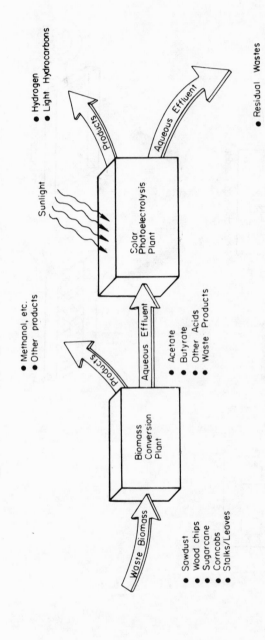

FIGURE 17. "TANDEM" PHOTOELECTROLYSIS PLANT CONCEPT

cellulose and lignin, although with only modest efficiency at present.[26]

Thus, we are led to the concept of a "tandem" photoelectrolysis plant, in which a solar photoelectrolysis device for the production of gaseous fuels as described above would be operated in tandem with a biomass conversion plant (for the production of acid-containing aqueous input to the photoelectrolysis plant). This is illustrated below in Figure 17. In principle, it should be possible to design a biomass conversion unit which could accept as input a variety of waste biomass materials (such as sawdust, woodchips, corncobs, stalks, and so on) which would otherwise be burned or discarded, and convert them to an aqueous mixture of simple organic acids and other products. This aqueous effluent would then serve as the input for the solar photoelectrolysis plant, which would carry out the conversion of the acids to gaseous fuel and discard the remaining biomass conversion products.

A scheme of this type offers a number of major advantages over other approaches, in that it represents a way of obtaining a "value added" from waste biomass resources which are renewable, and thus available in abundant supply. The production of simple acids by anaerobic digestion is faster and more efficient than the production of methane by fermentation,[26] so higher overall sunlight-to-fuel conversion efficiencies may be possible using this approach. Also, the design of the solar photoelectrolysis plant should be simplified significantly by the removal of the necessity for physically separating the anodic and cathodic gases. Finally, the use of a naturally-derived input redox electrolyte would improve both the economic and environmental costs of operating the system. Our results are particularly noteworthy in this context, as the concentration of acetic acid which was used (0.5 $\underline{M}$, or $\underline{ca.}$ 4% by volume) in our experiments is similar to that formed in certain types of microbiological fermentation. Thus, we have established that the photoelectrolytic Kolbe oxidation of acetic acid can occur efficiently at concentrations equivalent to those obtained in at least some fermentation systems.

CONCLUSIONS

In conclusion, we have shown that the photoelectrolysis

111

diverted from the production of oxygen and
ᵣoduction of alkanes, carbon dioxide, and
ᵤy the addition of simple organic acids to
. we have described a strategy, the "tandem
ᵣolysis plant" concept, for the large scale
ᵢon of gaseous fuels by means of the solar photo-
ᵣrolysis of the acid-rich aqueous effluent from a
ᵢomass conversion plant. It is important to note that in
order for this strategy to be useful, however, a number of
problems must first be solved. These include the problems
of electrode instability and poor sunlight absorption which
have plagued the field of photoelectrolysis since its
inception, and the problems associated with improving the
efficiency of acid production by means of microbiological
hydrolysis procedures. Further, and perhaps most
importantly, it is not yet known what effect the many
other materials produced by biomass conversion might have
on the course of the desired photoelectrochemical process.

We have avoided any discussion of comparative cost
estimates or projected overall efficiencies of a "tandem"
photoelectrolysis plant at this time because of the large
uncertainities associated with the successful resolution of
the above problems. Only after it has been demonstrated
that the photocatalytic Kolbe reaction can be made to
occur by a stable, visible-absorbing semiconductor material
will it be appropriate to consider cost estimates in a
preliminary way.

One can hope, though, that this problem, and the
microbiological problems, will be solved quickly. At
least, the worldwide research efforts in both photoelectrochemistry and biomass conversion have grown dramatically
over the past several years. In the meantime, as Ciamician
said so long ago[1], "I believe that industry will do well
in using from this very day all the energies that nature
puts at its disposal. So far human civilization has made
use almost exclusively of fossil solar energy. Would it not
be advantageous to make better use of radiant energy?" This
goal remains elusive, but well worth striving for.

ACKNOWLEDGEMENTS

This research was supported in part by the U.S. Department of Energy, under Contract No. W-7405-Eng-92 (Task 95). We also thank B. R. Allen and R. B. Iden (Battelle, Columbus Laboratories) for helpful comments and assistance with this work, and Ohio Semitronics, Inc. (Columbus, Ohio) for a gift of materials used in the fabrication of several photoelectrolysis cells.

REFERENCES

1. G. Ciamician, Science, 36, 385 (1912).

2. A. Fujishima, K. Honda, and S. Kikuchi, J. Chem. Soc. Japan, 72, 108 (1969).

3. A. Fujishima and K. Honda, J. Chem. Soc. Japan, 74, 355 (1971).

4. (a) A. Fujishima, E. Sugiyama, and K. Honda, Bull. Chem. Soc. Japan, 44, 304 (1971); (b) A. Fujishima and K. Honda, ibid, 44, 1148 (1971).

5. A. Fujishima and K. Honda, Nature, 238, 37 (1972).

6. A. Fujishima, K. Kobayakawa, and K. Honda, Bull. Chem. Soc. Japan, 48, 1041 (1975).

7. A. J. Nozik, Appl. Phys. Lett., 29, 150 (1976).

8. H. Yoneyama, H. Sakamoto, and H. Tamura, Electrochim. Acta, 20, 341 (1975).

9. H. Gerischer, in "Advances in Electrochemistry and Electrochemical Engineering", Vol. I, P. Delahay and C. W. Tobias, Eds., Interscience, New York (1961), Chapter 4.

10. See, for instance: (a) H. Tributsch and H. Gerischer, Ber. Bunsenges. Phys. Chem., 73, 851 (1969); (b) R. Memming and H. Tributsch, J. Phys. Chem., 75, 562 (1971), and references cited therein.

11. (a) A. B. Ellis, S. W. Kaiser, and M. S. Wrighton, J.
    Amer. Chem. Soc., 98, 1635 (1976); (b) A. B. Ellis,
    S. W. Kaiser, and M. S. Wrighton, ibid., 98, 6418
    (1976); (c) A. B. Ellis, S. W. Kaiser, and M. S. Wrighton,
    ibid., 98, 6855 (1976).

12. (a) J. Manassen, G. Hodes, and D. Cahen, J. Electrochem.
    Soc., 124, 532 (1977); (b) J. Manassen, D. Cahen,
    and G. Hodes, Nature, 263, 97 (1976); (c) G. Hodes,
    J. Manassen, and D. Cahen, ibid., 261, 403 (1976),
    and references cited therein.

13. See the comment by G. C. Barker in the Discussion follow-
    ing the paper: H. Gerischer, J. Electrochem. Soc., 113.
    1174 (1966).

14. A. J. Nozik, Appl. Phys. Letters, 30, 567 (1977).

15. A. J. Nozik (Materials Research Center, Allied Chemical
    Company), personal communication to R. E. Schwerzel
    (February 1977).

16. R. Williams (David Sarnoff Research Center, RCA
    Laboratories), personal communication to R. E. Schwerzel
    (February 1977).

17. M. S. Wrighton, P. J. Wolczanski, and A. B. Ellis, J.
    Solid State Chem., 22, 17 (1977).

18. (a) B. Kraeutler and A. J. Bard, J. Amer. Chem. Soc.,
    99, 7729 (1977); (b) B. Kraeutler and A. J. Bard, ibid.,
    100, 2239 (1978); (c) B. Krautler and A. J. Bard, ibid.,
    100, 5985 (1978).

19. (a) M. A. Butler, R. D. Nasby, and R. K. Quinn, Solid
    State Commun., 19, 1011 (1976); (b) G. Hodes, D.
    Cahen, and J. Manassen, Nature, 260, 312 (1976);
    (c) A. B. Ellis, S. W. Kaiser, and M. S. Wrighton, J.
    Phys. Chem., 80, 1325 (1976); (d) J. H. Kennedy and
    K. W. Frese, Jr., J. Electrochem. Soc., 123, 1683 (1976);
    (e) H. Tributsch, J. Electrochem. Soc., 125, 1086 (1978).

20. W. W. Anderson and Y. G. Chai, Energy Conversion, 15,
    85 (1976).

21. G. Kortüm and J. O'M. Bockris, "Textbook of Electro-Chemistry", Vol. I., Elsevier Publishing Co., New York (1951) Chapter 8.

22. J. M. Bolts and M. S. Wrighton, J. Phys. Chem., 80, 2641 (1976).

23. R. E. Schwerzel, "Methods for the Photochemical Utilization of Solar Energy", paper presented at the 29th Southeast Regional Meeting of the American Chemical Society, Tampa, Florida (November 9-11, 1977).

24. R. E. Schwerzel, in "Radiation Energy Conversion in Space", K. W. Billman, Ed, American Institute of Aeronautics and Astronautics, New York (1978), pp. 626-657.

25. A. K. Vijh and B. E. Conway, Chemical Reviews, 67, 623 (1967).

26. See, for instance, S. C. Prescott and C. G. Dunn, "Industrial Microbiology", 3rd Ed., McGraw-Hill, New York (1959); (b) J. E. Sanderson, D. L. Wise and D. G. Augenstein, "Liquid Hydrocarbon Fuels from Aquatic Biomass", Paper No. 27 presented at the Second Annual Fuels from Biomass Symposium, Rensselaer Polytechnic Institute, Troy, New York (June 20-22, 1978).

27. S. N. Frank and A. J. Bard, J. Phys. Chem., 81, 1484 (1977).

28. "CRC Handbook of Chemistry and Physics", 53rd Ed., Chemical Rubber Publishing Company, Cleveland, Ohio (1972).

LIGHT-INDUCED ELECTRON TRANSFER REACTIONS IN SOLUTION, ORGANIZED ASSEMBLIES AND AT INTERFACES: SCOPE AND POTENTIAL APPLICATIONS.[1]

David G. Whitten*, Patricia J. DeLaive, Thomas K. Foreman, Janet A. Mercer-Smith, Russell H. Schmehl and Charles Giannotti**

Department of Chemistry, University of North Carolina, Chapel Hill, N.C. 27514

INTRODUCTION

Light-induced electron transfer reactions and redox chemistry resulting from these processes have been the subject of extensive investigations during the past several years. Quenching of a very wide variety of excited states by electron transfer or by formation of exciplexes in which there is considerable charge separation has been shown to be quite general; the chemical consequences of these quenching events can range from no net chemistry to such diverse processes as cycloadditions, free radical or radical ion reactions and polar substitutions or additions.[2-27] Much study in this area has focused on excited states of transition metal complexes; the rich array of accessible redox states available in many complexes coupled with low excitation energies and relative photostability makes them potentially attractive candidates for a number of applications. For several complexes it has been found possible to quench the same excited state by both electron donors (eq 1) and electron acceptors (eq 3):

$$MC^{n+*} + D \rightarrow MC^{(n-1)^{+}} + D^{+} \qquad (1)$$

*Author to whom correspondence should be directed.
**Institut de Chimie des Substances Naturelles CNRS
91190 Gif-sur-Yvette, France

$$MC^{(n-1)^+} + D^+ \rightarrow M^{n^+} + D \qquad (2)$$

$$MC^{n^{+*}} + A \rightarrow MC^{n+1} + A^- \qquad (3)$$

$$MC^{n+1} + A^- \rightarrow MC^{n^+} + A \qquad (4)$$

While rates and efficiencies for the quenching processes are often very high, the energy wasting back
reactions (eqs 2 and 4) also occur rapidly resulting in no
net chemistry in many of the systems investigated thus far.
The present paper will review some recent investigations of
these processes occurring in solutions and in various molecular organizates. The first section, dealing with solution-phase photoredox processes, will focus on ways to
prevent energy wasting back reactions such that the high
energy products produced in reactions such as 1 and 3 can
be diverted to provide usable reagents and/or energy storage.
The second part will review some recent studies of electron
transfer reactions in organizates and at interfaces which
suggest an expanded scope and possible new applications for
these processes.

PHOTOINDUCED REDOX REACTIONS IN SOLUTION

Several recent reviews have indicated the wide range
and scope of light-induced electron transfer reactions with
complexes and other substrates.[25,27,29-31] Therefore the
current paper will focus on only a few of these processes
with what might be regarded as prototype systems and the
emphasis will be on possible applications of these specific
reactions to energy conversion and other uses. Metal complexes used as excited substrates in these studies include
compounds 1-7. All of these complexes have been found to
participate in electron transfer quenching processes such
as 1 or 3 with a variety of donors or acceptors in reactions
where no net chemistry results as a consequence of the
quenching.[13-19] More recently we have found it possible
in several cases to use the same complexes to accomplish
permanent conversions.[28,29,23,33] It is instructive to
examine some of the latter cases in detail.

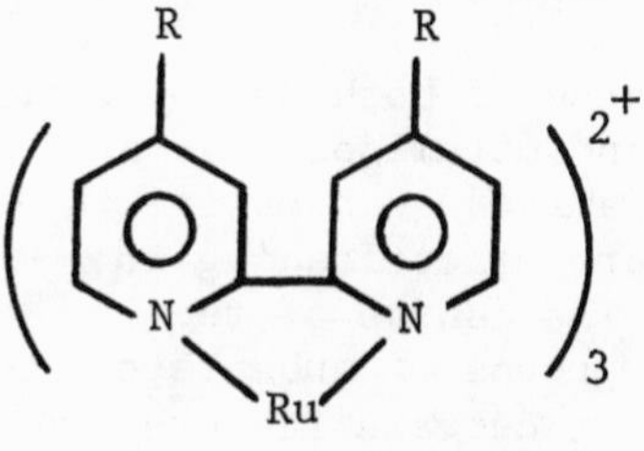

1 R = H

2 $R = -CO_2CH(CH_3)_2$

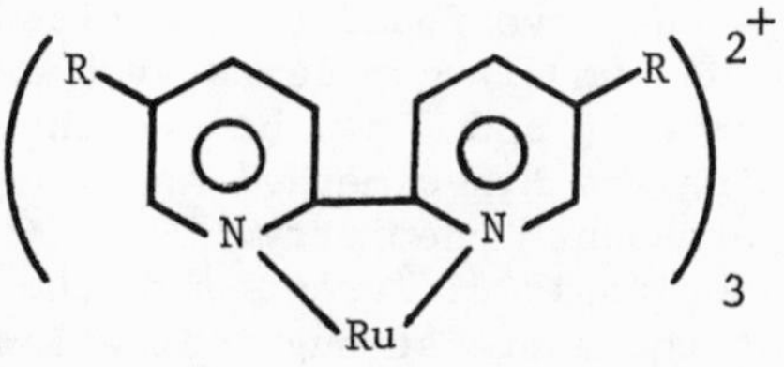

3 $R = -CO_2CH(CH_3)_2$

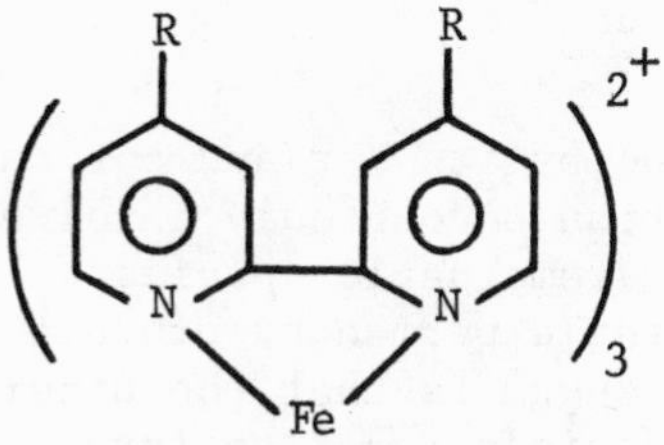

4 R = H

5 $R = -CO_2-CH(CH_3)_2$

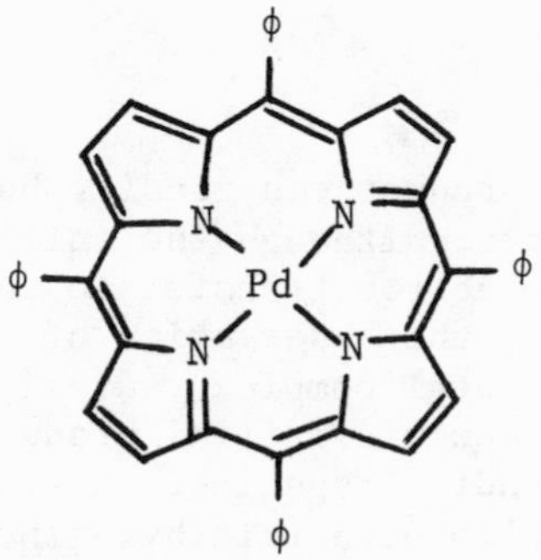

6

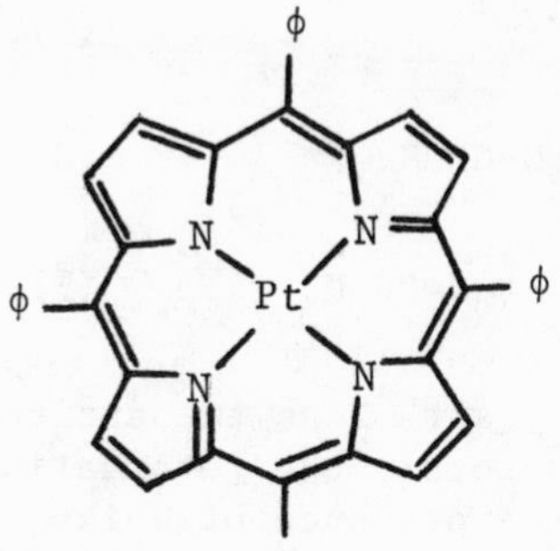

7

Reductive Quenching of Metal Complex
Excited States by Amines

We have found that excited states of both hydrophobic ruthenium (II) complexes such as 2 and metalloporphyrins such as 6 and 7 can be quenched by amines such as triethylamine and N,N-dimethyl aniline in processes leading to "permanent" chemistry.[28, 29, 32, 34] The course of the reaction is different with the two groups of substrates but there are strong indications that certain factors are of importance in both cases. For the ruthenium (II) poly-pyridine complexes the net chemistry readily observable upon irradiation of the complex in the presence of the amines is one-electron reduction of the metal complex to the paramagnetic monocation (eq 5):[28]

$$RuL_3^{2+} \xrightarrow[R_3N]{h\nu} RuL_3^+ \tag{5}$$

This product can readily be detected by its esr and optical spectra; although the initial reaction undoubtedly involves one electron transfer to form two paramagnetic species (eq 6) the observation of only a single permanent product (the metal complex radical cation) suggests that the other paramagnetic initial product must rapidly react to form non-radical species. A study of products formed in the reaction with triethylamine indicates the overall sequence (at high $[Et_3N]$) to be that given by eq 6-9. In this case

$$RuL_3^{2+*} + Et_3N: \rightarrow RuL_3^+ + Et_3N \cdot^+ \tag{6}$$

$$RuL_3^+ + Et_3N \cdot^+ \rightarrow RuL_3^{2+} + Et_3N: \tag{7}$$

$$Et_3N \cdot^+ + Et_3N: \rightarrow Et_3NH^+ + Et_2N - \underset{\cdot}{C}HCH_3 \tag{8}$$

$$Et_2N - \underset{\cdot}{C}HCH_3 + RuL_3^{2+} \rightarrow RuL_3^+ + Et_2\overset{+}{N} = CHCH_3 \tag{9}$$

the occurrence of permanent chemistry with 2 as the excited substrate is due primarily to two factors: the retardation of the back reaction (eq 7) due to the presence of bulky hydrophobic groups in 2 and the rapid subsequent reactions of one of the initial products (in this case the reactive radical cation $Et_3N \cdot^+$) to give products which will not readily participate in energy wasting back reactions.

Evidence that both these factors are critical comes from our finding that irradiation of "unhindered" 1 in the presence of Et_3N: gives only transient reduction detectable by flash spectrscopy;[32] here the back reaction is fast enough to compete with subsequent processes destroying the reactive amine radical cation. In contrast, irradiation of the hindered complex 2 in the presence of triphenylamine leads to only transient formation of RuL_3^+; in this case the amine radical cation is sufficiently stable such that it survives long enough for even a retarded back reaction to be the dominant process subsequent to the quenching event.

An analysis of the triethylamine induced photoreduction of 2 indicates the reaction involves considerable storage of energy. Thus in the thermochemical scheme given by eq 10-13 it is clear that steps 10 and 12 are energetically

$$2(RuL_3^{2+} + e \rightarrow RuL_3^+) \tag{10}$$

$$2(Et_3N + H^+ \rightarrow Et_3NH^+) \tag{11}$$

$$(Et_3NH^+ \rightarrow H_2 + Et_2\overset{+}{N}=CHCH_3) \tag{12}$$

$$(H_2 \rightarrow 2H^+ + 2e^-) \tag{13}$$

uphill while 11 is downhill. It appears that steps 11–13 approximately balance one another so that the net energy stored in the process is given by at least the potential of the $Ru^{2+}/1+$ complex which is −0.9V or 20.7 kcal/mole.[29] The "efficiency" (Q) of the process can be thus estimated using the equation:[35]

$$Q = \frac{100 \cdot \Delta H \cdot \phi}{E_\lambda} \tag{14}$$

For light of 540 nm (near the excitation threshold) the calculated Q for a quantum efficiency, ϕ = 0.35 (acetonitrile) is 14%. This compares very favorably with several organic systems which can be activated only with much shorter wavelength light.[35-38] From the mechanism given by eqs 6-9 it can be estimated that a quantum yield of 2 should be attainable. If this were the case the efficiency could

be as high as 75%. With the particular system $\underset{\sim}{2}$-triethyl-
amine we have not yet obtained efficiencies higher than
0.35; in less polar solvents such as isobutyronitrile and
tetrahydrofuran the efficiency is reduced to values of 0.2
and 0.05 respectively. The source of the inefficiencies
in the process are not completely clear but it appears
likely that the major loss of efficiency may occur during
the initial electron transfer to form separated ions (eq 6).
While we have no data for eq 6, in other instances where
excited states of polypyridyl ruthenium (II) complexes are
quenched, efficiencies in the range of 0.1-0.3 have been
estimated.[39,40] However in the case of quenching of 1 by
Fe^{3+} in aqueous perchloric acid efficiencies in the range
0.7-0.9 are indicated.[41]

While the scheme outlined in eqs 6-9 involves fairly
efficient energy conversion, it is clear that any viable
scheme for utilizing complexes such as $\underset{\sim}{2}$ would depend on
its clean regeneration concurrent with efficient recovery
and/or utilization of the energy stored in the reductive
process. We have in fact observed a slow dark reaction in
several solvent systems whereby light-reduced $\underset{\sim}{2}$ is converted back to the starting complex. One possible method for
accomplishing a regeneration of the dication involves reduction of water (eq 15). This reaction is energetically

$$2RuL_3^+ + 2H^+ \rightarrow 2RuL_3^{2+} + H_2 \qquad (15)$$

favorable and in fact we find that the reduced complex
reacts rapidly with water to regenerate the starting
RuL_3^{2+}.[28,32] However when the reaction is carried out in
dilute aqueous acetonitrile solutions there appears to be
little, if any, generation of hydrogen. We have recently
found that the use of a heterogeneous catalyst such as
PtO_2 (vide infra) can bring about what appears to be a
fairly efficient generation of hydrogen.[42] Thus irradiation
of $\underset{\sim}{2}$ in the presence of triethylamine and a suspension of
PtO_2 in aqueous acetonitrile leads to generation of hydrogen without consumption of the ruthenium complex catalyst.
While the hydrogen evidently comes from reduction of water,
the net reaction involves consumption of triethylamine
which is used as the reductant. While this and other
reactions in which hydrogen can be generated from water
using visible light[43,45] are of obvious interest, it is
clear that utilization of the high energy RuL_3^+ in this

manner is an energy wasting process for which much better
substitutes might be found.

A second photoreduction process involving amines and
excited states of metal complexes has been observed upon
irradiation of metalloporphyrins with visible light. In
this case there is also an electron transfer step but the
overall process involves a transformation in the porphyrin
ring. Our studies have focused on reactions of palladium
(II) and platinum (II) porphyrins such as 6 and 7.[34] On
irradiation in the presence of amines such~as triethylamine
or N,N-dimethylaniline these porphyrins react to give
characteristic spectral changes associated with consecutive
reduction of two double bonds in the porphyrin macrocycle
to yield first chlorin and subsequently isobacteriochlorin
(dihydrochlorin) structures (eq 16).[46-48] A detailed study
of the reaction of excited 6 with N,N-dimethylaniline indicates that in this case the reaction involves exclusively
reductive addition to give 8b and 9b rather than the simple
reduction products 8a and 9a.[34] Mechanistically the formation of 8b can be rationalized by eqs 17-24; the reaction
observed here appears closely related to other amine-
mediated photoreductions.[49-51] Quenching studies indicate
an exciplex is intermediate in the process; however a
flash photolysis study indicates little, if any, buildup
of "free" radical ion intermediates. Studies with the
spin trap 5,5-dimethyl-1-pyroline-1-oxide (DMPO) have

led to the trapping of the radical $\cdot CH_2-\overset{\overset{\displaystyle CH_3}{|}}{N}-C_6H_5$ and a

hydrogen atom adduct.[52] These species presumably arise
from reaction of DMPO with the radicals or radical pairs
produced in eq 20.

The reaction between excited 6 and amines such as
N,N-dimethylaniline or triethylamine proceeds with low
efficiencies (0.0001-0.001) in all solvents studied thus
far. That the quantum yields are low is not surprising
since electron transfer (eqs 18-19) is energetically
uphill by ~6-7 kcal/mole for both amines and rapid recombination of the oppositely charged ions produced in eq 19
would be expected to be rapid and probably the dominant
path. That some reaction actually occurs is apparently
due to the high reactivity of the amine radical cation via
eq 20. A rather instructive indication that a high
reactivity is prerequisite to avoiding the back reaction

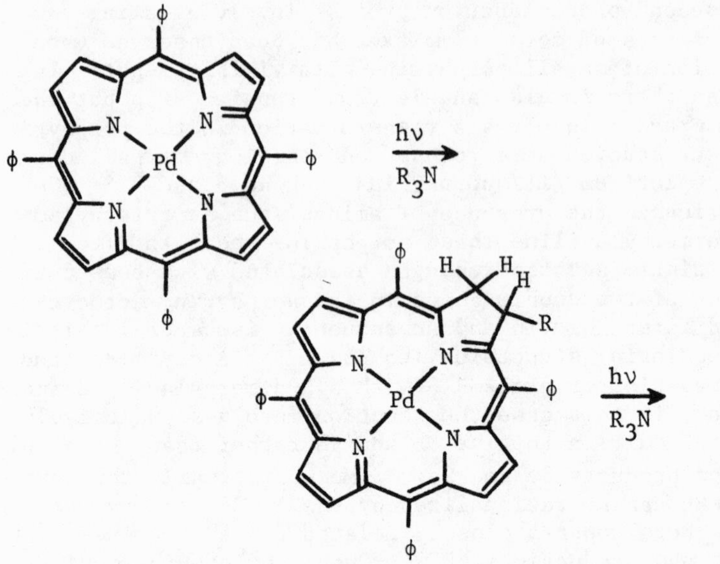

$$\qquad\qquad\qquad (16)$$

8a   R = H

8b   R = CH$_2$-N-C$_6$H$_5$
         |
         CH$_3$

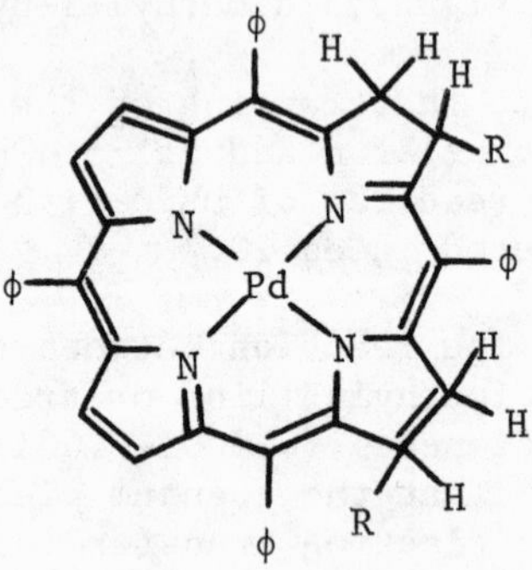

9a   R = H

9b   R = CH$_2$-N-C$_6$H$_5$
         |
         CH$_3$

$$PdP \xrightarrow{h\nu} PdP^{1*} \longrightarrow PdP^{3*} \tag{17}$$

$$PdP^{3*} + (CH_3)_2\ddot{N}\phi \rightarrow [PdP \cdots (CH_3)_2\ddot{N}\phi]^* \tag{18}$$

$$[PdP \cdots (CH_3)_2\ddot{N}\phi]^* \longrightarrow PdP\bar{\cdot} \cdots (CH_3)_2\overset{+}{\dot{N}}\phi \tag{19}$$

$$PdP\bar{\cdot} \cdots (CH_3)_2\overset{+}{\dot{N}}\phi \longrightarrow PdPH\cdot \; \cdots \; \cdot CH_2\overset{CH_3}{N}\phi \tag{20}$$

$$PdPH\cdot \; \cdots \; \cdot CH_2\overset{CH_3}{N}\phi \longrightarrow \underset{\sim\sim}{8b} \tag{21}$$

in this case is given by comparison of the reactivity of 6 with DMA and 4-methoxy-N,N-dimethyl aniline (MDMA).[52] The latter amine is a better reductant by 0.22 V; consequently electron transfer to triplet 6 from MDMA is still endothermic (by ca. 2 kcal/mole) but much more favorable than with DMA. Not surprisingly it is found that MDMA quenches the phosphorescence of 6 with a rate constant two orders of magnitude faster than that obtained for DMA. However no products are produced in the quenching process and no photoreduction is observed even on long-term irradiation. No transients are detected by flash photolysis and we have been unable to trap any radicals with DMPO under conditions where products are readily obtained with DMA.[52] The most consistent explanation for the high quenching but lack of products is that for MDMA the analog of steps 17-19 are facilitated while the reaction corresponding to 20 is returned due to the "stability" of the MDMA radical cation. Thus for this cation radical reverse electron transfer completely dominates proton transfer (eq 20) and no net chemistry occurs.

Results similar to those observed for these systems with amines as reducing agents should probably be extendable to other potential reductants. Particularly attractive is the possibility that the "tuning" of the redox properties of the metal complex (vide infra) will permit the use of less powerful reductants.

Oxidative Quenching of Metal
Complex Excited States

Oxidative quenching of metal complex excited states has been observed for a wide variety of substrate quencher combinations.[12-16,41] The usual pattern of quenching-back reactions as outlined in eqs 3 and 4 is observed in several cases where separated ions of like charge can be formed; in cases where ions of unlike charge are generated it is frequently observed that no long-lived transients are detected due apparently to rapid decay of the ion-pair without separation.[14,53] A quencher-oxidant that has been used in several studies is the organic dication paraquat or methyl viologen (PQ^{2+}) (N,N'-dimethyl-4,4'-bipyridine). This is an attractive quencher since it is readily reduced (-0.446V vs. SCE)[54] but it does not absorb light in the near uv or visible regions and has no low-lying excited states; furthermore the radical cation formed on reduction is easily detectable by its characteristic esr and absorption spectra.[54,55]

We and others have recently found that it is possible to quench excited states of 1, 2, 6, 7 and other compounds with PQ^{2+} in processes leading to net chemical reaction in relatively high efficiency.[29,22,56,57] The key to accomplishing net chemical conversion in these cases is the use of an oxidizable substrate (S) to rapidly consume the oxidized metal complex. Thus in several cases a sequence given by eqs 1', 22, 23 occurs and the production of the

$$MC^{n+*} + PQ^{2+} \rightarrow MC^{n+1} + PQ^{+} \quad (1')$$

$$MC^{n+1} + S \rightarrow MC^{n+} + S^{+} \quad (22)$$

$$S^{+} \rightarrow S_{ox} \text{ or Products} \quad (23)$$

$$MC^{n+1} + PQ^{+} \rightarrow MC^{n+} + PQ^{2+} \quad (24)$$

$$S^{+} + PQ^{+} \rightarrow S + PQ^{2+} \quad (25)$$

radical cation PQ^{+} as a permanent product can be observed.[58] The observation of permanent PQ^{+} production in these cases indicates both that the oxidized metal complex is consumed in competition with back electron transfer (eq 24) and that the product of one-electron oxidation of the substrate S^+ must also react rapidly to yield stable products since in most cases it would be expected that S^+ should rapidly oxidize PQ^+ (eq 25). In fact it is easy to show that for substrates such as triphenylamine which form reasonably stable one-electron oxidation products, the combination of reactions 1', 22 and 25 occurs producing no net chemical change.[18] Substrates which we have found reactive with the excited states of 2 include pyridine, 2,6-lutidine, N,N-dimethylformamide, triethylamine and water. In related studies using 1 and other excited substrates it has been found that thiol containing substrates such as cysteine can effect the same changes.[59] For the studies with 2 as the excited complex it is not surprising that rapid oxidation of a variety of substrates occurs since the potential for the $RuL_3^{3+/2+}$ couple is 1.59V indicating that oxidized 2 should be a powerful oxidant.[28] The overall quantum efficiencies observed with 1 and 2 in nitrile solvents with amines as oxidizable substrates range from 0.01-0.4, depending on the precise system investigated.[33] That relative rates of the back reaction (eq 24) are important is suggested by the finding that efficiencies obtained with 2 are generally two to three times those obtained with 1 for the same substrate-solvent combinations.[33]

The nature of the permanent product-forming step (eq 23) has been investigated for some of the substrate-solvent systems studied. In several cases it appears that processes somewhat analogous to those observed in reductive quenching occur. Thus for reactions in acetonitrile or isobutyronitrile it appears that hydrogen abstraction from the solvent (eq 26) is a major path. Analogous processes have

$$S \cdot^+ + H\text{-}Solv \rightarrow SH^+ + \cdot Solv \qquad (26)$$

been reported for anodic oxidation of amines and related substrates[60-62] and we have been able to trap the solvent radicals by irradiation in the presence of spin traps.[33,52] Hydrogen atom abstraction also occurs when oxidized substrate is generated in acetone solution; in this case we detect the same products formed from generation of the $\cdot CH_2\overset{O}{\overset{\|}{C}}\text{-}CH_3$ radical by other methods.[33]

The potential for utilization of reactions of this sort remains as yet undetermined principally because the full scope of usable substrate-solvent combinations and

their reactions have not been investigated. While the PQ^+
produced as a permanent product in the sequence 1', 22 and
33 is itself quite possibly a usable material (vide infra),
the reagent with the most potential for useful chemistry is
the oxidized metal complex. For example, as pointed out
earlier, the oxidized 2 generated in reaction 1' is a powerful oxidant which is in fact as strong an oxidizing agent
as Ce (IV) in nitric acid.[63] However under the conditions
of irradiation with visible light oxidized 2 is generated
in a neutral nonaqueous medium. Thus the reaction offers
the possibility of carrying out selective oxidations of a
number of substrates under extremely mild conditions. The
problem remains however that unless the oxidized substrate
is rapidly diverted to nonoxidizing products, no net chemistry will result due to the occurrence of reaction 25. An
example where reaction 25 (or analogous processes) is dominant involves the use of water as a substrate. Here we
find only a very low steady state buildup of PQ^+ even though
it is clear that oxidized 2 should oxidize water. In fact
we find that addition of water to solutions containing
reactive substrates such as 2,6 lutidine suppresses the
buildup of PQ^+ suggesting competition between the amine and
water for oxidized 2. Thus in terms of potential energy
application the most promising approaches would appear to
involve the use of either (a) reactive oxidizable substrates
which rapidly generate non-oxidizing but usable products or
(b) other excited state quencher-oxidants which form products that are not good reducing agents. If the latter
could be accomplished it should be possible to use the oxidized metal complex either directly or in combination with
a catalyst (vide infra) to effect the oxidation of water to
oxygen.

Utilization of Non-Emitting Metal
Complex Excited States

Although electron transfer quenching of metal complex
excited states has been most thoroughly studied with complexes having long-lived intensely luminescent excited
states such as 1, 2, 6 and 7, there have been a few cases
reported where relatively shorter-lived non-luminescent
excited states participate in analogous processes.[46, 53, 64]
For example relatively long-lived triplet states of zinc
and magnesium porphyrins can serve as electron donors to a
number of acceptors in polar solvents; this process can be

studied by both esr and flash spectroscopy. More recently
it has been shown that even relatively short-lived excited
states can be intercepted in processes completely analogous
to those reported for 1.[64] Among non-luminescent excited
states that would be particularly attractive from the point
of view of their light absorption, availability, and redox
properties are the polypyridyl iron (II) complexes. These
complexes have absorption spectra and stability comparable
to their better studied ruthenium (II) counterparts but
have not until recently been found photochemically active.
It has recently been shown indirectly that the simple tris-
(2,2'-bipyridine) iron (II) can be quenched by electron
transfer to Fe^{3+} in aqueous solution; from these studies
an estimate of the lifetime for the triplet of 4 of 23 nsec
has been made.[65] Thus the excited state of 4, although
non-luminescent, should be sufficiently long lived to participate efficiently in bimolecular reactions in the diffusion-controlled range.

We have recently found that the iron (II) analogs of
1 and 2, 4 and 5, undergo reactions analogous to the ruthenium (II) complexes in both oxidative and reductive quenching
processes.[66] Both complexes are stable to long-term irradiation in the absence of electron donors or acceptors;
irradiation of 4 in the presence of PQ^{2+} in acetonitrile
leads to no permanent chemistry but the PQ^+ radical cation
can be easily detected by flash techniques and the back
reaction (eq 2) ($k = 3.3 \times 10^9$ M^{-1} sec^{-1}) readily observed.
Addition of substrates such as 2,6-lutidine or triethyl-
amine results in "permanent" generation of the PQ^+ but with
lower efficiencies than observed with the ruthenium complexes. Reductive quenching of 5 by triethylamine apparently is more complicated than the corresponding reaction
with 2 since irradiation results in a net bleaching process.
However a study of the process by spin trapping and esr
indicates the same radical and radical ion intermediates
occurring with the ruthenium complex 2 and the triethylamine are formed. The indications that non-emitting first
row metal complexes such as these iron (II) derivatives can
participate in the same sort of electron transfer reactions
as the luminescent ruthenium, platinum and palladium complexes suggest that reagent expense need not be a major
factor in limiting the practical usefulness of these
processes.

Potential Applications of Solution
Phase Reactions

As discussed in the previous sections, light driven one-electron transfer reactions of metal complexes (and other substrates) can be frequently channeled so that one of the reactive initial products is stored and the other diverted to a permanent product. While one or both of these products may themselves be utilizable in several cases, it would be most attractive to have available means whereby the original reagents can be regenerated concurrent with release of energy or, perhaps more desirable, the formation of a stable "fuel". One such example of the latter case would be the production of hydrogen from water. In actuality it has been found possible to use even relatively weak reductants such as the radical cation $PQ.^+$ to generate hydrogen in aqueous solutions using catalysts such as platinum oxide[43-45] or enzymes.[56] There have been several recent reports of experiments in which a light activated one-electron transfer process in aqueous medium is coupled by use of heterogeneous catalysts such as PtO_2 or colloidal platinum to generate molecular hydrogen; while in some cases it is uncertain as to what the reducing species is, in others it appears that either a reduced metal complex or organic reductant is active.[43-45,67] For example, it has recently been reported that photogeneration of reduced paraquat can be coupled with heterogeneous catalysis using PtO_2 to generate hydrogen from water in aqueous solution.[67] In our own work we have found that the reduced 2 formed by photolysis in the presence of amines can generate hydrogen when the irradiation is carried out in the presence of water and a suspension of PtO_2.[42] In all of these cases it is fairly clear that the hydrogen comes from water; however these reactions do not appear to involve a true "water splitting" since some reductant is consumed irreversibly in each case. In fact it is clear that for reactions of this sort to have any practical utility it will be necessary to use water or some other inexpensive weak reductant which can be consumed in the overall process. In reality the coupling of oxidation of water to a light induced one-electron transfer process seems to be a more difficult process; nevertheless it appears that the use of appropriate catalysts with some systems already developed could possibly succeed. It has recently been reported that one-electron oxidants such as Ce^{+4} can be used in conjunction with PtO_2

or IrO_2 to mediate the oxidation of water to molecular oxygen.[68] Under similar conditions $Ru(bipy)_3{}^{3+}$ is also reduced in the presence of the same catalysts.[68] Since our work has indicated oxidants as powerful as Ce^{+4} can be generated from irradiation of metal complexes with visible light it does not appear unreasonable that catalysts could mediate water to oxygen oxidation in some of these systems upon photolysis. Experiments in this direction are currently in progress in several laboratories.

One of the most positive features of the one-electron transfer reactions discussed in this report is the versatility that can be achieved by variation of the excited substrate, quencher, solvent or added substrate. Our own investigations have indicated major changes have been possible by a simple tuning of redox properties of the metal complex by minor variation of the ligand structure. Thus in the series of ruthenium (II) complexes represented by 1, 2 and 3 the excitation energy remains nearly constant as the ligand is varied but the couples for both the +2/+1 and +3/+2 reductions are shifted to more anodic values. The ease of reduction of the dication 3 compared with 1 (potential shifted by >0.6V) is particularly noteworthy and suggests that this excited state should be a powerful oxidant. Corresponding variations in redox potentials have also been observed for the iron complexes; thus there appears to be the possibility using these or other metal complexes of constructing reagents or excited complex-quencher systems "tailor-made" for carrying out specific substrate redox chemistry.

ELECTRON TRANSFER IN ORGANIZATES

There has been much recent interest in photoprocesses occurring with molecular organizates such as micelles, vesicles and monolayers or at solution-organizate interfaces. Not surprisingly much of this interest has focused on electron transfer phenomena. Several elegant studies by Thomas, Grätzel and others have demonstrated that incorporation of either excited substrate or quencher in a charged micellar environment can result in pronounced changes in rates of both the quenching and back electron transfer processes.[40,68-71] It has been suggested that the use of micelles or other organizates to control or modify

rates of these processes should furnish a viable path to
energy storage.[72] Our work in this area has involved both
micelles and monolayer assemblies formed by successive
transfer of several monolayer films from an air-water interface to a rigid support. In the present discussion we will
outline a few of our recent findings with the latter and
suggest some possible applications of these to energy conversion.

Our studies of light-induced electron transfer processes in monolayer assemblies have included systems in
which the electron transfer partners are both immobilized
in assemblies as well as those in which the light-absorbing
substrate is immobilized in an assembly such that the
quencher-reactant penetrates the assembly via interfacial
contact from an aqueous or hydrocarbon solution. In both
situations we have observed electron transfer quenching
processes analogous to those observed in solution. Examples
of the former include studies with surfactant analogs of
ruthenium (II) complexes such as 1 and 2 and a surfactant
paraquat.[73] Here we find efficient electron transfer
quenching of the ruthenium complex luminescence can occur
provided contact between the complex and quencher exists.
Thus assemblies with complex and quencher in the same layer
or in adjacent layers having hydrophobic contact results in
efficient quenching but no net chemical reaction; here as
in solution the back electron reaction is evidently very
rapid and efficient.[73] We have also studied the behavior
of assemblies containing surfactant metalloporphyrins
including analogs of 6.[74] Irradiation of the assembly incorporated palladium porphyrin in the presence of the surfactant amine 10, leads to the formation of reductive

$$\underset{\sim\sim}{\underline{10}} \qquad \text{(structure: phenyl ring attached to } N \text{ with two } C_{18}H_{37} \text{ groups)}$$

adducts analogous to the reaction (eq 16) observed in solution. Although the process is not extremely efficient in
the assemblies, the finding that permanent chemistry can
compete with back electron transfer even when both reagents
are relatively immobilized suggests the possibility that a
complex immobilized in an assembly or some other organizate
might function usefully to mediate permanent redox chemistry.

We have recently studied interfacial quenching phenomena using a number of different chromophores and quenchers. Our results indicate that access and in some cases selective binding of the quencher to sites within the assembly plays a major role. Before discussing these results it is useful to summarize findings obtained in related studies from these and other laboratories regarding penetration of multilayer assemblies. A number of investigations have established that small neutral molecules such as O_2, CO and NO can readily penetrate through several multilayers and react at both hydrophilic and hydrophobic sites.[75-77] Small neutral organic molecules can likewise penetrate both hydrophobic and hydrophilic regions of the multilayer assembly and there have been a number of cases observed where rapid reaction takes place.[78,79] In contrast ionic species are in general unable to penetrate hydrophobic regions of the assemblies. In some cases, depending both on the nature of the assembly and the size or mass/charge of the solute, there is no penetration but only an adsorption on the outer surface of the assembly.[78-80] However most frequently metal ions and small organic ions can penetrate to hydrophilic regions inside the assembly and move rapidly at a hydrophilic-hydrophilic interface.[80]

We have studied the reductive photoaddition of two surfactant palladium tetraphenyl porphyrin derivatives whose basic chromophore is the same as 6. In one case the porphyrin resides at a hydrophilic interface while in the other the porphyrin is hydrophobic and probably resides buried in a mostly hydrocarbon region. In the latter case we can observe phosphorescence from the palladium complex in dry assemblies or those contacted with water but the hydrophilic palladium porphyrin unfortunately exhibits no phosphorescence. For the assemblies with the porphyrin in a hydrophobic site addition of N,N-dimethylaniline solvated in aqueous solution results in a quenching of the porphyrin phosphorescence; concurrent with quenching of the luminescence we observe a reductive photoaddition of N,N-dimethylaniline to the porphyrin analogous to the solution reaction.[74] Thus it is clear in this case that the N,N-dimethylaniline can diffuse to the hydrophobic reaction site and participate in the "normal" solution-phase photoreactions. In contrast to the above-discussed reaction where surfactant porphyrin and surfactant are both immobilized in the assembly, the interfacial process proceeds much further towards

completion and a considerable amount of doubly-reduced product (eq 16) can be detected. Similar results are observed
when assemblies containing the other surfactant palladium
porphyrin which resides at a hydrophilic site, are irradiated in the presence of N,N-dimethylaniline; relatively
rapid reaction analogous to that observed in solution occurs
indicating that the N,N-dimethylaniline rapidly penetrates
to hydrophilic regions and can react in analogous fashion.
The observation of comparable reactivity in this reaction
at both hydrophobic and hydrophilic sites is reasonable in
view of our findings in solution investigations that the
reaction is relatively little affected by changes in solvent polarity.

We are currently studying electron transfer quenching
phenomena in hydrophilic sites with neutral and charged
potential quenchers. In these studies we are using the
surfactant ruthenium (II) complex 11[81] and the organic dye
12.[82] Although our study here is not yet complete, we find

11   12

strong indications that access and binding of the quencher
can play a dominant role in controlling photophysical
phenomena. The strongly fluorescent 12 is readily incorporated into assemblies using hosts such as fatty acids
(anionic, since the assemblies are usually prepared at near
neutral pH) or surfactant tetralkylammonium (cationic)
salts. It has previously been found that the fluorescence
of 12 can be quenched efficiently by introduction of a
surfactant paraquat derivative; the quenching has been

shown to be an electron transfer process producing reversibly the paraquat cation radical.[83-85] While the "static"
quenching in the assemblies is quite pronounced -- nearly all
of the fluorescence from 12 can be quenched even with
fairly low paraquat "concentrations"[83] -- the fluorescence
of 12 in solution is relatively difficult to quench because
of the relatively low solubility of paraquat and the short
fluorescence lifetime for 12. Thus at the highest concentration of paraquat in solution (0.02 M) only about 6%
quenching of 12 occurs (K_{SV} = 3.2 M^{-1}).

When 12 is incorporated into monolayer assemblies
using a cationic host and contacted with an aqueous solution containing paraquat, there is little, if any, fluorescence quenching either immediately or on standing in contact with the paraquat solution. In contrast, we find
that assemblies containing 12 and an anionic host exhibit
strong fluorescence quenching upon exposure to aqueous
solutions containing even relatively small amounts of paraquat; here then is little, if any, prompt quenching but a
relatively slow (τ ~ minutes) quenching occurs upon standing.
This slow quenching can be almost completely reversed by
replacing the aqueous paraquat solution with an aqueous
solution of Cd^{2+}, an ion which does not quench the fluorescence of 12. In contrast, exposure of "quenched" asemblies
of 12 to pure water reacts in no reversal of the quenching.
A quenching of 12 similar to that produced by paraquat can
be produced by Cu^{2+} ions. The results indicate that both
the paraquat dication and Cu^{2+} quench as a result of their
penetration and binding to a site in the assemblies near
the immobilized dye 12. Thus in the case of an anionic
host a substantial interfacial quenching is obtained at
concentrations of quencher which would not give rise to
much quenching in homogeneous solution. The pronounced
effects observed can be attributed to a "concentrating" of
the solute-quencher (paraquat or Cu^{2+}) in the assemblies
by binding to the anionic host. In some ways these results
are similar to those obtained by Matheson and Meisel[86,87]
where addition of an anionic polymer (polyvinylsulfate)
to aqueous solutions of cationic substrates and quenchers
results in accelerated quenching phenomena.

We have also observed pronounced environmental effects
in our preliminary studies with 11. Here the excited state
lifetime is much longer (ca. 1 μsec in solution) than for
12 and paraquat can very effectively quench the luminescence

in solution even at low concentrations. In contrast we
find almost no quenching of the luminescence in assemblies
constructed from pure 11 either upon surface contact with
aqueous paraquat solutions or upon interfacial contact of
the solution with assemblies containing "buried" layers of
11. While the dication paraquat is unable to quench
assemblies produced from dication 11, the neutral acceptor
p-benzoquinone is an efficient quencher of the same assemblies; it is not yet clear with the quinone whether there
is simply penetration or a selective binding. In any case
the initial results indicate that in these interfacial
processes using an immobilized excited substrate access
and binding of the quencher can play a major role such that
dramatically enhanced or reduced quenching processes compared to solution can be obtained. As more results are
obtained with different hosts, substrates and quenchers it
should be possible to construct systems where efficient and
selective processes not obtainable in solution occur.
Perhaps the most exciting possibilities for the use of
organizates involve the construction of three (or more)
component functionalized assemblies in which cooperation
of a one-electron transfer event with a catalytic dark
reaction results in efficient net conversion of solute or
solvent molecules to stable products. A potential example
of this might involve monolayer deposition or adsorption
of 11, 12 or other substrate (with a suitable host) on a
catalyst surface such as PtO_2; irradiation of such assemblies in the presence of an aqueous solution with the appropriate quencher should lead to efficient and sustained net
oxidation as reduction of water, depending upon the sub-
strate-quencher combination selected. Other examples could
involve more complicated functionalized assemblies including
the use of surfactant catalysts. These and related applications involving the use of organizates to modify and control light driven one-electron transfer processes are
currently under investigation.

Acknowledgment. The authors are grateful to the National
Science Foundation (Grant No. CHE76-01074), Army Research
Office (Grant No. DAAG29-77-G-0063) and National Institutes
of Health (Grant No. GM15238-11) for support of this work.
We are also especially grateful to the W. R. Grace Company
for an award which assisted investigations by R. H. Schmehl
and P. J. DeLaive and to the North Atlantic Treaty Organization for a grant which made possible collaborations between Dr. C. Giannotti and the North Carolina group.

REFERENCES

1. Photochemical Reactivity in Organized Assemblies XIV.
    Paper XIII, D. G. Whitten, J. A. Mercer-Smith, R. H.
    Schmehl and P. R. Worsham, submitted for publication.
2. H. Leonhardt and A. Weller, Ber. Bunsenges. Phys.
    Chem., 67, 791 (1963).
3. H. Knibbe, D. Rehm and A. Weller, ibid., 72, 257 (1968).
4. D. Rehm and A. Weller, ibid., 73, 834 (1969).
5. D. Rehm and A. Weller, Israel. J. Chem., 8, 259 (1970).
6. K. H. Grellmann, A. R. Watkins and A. Weller, J. Phys.
    Chem., 76, 469 (1972).
7. K. H. Grellmann, A. R. Watkins and A. Weller, ibid.,
    76, 3132 (1972).
8. K. Kawai, N. Yamamoto and T. Tsubomura, Bull. Chem.
    Soc. Japan, 42, 369 (1969).
9. H. Yamashita, H. Kokubun and M. Kuizumi, ibid., 41,
    2312 (1968).
10. R. Bonneau, P. Fornier-de-Violet and J. Joussot-Dubien,
    Photochem. Photobiol., 19, 129 (1974).
11. E. Vogelmann and H. E. A. Kramer, ibid., 23, 383 (1976).
12. H. D. Gafney and A. W. Adamson, J. Am. Chem. Soc., 94,
    8238 (1972).
13. C. R. Bock, T. J. Meyer and D. G. Whitten, ibid., 96,
    4710 (1974).
14. C. R. Bock, T. J. Meyer and D. G. Whitten, ibid., 97,
    2909 (1975).
15. G. Navon and N. Sutin, Inorg. Chem., 13, 2159 (1974).
16. G. S. Lawrence and V. Balzani, ibid., 13, 2976 (1974).
17. J. R. Harbour and G. Tollin, Photochem. Photobiol.,
    19, 147 (1974).
18. R. C. Young, T. J. Meyer and D. G. Whitten, J. Am.
    Chem. Soc., 97, 4781 (1975).
19. R. C. Young, T. J. Meyer and D. G. Whitten, ibid., 98,
    286 (1976).
20. C. Creutz and N. Sutin, Inorg. Chem., 15, 496 (1976).
21. C. T. Lin, W. Böttcher, M. Chou, C. Creutz and N.
    Sutin, J. Am. Chem. Soc., 98, 6536 (1976).
22. C. T. Lin and N. Sutin, ibid., 97, 3543 (1975).
23. C. Creutz and N. Sutin, ibid., 99, 241 (1977).
24. H. E. Toma and C. Creutz, Inorg. Chem., 16, 545 (1977).
25. V. Balzani, L. Moggi, M. F. Manfrin, F. Bolletta and
    G. S. Lawrence, Coord. Chem. Revs., 15, 321 (1975).
26. A. Juris, M. T. Gandolfi, M. F. Manfrin and V. Balzani,
    J. Am. Chem. Soc., 98, 1047 (1976).

27. N. Sutin and C. Creutz, Adv. Chem. Ser., 168, 1 (1978).
28. P. J. DeLaive, J. T. Lee, H. Abruña, H. W.
     Sprintschnik, T. J. Meyer and D. G. Whitten, ibid.,
     168, 28 (1978).
29. P. J. DeLaive, C. Giannotti and D. G. Whitten, ibid.,
     in press.
30. R. Ballardini, G. Varani, M. T. Indelli, F. Scandola
     and V. Balzani, J. Am. Chem. Soc., in press.
31. C. R. Bock. J. A. Connor, A. R. Gutierrez, T. J.
     Meyer, D. G. Whitten, B. P. Sullivan and J. K. Nagle,
     submitted for publication.
32. P. J. DeLaive, J. T. Lee, H. W. Sprintschnik, H.
     Abruña, T. J. Meyer and D. G. Whitten, J. Am. Chem.
     Soc., 99, 7094 (1977).
33. P. J. DeLaive, C. Giannotti and D. G. Whitten, ibid.,
     100, 7413 (1978).
34. J. A. Mercer-Smith and D. G. Whitten, submitted for
     publication.
35. G. Jones, II, T. E. Reinhardt and W. R. Bergmark,
     Solar Energy, 20, 241 (1978).
36. C. Kutal, Adv. Chem. Ser., 168, 158 (1978).
37. P. A. Grutsch and C. Kutal, J. Am. Chem. Soc., 99,
     6460 (1977).
38. D. P. Schwendiman and C. Kutal, ibid., 99, 5677 (1977).
39. J. K. Nagle, unpublished results.
40. M. Grätzel, private communication.
41. C.-T. Lin and N. Sutin, J. Phys. Chem., 80, 97 (1976).
42. P. J. DeLaive and B. P. Sullivan, unpublished results.
43. J.-M. Lehn and J. P. Sauvage, Nouveau J. de Chim., 1,
     449 (1977).
44. B. V. Karyakin, T. S. Dzhabiev and A. E. Shilov,
     Dokl. Akad. Nauk SSSR, 233, 620 (1977).
45. K. Kalyanasundaram, J. Kiwi and M. Grätzel, Helv.
     Chim. Acta, in press.
46. D. G. Whitten, J. C. Yau and F. A. Carroll, J. Am.
     Chem. Soc., 93, 2291 (1971).
47. G. R. Seely and M. Calvin, J. Chem. Phys., 23, 1078
     (1955).
48. W. P. Suboch, A. P. Losev and G. P. Gurinovitch,
     Photochem. Photobiol., 20, 183 (1974).
49. S. G. Cohen, A. Parola and G. H. Parsons, Jr., Chem.
     Revs., 73, 141 (1973).
50. S. G. Cohen and R. J. Baumgarten, J. Am. Chem. Soc.,
     87, 2996 (1965).

51. F. D. Lewis and T. J. Ho, ibid., 99, 7991 (1977).
52. C. Sutcliffe, P. DeLaive and C. Giannotti, unpublished.
53. D. G. Whitten, J. K. Roy and F. A. Carroll, in "The Exciplex," Ed. M. Gordon and W. R. Ware, Academic Press, New York, N. Y., 1975, p. 247.
54. A. Ledwith, Accts. Chem. Res., 5, 133 (1972).
55. E. M. Kosower and J. L. Cotten, J. Am. Chem. Soc., 86, 5524 (1964).
56. S. Markiewicz, M. S. Chang, R. H. Sparks, C. A. Evans and J. R. Bolton, Int'l Conf. on Photochem. Conversion and Storage of Solar Energy, London, Canada, 1976, Abstract E-7.
57. K. Takuma, M. Tajiwara and T. Matsuo, Chem. Lett., 1199 (1977).
58. These reactions are only observed in oxygen-free solutions.
59. K. Kalyanasundaran, J. Chem. Soc., Chem. Comm., 628 (1978).
60. R. F. Dapso and C. K. Mann, Anal. Chem., 35, 677 (1963).
61. C. K. Mann and K. K. Barnes, "Electrochemical Reactions in Nonaqueous Systems," Marcel Dekker, Inc., New York, N. Y., 1970, p. 268.
62. C. P. Russell, Anal. Chem., 35, 1291 (1963).
63. W. H. Richardson, "Oxidation in Organic Chemistry, Part A," Ed. K. B. Wiberg, Academic Press, New York, N. Y., 1965, p. 244.
64. R. C. Young, J. K. Nagle, T. J. Meyer and D. G. Whitten, J. Am. Chem. Soc., 100, 4773 (1978).
65. J. Phillips, J. A. Koningstein, C. H. Langford and R. Sasseville, J. Phys. Chem., 82, 622 (1978).
66. T. K. Foreman, P. J. DeLaive and D. G. Whitten, unpublished results.
67. A. Moradpour, E. Amouyal, P. Keller, and H. Kagan, Abstr. Solar Energy Meeting, Cambridge, England, 1978, p. 31.
68. J. Kiwi and M. Grätzel, unpublished manuscript.
69. M. Grätzel in "Micellization, Solubilization and Microemulsion Vol II," Ed., K. L. Mittal, Plenum Press, 1977, p. 531.
70. J. K. Thomas, F. Frieser and M. Wong, Ber. Bunsenges. Phys. Chem., 82, 937 (1978).
71. Y. Moroi, A. Braun and M. Grätzel, ibid., 82, 950 (1978).
72. A. Henglein and M. Grätzel in "Solar Power and Fuels," Ed., J. R. Bolton, Academic Press, New York, N. Y., 1977, p. 53.

73. G. Sprintschnik, H. W. Sprintschnik, P. R. Kirsch and D. G. Whitten, J. Am. Chem. Soc., 99, 4947 (1977).

74. J. A. Mercer-Smith and D. G. Whitten, unpublished results.

75. F. R. Hopf and D. G. Whitten, J. Am. Chem. Soc., 98, 7422 (1976).

76. B. E. Horsey and D. G. Whitten, ibid., 100, 1293 (1978).

77. B. E. Horsey, F. R. Hopf, R. H. Schmehl and D. G. Whitten, Porphyrin Sympos. Proc., in press.

78. H. Kuhn and D. Möbius, Angew Chem., Int'l Ed. Engl., 10, 620 (1971).

79. H. Kuhn, D. Möbius and H. Bücher, in "Physical Methods of Chemistry," Vol. 1, Part 3b, Ed., A. Weissburger and B. Rossiter, John Wiley and Sons, Inc., New York, N. Y., 1972, p. 577.

80. R. H. Schmehl, G. Shaw and D. G. Whitten, Chem. Phys. Lett., in press.

81. We thank Dr. W. Sasse for supplying us a sample of this compound.

82. We thank Dr. D. Möbius for a sample.

83. K.-P. Seefeld, D. Möbius and H. Kuhn, Helv. Chim. Acta., 60, 2608 (1977).

84. H. Kuhn, Pure Appl. Chem., in press.

85. H. Kuhn, Proc. Solar Energy Conference, Cambridge, England, 1978, in press.

86. D. Meisel, J. Rabani, D. Meyerstein and M. S. Matheson, J. Phys. Chem., 82, 985 (1978).

87. D. Meyerstein, J. Rabani, M. S. Matheson and D. Meisel, ibid., 82, 1879 (1978).

EFFECT OF THE MICELLAR PHASE ON PHOTO-INDUCED REACTIONS(1)

J. K. Thomas

Chemistry & Radiation Laboratory

Univ. of Notre Dame, Notre Dame, IN 46556

ABSTRACT

The effect of phase on electron ejection and electron transfer processes is discussed. In particular the effect of micellar systems on the efficiency of these processes and the decreased energy for threshold is illustrated and evaluated for the possible application to storage of solar energy.

INTRODUCTION

Events (2) of the last decade have urged the scientific community to improve conventional and to seek alternate sources of energy. One alternative that has appealed to chemists is the storage of solar energy and its subsequent conversion to heat, chemical compounds of high energy, or electrical power. The task is particularly appropriate to photochemistry, and this manuscript indicates several different photochemical avenues that might be explored. This book contains work by Hautala(3) and Jones(4) where light is used to isomerize a particular molecule to a high energy form. The light energy is stored as chemical energy in this process and is subsequently released for use at some later date when the original compound is reformed and heat is released. Another technique originally suggested by Rabinowitch(5) and extended by Lichtin(6) utilizes the re-

141

versible light driven reaction of ferrous ions and thionine
to give measurable yields of electrical power. Much of the
basic work in these systems is now complete.

The present paper discusses several possible avenues
of light driven reactors leading to the production of ions
and possibly to the breakdown of water into H_2 and O_2. In
particular the effect of medium on the promotion of ionization events is discussed. This work is not near
completion, but the promise of the initial data and the
intriguing details of the system are very appealing. For
example, micellar systems, which will occupy the bulk of the
discussion, provide the photochemist with a means of organizing his chemical reactants on a molecular scale, thereby
promoting the particular reaction of interest.

The experimental section for this manuscript should
cover well nigh all techniques of photochemistry and in
particular laser flash photolysis. The reader is referred
to the original papers for experimental details.

TYPES OF REACTION

Two photo-induced reactions in particular will be
discussed in this manuscript, simple photo-ionization of the
type,

$$A \xrightarrow{h\nu} A^+ + e^-$$

and photo-induced electron transfer,

$$A + B \xrightarrow{h\nu} A^+ + B^-$$

Photo-ionization processes of the type indicated above
require upwards of 6.5 ev in the gas phase and are operative
in the far uv spectral region. Both photo-ionization (PI),
and electron transfer (ET), tend to be inefficient due to
back neutralization of the ions producing excited states.
It is at this stage that operation of these systems in the
condensed phase leads to several advantages for photochemists.

PHOTO-IONIZATION IN LIQUIDS

Simple photo-induced conduction methods (7,8,9) have been used to measure onsets of photo-ionization in insulating liquids such as alkanes. Fig. 1 shows typical data. The onset of conduction on irradiation of the system with light of appropriate wavelength is an indication of photo-ionization. It is noted immediately that the ionization potential (IP) of the molecule in an alkane is some 2 ev lower than that in the gas phase. This is interpreted as being due to the polarization energy of the cation P_+, and the energy of the electron in the medium V_0. If the gas phase IP is I_g and that of the liquid I_s, then $I_s = I_g + P_+ + V_0$. As P_+ is $\sim-$ 1.5 ev and V_0 varies from 0 in hexane to as much as -0.5 in tetramethylsilane, TMS, then $I_s - I_g \sim-2.0$ ev.

Although significant, this lowering of IP by the condensed phase still requires light of $\lambda < 3000\overset{o}{A}$ for photo-ionization. However, the V_0 for liquid water is stated to be -1.6 ev(10) and IP would be reduced in aqueous systems by more than 3.0 ev, thus enabling violet light to produce photo-ionization. Unfortunately the molecules used in these experiments have limited solubility in water. This drawback can be overcome by utilizing micellar solutions where the solute to be photo-ionized is solubilized by an aggregated surfactant such as sodium lauryl sulfate, NaLS, cetyltrimethylammonium bromide, CTAB, etc. Dissolution of these molecules in water beyond a critical concentration called the critical micelle concentration (CMC) leads to the formation of aggregates of the surfactant called micelles (Fig. 2). About 50 to 100 monomer surfactant units aggregate together giving a roughly spherical entity of radius 15 to 20Å with a net charge of 20 to 30 units. This charge, formed only by ionic surfactants is due to dissociation of the head groups giving Na^+ ions (NaLS), or Br^- ions (CTAB). The micelle thus is surrounded by an atmosphere of counterions.

The micelles solubilize hydrophobic molecules of the type used in the present work, usually at the micelle/water interface. Laser photolysis of molecules such as phenothiazine, PTH, (11) tetramethylbenzidine, TMB, (12) aminopyrene (13) and aminoperylene (14) in NaLS micelles gives rise to a one photon photoionisation giving the solute cation and the hydrated electron e_{aq}^-. The cation remains

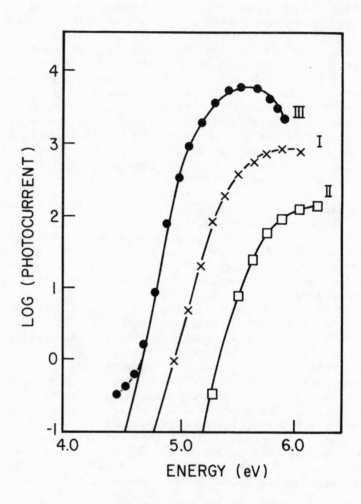

I   PHENOTHIAZINE IN   TMS

II   PHENOTHIAZINE IN   ISO OCTANE

III TMB IN TMS

FIGURE 1

Photo-conduction curves in non polar liquids.

$$CH_3 - (CH_2)_n - \overset{\overset{\displaystyle CH_3}{|}}{\underset{\underset{\displaystyle CH_3}{|}}{N^+}} - CH_3 \quad Br^-$$

CATIONIC

$$CH_3(CH_2)_n - SO_4^- \quad Na^+$$

ANIONIC

$$R - \langle \bigcirc \rangle - (OCH_2 - CH_2)_{\overline{n}} OH$$

NONIONIC

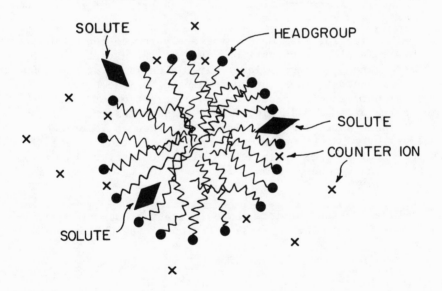

FIGURE 2

Surfactants and suggested structure of micelle.

associated with the anionic micelle while e^- exits into the
aqueous phase forming hydrated electrons, e_{aq}^-. In the case
of TMB the radical cation is stable for several days in
NaLS micelles provided e_{aq}^- is converted to a stable negative
species by reaction with a second solute in the aqueous
phase. Photo-ionization of molecules such as pyrene, anthracene, etc. proceeds via a two photon process. In all cases
the efficiency of photo-ionization of the above solutes is
much higher in micellar solution than in homogeneous solvents such as alcohol or hydrocarbons. This is interpreted
as being due to prevention of the back neutralization of the
cation and e_{aq}^- by the electrostatic repulsion of the negative micellar surface. (9)

$$S \xrightarrow{h\nu} S^+ + e_{aq}^- \qquad \text{homogeneous solvent}$$
$$\longrightarrow S \text{ or } S^*$$

$$(\text{micelle})\ S \xrightarrow{h\nu} (\text{micelle})\ S^+ + e_{aq}^- \longrightarrow \begin{array}{l} \text{Products of} \\ \text{ion escape} \end{array}$$

The micellar systems also severely reduce the IP of the
solute molecules below that observed in the gas phase. The
extreme case is aminoperylene where green light of $\lambda = 5300\overset{\circ}{A}$
(2.4 ev) photo-ionizes this molecule in NaLS solutions.
Typical data showing onsets of photoionization of several
molecules are shown in Fig. 3. It is seen that a non-polar
liquid such as tetramethylsilane, TMS, reduces IP of the
solutes by ~ 2.0 ev while micellar NaLS solution reduces it
even further. This is in accord with the earlier suggestion
that P_+ and V_0 change from ~ -2.0 ev in TMS to ~ -3.6 ev in
water.

Micelles thus promote photo-ionization both by reducing
IP and by decreasing the back reaction of the photoproduced
ions.

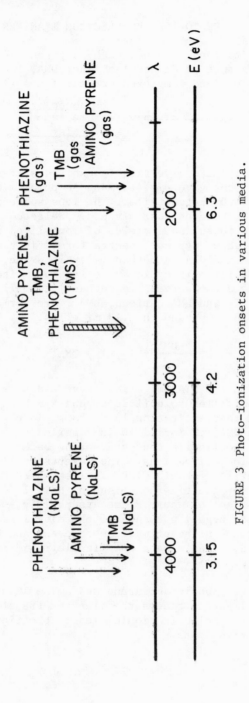

FIGURE 3 Photo-ionization onsets in various media.

PHOTO-INDUCED ELECTRON TRANSFER

Excitation of arenes such as pyrene, P, in the presence
of amines such as dimethylaniline, DMA, leads to the formation of excited complexes or exciplexes.(15)

$$P \xrightarrow{h\nu} P* \xrightarrow{DMA} (PDMA)* \begin{cases} \xrightarrow{\text{nonpolar medium}} \text{other excited states} \\ \xrightarrow{\text{polar medium}} \text{ion pairs} \end{cases}$$

These exciplexes give rise to excited states in non-polar
solvents, but give rise to ions in polar media such as
alcohols and acetonitrile. The ions subsequently recombine
rapidly in homogeneous solvents. Selected micellar systems
give rise to enhanced yields of long lived ions (16,17).
Figure 4 shows typical spectra formed in the laser photolysis
of pyrene and DMA in methanol, cyclohexane, C_6H_{12}, and in
CTAB micelles. In cyclohexane mainly excited triplet states
of pyrene are observed, in methanol a small yield of pyrene
anions, P^-, and DMA cations, DMA^+, are formed, this latter
yield being enhanced in CTAB micelles.

$$P \xrightarrow{h\nu} P* \xrightarrow{DMA} P^- + DMA^+$$

Again the CTAB micelles serve a double purpose:
(a) the micelles solubilize pyrene and DMA in close proximity so that excitation leads to rapid e^- transfer, and
(b) when DMA^+ is formed it is repelled by the positive surface of the micelle thus preventing back reaction and giving
rise to long lived ions. The inserts on Figure 4 show the
time profile of P^- in CTAB micelles where no decay is observed over the time scale shown (half-life $\approx$ 5 msec), and
in NaLS micelles where P^- decays rapidly. In this latter
system the negative change of the micelle attracts the
DMA^+ formed in the primary photo-event and because this
species does not escape the micelle, rapid neutralization with
P^- ensues.

Correct choice of arene and amine leads to systems where
visible light initiates the electron transfer. Correct
choice of micelle (cationic) leads to efficient production
of long lived ions.

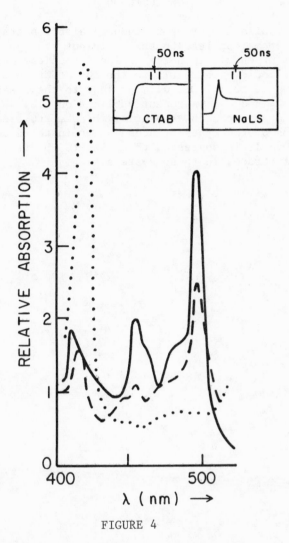

FIGURE 4

Short lived species in the laser photolysis of pyrene and
dimethylaniline in C_6H_{12}(.....); methanol (------); and
CTAB micelles (————). Inserts indicate the time
dependence for formation and decay of pyrene anion in NaLS
and CTAB micelles.

CONCLUSION

Photo-ionization or photo-induced electron transfer in micellar systems can lead to the efficient storage of light in the form of chemical energy of ionic species. Future development could either utilize the ions to form electrical current in an external circuit, or lead to the breakdown of water into H_2 and O_2. A recent publication (18) shows that colloidal catalysts such as platinum black, PtO_2, can utilize ions of the types produced above in the decomposition of H_2O. It is suggested that the PtO_2 particles act as micro-electrodes in these systems.

REFERENCES

(1) The research described herein was supported by the Office of Basic Energy Sciences of the Department of Energy. This is Document No. NDRL-1947 from the Notre Dame Radiation Laboratory.

(2) Science, 184, 247-386 (1974).

(3) R. Hautala, et al., this book.

(4) G. Jones, et al., this book.

(5) E. Rabinowitch, J. Chem. Phys., 8, 551 (1940).

(6) N. Lichtin and M. Hoffman, this book.

(7) R. C. Jarnagin, Acc. Chem. Res., 4, 420 (1971).

(8) R. A. Holroyd and R. L. Russell, J. Phys. Chem., 78, 2128 (1974).

(9) P. Piciulo and J. K. Thomas, J. Chem. Phys., 68, 3260 (1978).

(10) G. C. Barker, G. Bottura, G. Cloke, A. W. Gardner and M. J. Williams, Electroanal Chem. Interfacial Electrochem, 50, 323 (1974).

(11) S. A. Alkaitis, G. Beck and M. Grätzel, J. Am. Chem. Soc., 97, 5723 (1975).

(12) S. A. Alkaitis and M. Grätzel, J. Am. Chem. Soc., 98, 3549 (1976).

(13) J. K. Thomas and P. Piciulo in "Interfacial Photoprocesses: Applications to Energy Conversion and Synthesis," ACS Advances in Chemistry Series, ed. by M. S. Wrighton, American Chemical Society, Washington, D.C., in press.

(14) J. K. Thomas and P. Piciulo, J. Am. Chem. Soc., 100, 3239 (1978).

(15) H. Leon Hart and A. Weller, Ber. Bun. Phys. Chem., 67, 791 (1963).

(16) B. Razem, M. Wong, and J. K. Thomas, J. Am. Chem. Soc., 100, 1679 (1978).

(17) H. Masuhara, K. Kaji,and N. Mataga, Bull. Chem. Soc. Japan, 50, 2084 (1977).

(18) B. V. Korvakin, T. S. Dzhabiev and A. E. Shilov, Dokl. Phys. Chem., 233, 359 (1977).

PHOTOCHEMICAL DETERMINANTS OF THE EFFICIENCY

OF PHOTOGALVANIC CONVERSION OF SOLAR ENERGY

Morton Z. Hoffman and Norman N. Lichtin

Department of Chemistry, Boston University

Boston, Massachusetts 02215

ABSTRACT

A photogalvanic cell is a battery in which the cell
solution absorbs light directly to generate species which,
upon back reaction through an external circuit with the
aid of suitable electrodes, produces electric power;
photoactivation of the electrodes is not involved. The
charge-carrying species have storage capacity if they are
long-lived and can be prevented from engaging in degradative back reactions in bulk solution. The efficiency of a
photogalvanic cell for the conversion of photon energy into
electrical energy is determined by photochemical and electrochemical factors. Among the latter are the choice of
electrode materials and the kinetics of electron transfer
at the heterogeneous surfaces. In this paper we examine
the photochemical determinants of the efficiency of photogalvanic cell operation: the absorption spectral characteristics of the cell solution, the efficiency of formation
of separated charge carriers, and the lifetimes of the
carriers toward back electron transfer. Modulation of bulk
solution dynamics can be achieved by variation of the
solution medium. The photochemical determinants are
discussed with particular reference to the use of thionine
or $Ru(bpy)_3^{2+}$ as the light absorbing species.

INTRODUCTION

It is obvious that the practical value of a device
that can effect the direct conversion of solar energy into
electricity with the capacity for storage would be
enormous. Such a rechargeable solar battery could have
applicability as a mobile source and in areas isolated
from an electric grid system. In principle, photogalvanic
devices can achieve these goals. A photogalvanic cell is
a battery in which the cell solution absorbs light directly to generate species which, upon back reaction through
an external circuit with the aid of suitable electrodes,
produces electric power; photoactivation of the electrodes
is not involved. In its simplest terms, photoexcitation
of the absorbing species leads to an excited state which
can undergo reductive or oxidative quenching to form
charge carriers. The conversion of charge carriers to the
original substrates at suitable electrodes provides electrons that can be driven through an external circuit to
perform useful work. The capacity for storage derives from
the lifetimes of the charge carriers; the longer lived the
charge carriers with respect to energy degradative back
electron transfer in bulk solution, the greater the opportunity for storage to pass from the microtime domain into
the macrotime. Efficient storage would be achieved if the
charge carriers were kinetically stable.

In this paper, the photogalvanic process is examined
in terms of the photochemical determinants which establish
the efficiency of such a cell with particular reference to
the use of thionine and $Ru(bpy)_3^{2+}$ as the light absorbing
species in solution. As will be seen, the current state
of the art does not permit the utilization of photogalvanic
cells as practical solar energy devices at the present time
but some research approaches for future development are
presented.

THE PHOTOGALVANIC PROCESS

A photogalvanic cell as defined in the Introduction
is a closed-cycle quantum device which operates without
the net consumption of any chemical component (except
through undesirable side reactions or other losses). In
a photogalvanic transducer, absorbed quanta drive a redox
reaction which, in the ground state, is endergic; the

prompt, spontaneous reversal of this reaction through the external electrical circuit restores the system to its original composition. In a photogalvanic storage cell, spontaneous reversal does not take place promptly. Products of the photoredox reaction are stored by some means for an essentially infinite period but ultimately are caused to react through the external circuit to restore the original composition. In a photocatalyzed galvanic cell, production of electricity is associated with a net change in the chemical composition of the system. As a result, the cell cannot be operated in multiple closed cycles.

There are five fundamental steps inherent in photochemical conversion (1): 1) absorption of light; 2) generation of charge carriers; 3) transport of charge carriers to electrodes; 4) electron exchange between charge carriers and electrodes; 5) delivery of current to load. Step 1 is simple in concept and involves matching the absorption spectrum of the cell solution as well as possible to the insolation spectrum by means of the design of absorbing species and sensitizers. Step 2 may involve one or more elementary reactions and must produce charge carriers which can store the energy derived from the solar photons long enough to allow the following step to occur. Step 3 requires the transport of the charge carriers to the electrodes to compete with their bulk back reactions in homogeneous solution; by means of device design and chemical tuning, the solution lifetime of charge carriers can be maximized. Options available in step 4 include diffusion-controlled electron transfer at reversible electrodes and efficient blocking of the charge carriers at selective electrodes. Manipulation of these options is essential to the ultimate achievement of practical photogalvanic devices. When the electrodes are conductors, step 5 is trivial. When a semiconductor is used as an electrode to achieve selectivity, resistive losses in the electrode can be substantial if suitable precautions are not taken.

Photogalvanic Devices and Electrical Output

A number of analyses of the general thermodynamic and kinetic aspects of the electrical output of photogalvanic cells have been published (2-4). It is inconvenient to separate such analyses completely from device specifics so that in this section we shall discuss the electrical

output of three types of photogalvanic converters utilizing the iron-thionine system: "primitive beaker" cell,
the totally illuminated thin-layer (TI-TL) transducer,
and storage cells.

After the discovery over forty years ago (5,6) that
thionine is reversibly photoreduced by Fe^{2+} in acidic
aqueous solution, Rabinowitch (7) explored the possibility
of utilizing the system in the construction of a photogalvanic cell for solar energy conversion. In the "primitive beaker" cell shown in Figure 1, the electrodes are
identical with one illuminated and the other in the dark.
The open circuit voltage, V_{oc}, is ideally the difference
in the reversible potentials at the two electrodes. For
the iron-thionine system, the concentration of leucothionine is negligible at the dark electrode so that the
exchange current is determined entirely by the iron couple.
The potential at the dark electrode is given by the Nernst
expression in equation 1 (8) where E^o_{Fe} is the standard

$$E_{dark} = E^o_{Fe} + 0.06 \log([Fe^{3+}]/[Fe^{2+}])_{dark} \qquad (1)$$

reduction potential of the Fe^{3+}/Fe^{2+} couple and the terms
in the square brackets represent the activities of the
species. The potential at the illuminated electrode is
a complex function and depends on the concentrations of
all the redox species at the electrode (8). The expression
for this potential can be simplified for an ideal electrode
by assuming that the electron transfer rate constants for
all the redox species are the same and further simplified
by assuming that the photostationary concentration of the
half-reduced dye, semithionine, is negligible. The result
is equation 2 where $f = F/RT$, F is the Faraday in coulombs,
E_{light} is the potential of the illuminated electrode, E^o_{Fe}

Thionine, T   Leucothionine, L

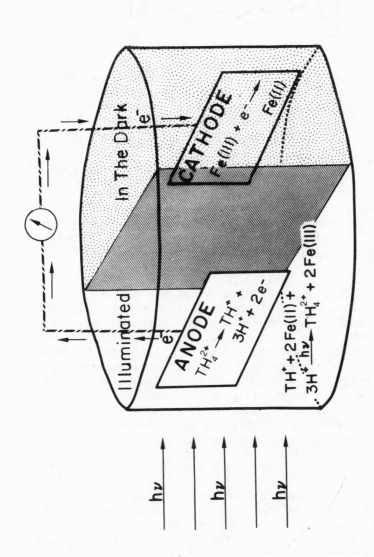

Figure 1. "Primitive-beaker" iron-thionine photogalvanic cell. TH^+ = thionine, TH_4^{2+} = leucothionine, both at pH 2. The overall light-driven cell reaction and the spontaneous energy-producing reactions at the electrodes are shown.

$$\left[\underset{H_2N}{\underbrace{}} \overset{S}{\underset{\underset{H}{N}}{}} \underset{NH_2}{} \right]^{+}$$

Semithionine, S

is the standard reduction potential of the Fe^{3+}/Fe^{2+}
couple, $E^o_{T \to S}$ is the standard potential for the one-electron reduction of thionine, and $E^o_{S \to L}$ is the standard
potential for the one-electron reduction of semithionine.
The terms in the square brackets represent the activities

$$\exp(fE_{light}) = \frac{[Fe^{3+}]\exp(fE^o_{Fe}/2) + 2[T]\exp(fE^o_{T \to S}/2)}{[Fe^{2+}]\exp(-fE^o_{Fe}/2) + 2[L]\exp(-fE^o_{S \to L}/2)} \quad (2)$$

of the various species in the photostationary state.
Analogous terms would be appropriate for other photoredox
systems. The potential of the "primitive beaker" photogalvanic cell under open circuit photostationary conditions
is equal to $E_{dark}-E_{light}$ as calculated from equations 1 and
2.

The current produced by a "primitive beaker" photogalvanic cell with ideally reversible electrodes depends
on the composition of the bulk solution in the photostationary state under short-circuit or other conditions of
current withdrawal and on the rate of diffusion of charge
carriers to the electrodes. This current is related to
various system parameters via Nernst's diffusion current
relationship (9), equation 3, where n is the number of
redox equivalents per mole of a given charge carrier, F is
the Faraday, A is the electrode area, D is the diffusion
coefficient of a given charge carrier, C_b is the concentra-

$$i = n \, F \, A \, D \, C_b/\delta \quad (3)$$

tion of a given charge carrier in bulk solution under
photostationary conditions, and δ is the thickness of the
diffusion layer. The situation is complicated when
electrodes are reversible to both redox couples (2) since
the current cannot be related to the bulk concentration of
a single charge carrier at each electrode. A detailed
treatment of this is beyond the scope of this presentation.
For "primitive beaker" cells, only very low sunlight
engineering efficiencies (S.E.E. $\leq 10^{-3}$%) have been

reported. The crucial figure of merit in evaluating the efficiency of solar energy conversion is S.E.E. as defined in equation 4 for conversion to electrical power or energy.

$$S.E.E. = \frac{\text{electrical power or energy delivered to load}}{\text{incident sunlight power or energy}} \times 100$$

(4)

The totally illuminated thin-layer (TI-TL) photogal-vanic cell was first described by Clark and Eckert (10) and a number of studies of the behavior of the iron-thionine photoredox system in such cells have been report-ed (8,11-17). In TI-TL cells, at least one electrode is selective; ideally, a selective electrode completely blocks one redox couple while it is completely reversible to the other. Greatest efficiency would be achieved with a TI-TL cell in which each electrode was perfectly selec-tive for a different couple. Figure 2 shows a schematic drawing of a TI-TL iron-thionine cell in which the anode is n-type SnO_2 which responds selectively to the thionine/reduced thionine couple while non-selective materials (such as Pt, indium tin oxide, or Cd_2SnO_4) are used as cathodes. TI-TL cells offer the possibility of develop-ment as practical area devices analogous to solid-state photovoltaic devices. The best S.E.E. value reported (12) for a TI-TL iron-thionine cell, $\sim 0.06\%$, is more than 100 times less than S.E.E. values readily achieved with both solid-state and liquid-junction (18) photovoltaic devices. The theoretical upper limit of S.E.E. of TI-TL iron-thionine photogalvanic cells is 2-3% and with sensitiza-tion to the blue is 5-10%.

Little research on photogalvanic storage cells has been reported. Such devices require long-term storage of potential energy in a galvanic cell which can be charged directly by light and recharged by light after each cycle during which its potential energy is withdrawn as electri-city. The ideal system for such storage would be one in which the photoredox reaction produced endergic products which would release their free energy efficiently upon back reaction at the electrodes but at a negligible rate by back reaction in bulk medium. The back reaction through electron transfer at suitable electrodes would take place only when the external electrical circuit was closed between the electrodes. No such system appears to

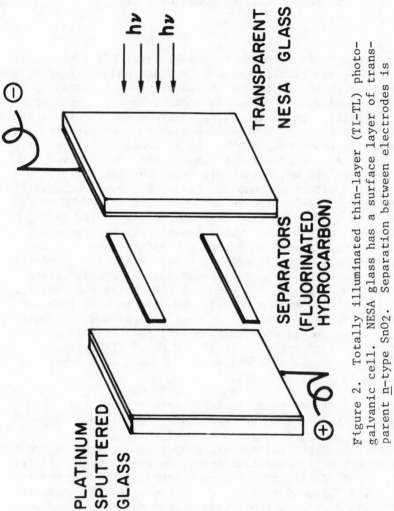

Figure 2. Totally illuminated thin-layer (TI-TL) photo-galvanic cell. NESA glass has a surface layer of transparent n-type SnO2. Separation between electrodes is typically 25-80 μm.

have been reported. The design of such systems is an important and exciting challenge to photochemists.

At least two approaches to photogalvanic storage have been developed which do not depend upon the inertness of photoredox products in bulk medium. Both approaches involve physical separation, one of reactants, the other of products. The process of charging a photogalvanic storage cell with reactants separated requires that light-driven electron transfer takes place through an external circuit or an asymmetric bilayer membrane. Storage via physical separation of photoredox products requires separation techniques, e.g., phase separation, which are fast compared to spontaneous back reaction in bulk medium. The photoreduction of thionine to leucothionine by Fe^{2+} in the presence of excess acetate ion, which is irreversible because of the precipitation of insoluble ferric acetate (19), could conceivably be the basis for a "products separated" photogalvanic storage cell; it does not appear to have been exploited for this purpose. The relatively slow rate of the bulk back reaction of leucothionine with the EDTA complex of Co(III) has been exploited in the extraction of leucothionine into ether from aqueous medium; the back reaction of the separated products via an external circuit has been demonstrated (20).

THE ELECTROCHEMICAL DETERMINANTS

Regardless of the type of photogalvanic cell, efficient electron transfer must occur between photo-chemically-generated separated charge carriers and appropriate electrodes. The determinants of the electrode processes in "primitive beaker" cells, in which electrodes are reversible to all the redox couples, have been discussed in the previous section. If practical photo-galvanic cells are ever achieved, it appears likely that electrodes selective to particular redox couples will be employed. Selectivity is not, however, the only require-ment. In addition, the electrode must not catalyze back reaction of charge carriers on its surface. Such back reaction is equivalent to shorting a circuit in parallel with load. Surface catalysis of the oxidation of leuco-thionine by Fe^{3+} on its n-type SnO_2 selective anode has been identified as a major loss factor in TI-TL iron-thionine cells (16).

Requirements for Electrode Selectivity

The theory of the phenomena at the interface between semiconductors and electrolytic solutions has been developed by Gerischer (21) and by Levich (22). This theory has been applied extensively to photoelectrochemical (liquid-junction) devices (23) in which quanta are absorbed by the semiconductor electrodes. Application to photo-galvanic devices is simpler because excitation of the semiconductor is not involved and key features of this application follow.

1) The efficiency of electron transfer across the interface between electrolyte and semiconductor is largely governed by relationships between the redox potentials of electrolyte couples and the energies of bands of the semiconductor. Reduction potentials on the NHE scale in aqueous solution can be converted to the energy scale used for semiconductor levels which takes the energy of a free electron in vacuum as zero, by reversing the sign of the reduction potential and subtracting 4.6.

2) Electron transfer between a redox couple and a semiconductor electrode is blocked, in whole or in part, if the redox potential of the couple falls within the gap between the highest energy level of the valence band of the semiconductor and the lowest energy level of its conduction band. In general, blocking is more complete the farther the redox potential of the couple is from the two band edges. The relationship between redox potential and semiconductor energy levels depends on the magnitude of the band gap, the energy of the valence (or conduction band edge), and the redox potential of the couple under these conditions of use. It should be noted that since the band gap determines the wavelength of light above which a semiconductor does not absorb light, band gaps of 3.5V or greater (354 nm or less) are particularly appropriate for photogalvanic solar devices because essentially 100% of the insolation spectrum at sea level is at wavelengths longer than this value.

3) n-Type semiconductors are best suited for use as selective anodes while p-type semiconductors are most appropriate as selective cathodes. The electron energy levels of an n-type semiconductor bends upwards near the interface with an electrolytic solution (see Figure 3) while band-bending is downward for a p-type semiconductor. Thus, electrons can flow into the conduction band of an n-type semiconductor from a redox couple with a potential

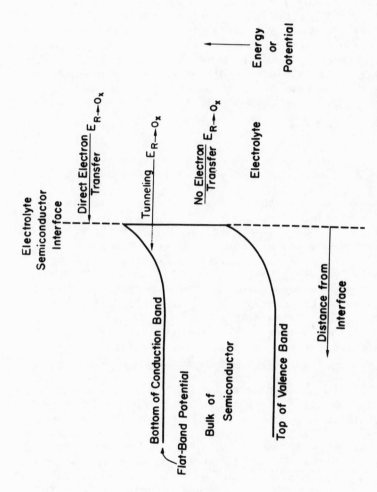

Figure 3. Diagram of energy relationships in the vicinity of the interface between an n-type semiconductor and redox electrolyte solutions.

at or above the bottom edge of the conduction band at the
interface. If the redox potential is below that level but
at least as high as the flat-band potential of the n-type
semiconductor (i.e., the lower edge of the conduction band
in the bulk of the semiconductor and outside the band-
bending region), electron transfer to the electrode may
still occur by tunneling as illustrated in Figure 3.
Similarly, electrons can flow from the conduction band of
a p-type semiconductor to a redox couple, the potential
of which lies below the bottom edge of the conduction band
at the interface or at least below the flat-band potential. Many n-type semiconductors are known which have
large enough band gaps to be transparent to sunlight
(>3.5V) but few, if any, transparent p-type semiconductors
are available.

Iron-Thionine TI-TL Cell with Selective
SnO_2 Anode and Poorly Selective Cathode

It has been shown that conversion of visible light to
electricity by the iron-thionine TI-TL photogalvanic transducer with n-type SnO_2 anode and Pt, indium tin oxide, or
Cd_2SnO_4 cathode is largely due to the preference of the
anode for accepting electrons from leucothionine rather
than from Fe^{2+} (14-16). Selectivity is limited with the
result that photogalvanic voltages are substantially
reduced because the anode potential reflects a mixed
exchange current. Exchange with the dye couple predominates but exchange with the iron couple is significant (8).
The cathodic voltage reflects a mixed potential in which
exchange by the iron couple predominates.

It has been shown that essentially all charge carriers
generated in the TI-TL iron-thionine photogalvanic cell
reach the electrodes which are separated by 25 or 80 μm.
Incompleteness of selectivity of electrode processes does
not fully account for deficiency of output compared to that
calculated from the photostationary compositions. A loss
factor of 4-8 appears to be due to inefficient transfer
of electrons between charge carriers and electrodes. This
inefficiency has been ascribed to complexation of Fe^{3+} by
leucothionine (24) adsorbed on SnO_2 followed by back
reaction of a portion of the adsorbed complexes, desorption of the other complexes, and diffusion back into the
bulk solution (16).

THE PHOTOCHEMICAL DETERMINANTS

Even if electron transfer at the electrodes were perfectly efficient and the maximum current and voltage from the photogenerated charge carriers could be achieved, the efficiency of photogalvanic conversion of solar energy would be limited by those factors that are part of the photochemical mechanism: the absorption spectral characteristics of the cell solution, the efficiency of formation of the separated charge carriers, and the lifetimes of the carriers toward back electron transfer. In this section of the paper we shall discuss the general nature of the photochemical determinants, their limitations with regard to thionine and $Ru(bpy)_3^{2+}$ (bpy = 2,2'-bipyridine) as the absorbing species, and some means which are available for varying and optimizing these parameters. The use of thionine in photogalvanic cells has been discussed in previous sections; $Ru(bpy)_3^{2+}$ has also been shown (25) to engage in photogalvanic action.

Absorption of Light

The essential requisite for any photochemical process is the absorption of light; the ideal solar energy absorber is one which could absorb all the photons from the sun that impinge upon it with the concomitant generation of the reactive excited states that lead to the separated charge carriers. In reality, not all the photons will be absorbed nor will they all, even if absorbed, lead to the reactive excited states and charge carriers. The absorption of light is a wavelength-dependent phenomenon governed by the Beer-Lambert law, $A_\lambda = \varepsilon_\lambda \ell c$, where A_λ is the absorbance of the solution at a particular wavelength, ε_λ is the molar absorptivity $(M^{-1}cm^{-1})$ of the solution at that wavelength, ℓ is the pathlength (cm) of the absorbing solution, and c is the concentration (moles liter^{-1}) of the absorbing species. If the absorbance of the solution were unity at all wavelengths, 90% of the sunlight would be absorbed; if A = 2, 99% of the sunlight would be absorbed. In order to achieve that high absorption of all the wavelengths of sunlight, the pathlength and/or concentration of the absorbing solution would have to be large enough to offset the low values of ε at some wavelengths. Since ε is not uniform across the wavelength range, the very high absor-

bancies that must occur at some wavelengths can be limiting factors; absorption of light would be virtually
complete in a very small depth (perhaps less than 1mm) at
the window. The resultant inhomogeneity of excited state
and charge carrier concentration across the bulk of the
solution would create kinetic and diffusional limitations.
Long pathlengths present the potential problem of the
diffusion of charge carriers through a large bulk of
solution. From the practical standpoint, large cell pathlengths will increase the volume and weight of solution
that must be contained and high concentrations of absorbers
can be equated with a high cost of expensive materials.

Assuming that a pathlength of $\sim$10 cm and an absorber
concentration of $\sim 10^{-3}$M probably represent upper limits,
an absorbance of unity is achieved for $\varepsilon_\lambda \geqslant 10^2$M^{-1}cm^{-1}.
In practice, ε_λ values of $<10^3$M^{-1}cm^{-1} are probably not
very useful for reasonable pathlengths. This is not to
say that weakly absorbing materials, due to low values of
ε or c cannot be used; cells could be stacked so as to
achieve a long effective pathlength. Such an approach,
however, may pose severe engineering problems.

Values of $\varepsilon_\lambda > 10^3$M^{-1}cm^{-1} are characteristic of highly
allowed electronic transitions (charge transfer, π-π*)
which will have threshold energies corresponding to the
0-0 transition. While there may be forbidden (d-d, S$\rightarrow$T)
transitions at energies lower than the highly allowed
transition, their forbiddenness causes ε_λ to be low and
renders them of limited practical utility. Clearly, the
most effective materials for use as solar harvesters will
be those that absorb strongly in the visible and near-uv
regions of the spectrum. Figure 4 shows the absorption
spectra of thionine and Ru(bpy)$_3^{2+}$ in aqueous solution. It
is easy to see why these materials have been popular solar
absorbers; yet, even under the most favorable conditions
of pathlength and concentration, only a small fraction of
the solar radiation at air mass 1 (Figure 5) could be
absorbed. Even if all the other characteristics of photogalvanic cells employing these materials were perfectly
efficient, such cells would be limited by the fraction of
the solar output absorbed by the solution.

It being a shame to waste all those potentially
perfectly good solar photons, approaches must be taken to
improve the absorption characteristics of the medium,
especially to the red; there is no lack of chromophores

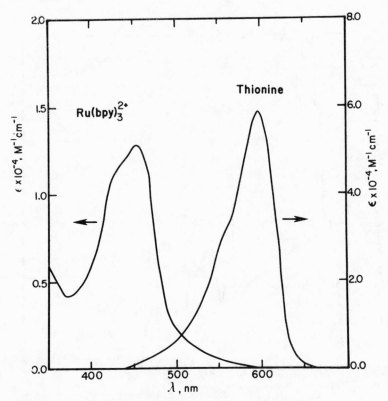

Figure 4. Absorption spectra of thionine (pH 2) and Ru(bpy)$_3^{2+}$ in aqueous solution.

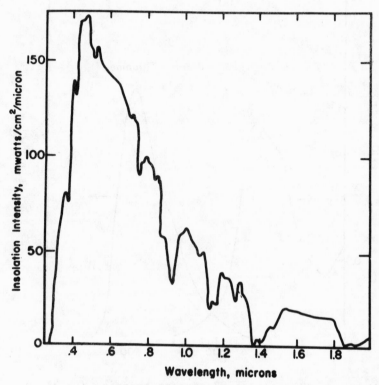

Figure 5.   Insolation spectrum at air mass 1 (sea level with sun at the zenith).

for the uv region. Molecular modification of thionine to
methylene blue results in a substantial shift of λ_{max} to
665 nm (ε_{max} 8.0 x $10^4 M^{-1} cm^{-1}$). For Ru(bpy)$_3^{2+}$, substitution of phen (=1,10-phenanthroline) for bpy or substitu-
tion on the polypyridyl rings has little effect on λ_{max} or
the shape of the visible absorption envelope (26,27).
Some red shift of λ_{max} to 473 and 501 nm is observed for
Ru(terpy)$_2^{2+}$ and Ru(TPTZ)$_2^{2+}$ (terpy = 2,2',2''-terpyridine;
TPTZ = 2,4,6-tri(2-pyridyl-s-triazine)) (26). These
effects are very modest but they show that alternation
of the ligand structure is a reasonable approach.
Recently, Dose and Wilson (28) prepared mono- and binu-
clear Ru(II) complexes containing N-bound aromatic ligands
which show λ_{max} close to 600 nm. Whether these or other
complexes yet to be synthesized will possess the other
qualities necessary for photogalvanic conversion remains
to be seen.

In most cases, the intense visible absorption bands
and those further to the blue are separated by regions of
significantly lower absorbance. In order to utilize those
solar photons, sensitizers can be employed to absorb light
and transfer energy from the excited state of the sensitizer to the photochemically active species. The requirement is for the energy of the excited state of the donor
to be higher than the energy required to reach the populated excited state of the acceptor. As a result, sensitization to the blue of the intense absorption band of the
acceptor can be achieved as has been demonstrated for
mixtures of dyes with thionine and methylene blue (29).
Sensitization to the red of the intense singlet absorption
band is limited by the energetics of the lowest, and
presumably reactive, excited state.

Formation of Separated Charge Carriers

There are many physical and chemical steps between
the formation of the initial excited state as a result of
the light-absorption process and the separated charge
carriers. In the first place, the initial excited state
may not be the reactive state that is the precursor of
the charge carriers. If the absorber is an organic
molecule, excitation in the intense absorption band produces the lowest excited singlet state, the lifetime of
which is of the order of 10^{-9}s. A lifetime that short

precludes efficient quenching; even if the quenching
rate constant were in the diffusion-controlled region
($\sim 10^{10} M^{-1} s^{-1}$), a quencher concentration of $\sim 1M$ would be
required in order to intercept $\sim 90\%$ of the excited state.
Intersystem crossing from the lowest singlet excited state
to the lowest triplet excited state, which, being longer
lived than the singlet is more likely to be the reactive
excited state that leads to the charge carriers, competes
with radiative and non-radiative internal conversion back
to the ground state and degradative reactive modes. Using
the symbols G, *E, and *R for the ground, initially
excited, and reactive excited states, respectively, the
scheme can be generalized with k_{rd}, k_{nr}, k_{rx}, and k_{isc}
representing the rate constants for radiative, non-radiative, reactive, and intersystem crossing decay of *E,
respectively. Inasmuch as the quantum yield of formation
of the excited state populated in the absorption process
must be unity, the quantum yield of formation of *R is
dictated by the efficiency of the intersystem crossing
step (η_{isc}) in competition with the energy degradation
modes: $\eta_{isc} = k_{isc}/(k_{rd} + k_{nr} + k_{rx} + k_{isc})$. For thionine
as for aromatic molecules in general, the major mode of
decay of the lowest excited singlet state is intersystem
crossing to the triplet state so that $\eta_{isc} = \phi_{*R} \to 1$.
For $Ru(bpy)_3^{2+}$, where *R corresponds to a metal-to-ligand
charge transfer excited state involving promotion of a
metal-centered $\underline{d}$ electron into highly delocalized π^*
ligand orbitals, population of *R occurs with nearly
unitary efficiency (30). In these systems, the quantum
yield of formation of *R cannot be viewed as a severe
limitation to the efficiency of photogalvanic conversion

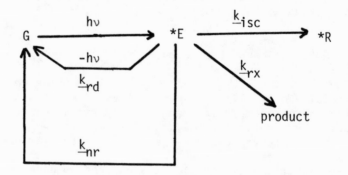

as long as the degradation of *E to some inactive product does not occur to any meaningful extent. Such degradation would preclude the use of the system in the required cyclic manner. However, if an interesting photogalvanic system were developed in which η_{isc} were not nearly unity, the internal or external heavy-atom effect could be used to enhance intersystem crossing.

With the quantum yield of formation of the reactive excited state, ϕ_{*R}, approximately unity, attention must be focused on the redox reactions that quench *R and generate the separated charge carriers. In the absence of redox quencher, the decay of *R is via radiative, non-radiative, and reactive modes with k'_{rd}, k'_{nr}, and k'_{rx} representing the rate constants of those processes. The intrinsic lifetime of *R is dictated by those values of k': $\tau^0_{*R} = 1/(k'_{rd} + k'_{nr} + k'_{rx})$; it is assumed that *R is the lowest electronically excited state of the absorber so that transitions to other excited states can be ruled out. In the presence of the quencher at a concentration [Q], the quenching reaction, with a second-order rate constant of k'_q, competes with the intrinsic modes so that the lifetime of *R is reduced: $\tau_{*R} = 1/(k'_{rd} + k'_{nr} + k'_{rx} + k'_q[Q])$. The efficiency of the quenching reaction, η_q, is given simply as $\tau^0_{*R}k'_q[Q]/\tau^0_{*R}k'_q[Q] + 1)$. It is easy to see that $\eta_q \to 1$ as [Q] increases so that $\tau^0_{*R}k'_q[Q] \gg 1$. Thus, if τ^0_{*R} is small, a high concentration of Q and a large value of k'_q must be combined in order for the quenching reaction to compete. For example, in order to quench $\sim$90% of *R, $\tau^0_{*R}k'_q[Q] \sim 10$; if $\tau^0_{*R} \sim 10^{-6}$s and k'_q were in the diffusion-controlled region ($\sim 10^{10} M^{-1}s^{-1}$), [Q] has to be of the

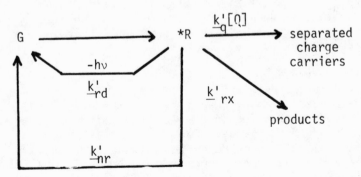

order of 10^{-3}M. In order to quench $\sim$99% of *R, [Q] would
have to be $\sim 10^{-2}$M; if $\underline{k}'_q$ were less than $\sim 10^{10}$M^{-1}s^{-1},
proportionally higher concentrations of Q would have to
be used.

It is clear that the parameters available to control
η_q are τ^o_{*R} and the product $\underline{k}'_q$[Q]. It is advantageous
to have τ^o_{*R} and $\underline{k}'_q$ as large as possible so as to be able
to quench *R without requiring the concentration of Q
to be so large as to exceed its solubility limit, interfere with the light absorption process, or interfere with
the electrochemical aspects of the cell.

The value of τ^o_{*R} for a particular excited state is
fixed by nature although molecular environment can have
an effect due to changes in the rate constants of non-
radiative and reactive modes; radiative decay, where it
exists at all in fluid media, is generally a small fraction
of the total decay paths and is relatively insensitive to
molecular environment. For thionine, the intrinsic lifetime of the lowest triplet state is 7-8 μs (31). For
Ru(bpy)$_3^{2+}$, the lifetime of the charge transfer excited
state is little affected by changes in the solution
medium; τ^o ranges from 0.60 μs in aqueous solution (32)
to 1.0 μs in D$_2$O (33) and is 0.76 μs in CH$_3$OH (32).
Variation of the substituent nature of the polypyridyl
ligands does result in modest changes in τ^o with 0.92 μs
for Ru(phen)$_3^{2+}$, 1.8 μs for Ru(5,6-(CH$_3$)$_2$phen)$_3^{2+}$ (34), and
$\sim$2 μs for Ru(II) complexes containing hydrophobic ligands
(27). Unfortunately, the more promising complexes from
the point of view of their absorption characteristics,
Ru(terpy)$_2^{2+}$ and Ru(TPTZ)$_2^{2+}$, have very short intrinsic
excited state lifetimes of <5 ns (26,35) which effectively
rule out their applicability as efficient generators of
separated charge carriers.

Any degradative reactions of *R that do not lead to
separated charge carriers and are competitive with the
desired redox quenching reaction are wasteful. These
include energy transfer quenching and interaction with the
solvent and other solutes. For excited triplet states of
aromatic molecules and the excited states of metal com-
plexes, rapid energy tranfer quenching by O$_2$ ($\underline{k} \sim 10^9$-10^{10}
M^{-1}s^{-1}) (36) must be avoided. While triplet thionine
appears to be insensitive to interaction with solvent
or other solutes in a degradative manner, although

detailed studies directed toward that issue have not been conducted, $*Ru(bpy)_3^{2+}$ has been found to engage in temperature-dependent photoanation reactions, albeit with low ($<10^{-3}$) quantum yields (37,38). Although the desirable redox quenching reaction can be made to overwhelm energy wasteage modes, the existence of even a very low residual quantum yield ($\sim 10^{-6}$) of these degradative reactions can render the photochemical system useless for the required operation over many million cycles.

The parameters affecting the values of $\underline{k}_q'$, the rate constant of redox quenching, have been the subject of increasing numbers of studies with the conversion of light energy into chemical energy as one of the underlying interests. Rehm and Weller (39) have discussed in detail the kinetic and thermodynamic aspects of the quenching process and have derived relationships between $\underline{k}_q'$ and the overall free energy change of the excited state reaction; $\underline{k}_q'$ increases as ΔG_q^o becomes more negative reaching a plateau value at very negative ΔG_q^o values corresponding to the diffusion-controlled limit. Other approaches have used the Marcus theory of outer-sphere electron transfer reactions (40) which predicts an inverted region where $\underline{k}_q'$ drops from the plateau value when ΔG_q^o becomes very negative. "Vestiges" of the inverted region have been found in the quenching of $*Ru(bpy)_3^{2+}$ (41) but not in the quenching of triplet thionine (42). Irrespective of the intricacies of these theories, the analysis performed above shows that for short-lived excited states ($\tau_{*R}^o \sim 10^{-6}$ s), reaction with $\underline{k}_q'$ at or near the diffusion-controlled limit is necessary for efficient conversion of $*R$ to separated charge carriers. Therefore, ΔG_q^o must be very negative and knowledge and control of the redox potential of $*R$ is required. One important consequence of electronic excitation is the increase of the electron affinity and decrease of the ionization potential of the molecule. Therefore, the electronically excited state is expected to be both a better reductant and a better oxidant than the ground state. By assuming that differences in shape, size, solution, and entropy of the ground and excited states are small, the redox potentials of the excited state are fairly well approximated by equations 5 and 6 (43).

$$E^o(M^+/*M) = E^o(M^+/M) - E_{o-o}(M-*M) \qquad (5)$$

$$E^o(^*M/M^-) = E^o(M/M^-) + E_{o-o}(M-^*M) \qquad (6)$$

Here $E^o(M^+M)$ and $E^o(M/M^-)$ are the standard reduction potentials of the one-electron oxidation product and the ground state, respectively, and $E_{o-o}(M-^*M)$ is the one-electron potential corresponding to the zero-zero spectroscopic energy of the excited state. Potentials are, of course, affected by acid-base reactions of the ground and excited states so that $\underline{k}_q'$ can have pH dependences of consequence. Excited state potentials can also be evaluated from quenching measurements using a series of quenchers of known potentials (44). For $Ru(bpy)_3^{2+}$ where $E_{o-o}(M-^*M) = 2.12V$, $E^o(M^+/^*M) = -0.84V$ and $E^o(^*M/M^-) = +0.84V$; in comparison, $E^o(M^+/M) = +1.26V$ and $E^o(M/M^-) = -1.28V$ (34). Variation of substituents on the polypyridyl rings, including hydrophobic groups, can alter the values of E^o by as much as 0.3V (27,34) which can result in variations of $\underline{k}_q'$ of more than a factor of 10 when below the diffusion-controlled plateau (45).

The value of $\underline{k}_q'$ for $^*Ru(bpy)_3^{2+}$ can also be varied by the solution medium. Meisel and coworkers (46) have shown that incorporation of $Ru(bpy)_3^{2+}$ into sodium dodecyl sulfate micelles increases the rate of $\underline{k}_q'$ for Cu^{2+} by a factor of two. This effect is modest compared to the factor of 2000 increase in $\underline{k}_q'$ when polyvinylsulfate is present in the solution (47). Here, very efficient quenching of $^*Ru(bpy)_3^{2+}$ by Cu^{2+} occurs in the potential field domains of the polyelectrolyte; a similar effect is seen for quenching by Fe^{3+} (48).

The redox potentials of both the ground and lowest triplet states of thionine are pH dependent because of the participation of these states, as well as the semi-reduced and fully-reduced species, in rapid, reversible protonic reactions. Relevant pK_a values in aqueous solution are given in equations 7-13. The dependence upon pH of the

<u>ground state thionine</u>

$$TH_2^{2+} \rightleftharpoons TH^+ + H^+ \qquad pK_a \ -0.7 \ (49) \qquad (7)$$

$$TH^+ \rightleftharpoons T + H^+ \qquad pK_a \ 11 \ (49) \qquad (8)$$

triplet thionine

$$^3TH_2^{2+} \rightleftharpoons {}^3TH^+ + H^+ \qquad pK_a \; 6.3 \; (50) \qquad (9)$$

$$^3TH^+ \rightleftharpoons {}^3T + H^+ \qquad pK_a \; 8.9 \; (51) \qquad (10)$$

semithionine

$$TH_2^{+\cdot} \rightleftharpoons TH\cdot + H^+ \qquad pK_a \; 8.2 \; (52) \qquad (11)$$

leucothionine

$$TH_4^{2+} \rightleftharpoons TH_3^+ + H^+ \qquad pK_a \; 4.4 \; (53) \qquad (12)$$

$$TH_3^+ \rightleftharpoons TH_2 + H^+ \qquad pK_a \; 5.3 \; (53) \qquad (13)$$

ground state two-electron reduction potential (vs NHE) in aqueous solution at 25°C is given by equation 14 (53) where T and L represent the sums of all forms of thionine and leucothionine, respectively. The equilibrium potentials

$$E^{o'}_{T \to L} = 0.563 - 0.09 \; pH + 0.03 \; log([T]/[L]) \qquad (14)$$

of the two one-electron couples at pH2 in water are $E^{o'}_{T \to S} = 0.192V$ and $E^{o'}_{S \to L} = 0.575V$ (54). The reduction potential for the $^3T/S$ couple at pH2 can be estimated from $E^{o'}_{T \to S}$ and $E_{o-o}(T-^3T)$; this latter quantity for $^3TH_2^{2+}$, the predominant form of triplet thionine at pH2, is $1.2 \pm 0.1V$ (55) (compared with E_{o-o} for $TH^+ = 1.7 \pm 0.1V$). The result is $E^{o'}_{^3T \to S} = 1.4 \pm 0.1$ V at pH2.

The intrinsic lifetime of triplet thionine is dependent on its state of protonation. In dilute aqueous solution at pH2, where $^3TH_2^{2+}$ predominates, $\tau^o = 7.7$ μs (31) while in neutral solution, where $^3TH^+$ predominates, $\tau^o \sim 60$ μs (56). The value of τ^o of $^3TH_2^{2+}$ remains essentially constant when the solvent is changed from water to 50 v/v% aqueous CH_3CN and the counter ions are changed from HSO_4^-/SO_4^{2-} to $CF_3SO_3^-$ (31). The lifetime of the protonated form of triplet methylene blue ($^3MBH^{2+}$) has been examined over a wider range of conditions (31) and has been found to be essentially constant (4.5 ± 0.5 μs) in water and 50 v/v% aqueous CH_3CN in the presence of 0.1 – 0.001 M acid, with HSO_4^-/SO_4^{2-}, $CF_3SO_3^-$, or Cl^- (57) as counter ions, and at ionic strengths between 0.1 and 1.0 M.

The values of $\underline{k}_q'$ for the quenching of $^3TH^{2+}$ and $^3MBH^{2+}$ by Fe_{aq}^{2+} are sharply dependent on the nature of the counter ions and, in the presence of HSO_4^-/SO_4^{2-} (but not $CF_3SO_3^-$) upon solvent (31). For $^3TH^{2+} + Fe_{aq}^{2+}$, $\underline{k}_q' = 6.0\pm0.5$ x 10^6 $M^{-1}s^{-1}$ in both water and 50 v/v% aqueous CH_3CN with $CF_3SO_3^-$ as the counter ion but is 6 x 10^7 $M^{-1}s^{-1}$ in water and 5 x 10^8 $M^{-1}s^{-1}$ in 50 v/v% aqueous CH_3CN with HSO_4^-/SO_4^{2-} as the counter ion. Values of $\underline{k}_q'$ for $^3MBH^{2+}$ $+ Fe_{aq}^{2+}$ are about 15% of the above values under identical conditions. The observed dependences of $\underline{k}_q'$ on solution medium are rationalized in terms of the electrostatic association of SO_4^{2-} with Fe_{aq}^{2+} and/or the dipositive dye cations (31). The high sensitivity of $\underline{k}_q'$ of the triplet dye molecules to the nature of the solution medium, coupled with the insensitivity of their intrinsic life-times, makes it possible to adjust conditions for optimum quenching of the reactive excited states. For example, in 50 v/v% aqueous CH_3CN in sulfate medium at pH2, 97.5% of $^3TH_2^+$ and 76% of $^3MBH^{2+}$ is quenched by 0.01M Fe_{aq}^{2+}. In order to take advantage of the much longer values of τ^0 for the deprotonated triplet states, $^3TH^+$ and $^3MB^+$, the use of quenchers are required which, unlike Fe_{aq}^{2+}, are compatible with neutral or alkaline solution. Such quenching reactions are under investigation in our labora-tory at the present time.

It is clear that conditions can easily be arranged such that the efficiency of quenching of the reactive excited state is virtually unity. The bimolecular quench-ing of an excited state can occur via electron transfer, energy transfer, or deactivation by some catalytic action and it is only the redox mode that can lead efficiently to the separated charge carriers. Catalytic deactivation processes are generally slow (58) so that the most impor-tant modes are electron and energy transfer. As long as the electronic energy requirement of the acceptor is less than the energy available from the donor, energy transfer is thermodynamically allowed; spin restrictions may cause the reaction to be kinetically hindered. It is only by the direct observation of the products of the reaction, redox or excited, that evaluation of the proportion of each process can be made. From the standpoint of energy conversion, it is essential that the efficiency of the redox process, η_{redox}, be as close to unity as possible. The reaction of triplet thionine with Fe_{aq}^{2+} yields semi-thionine, the one-electron reduction product, essentially quantitatively as determined by flash photolysis (56).

For $*Ru(bpy)_3^{2+}$, it is now well established (43) that oxidative and reductive quenching leads to the virtually quantitative generation of $Ru(bpy)_3^{3+}$ and $Ru(bpy)_3^{+}$, respectively, and the corresponding reduced and oxidized forms of the quencher. Depending upon the nature of the quencher, the derived product can be a stable oxidation state of a metal complex or an unstable free radical. The yield of these separated charge carriers depends upon the fraction of the redox component of the quenching step (η_{redox}) and the efficiency of separation of the species from their initially formed solvent cage (η_{sc}); for $*Ru(bpy)_3^{2+}$, $\eta_{redox} \sim 1$ (32). Because the products of electron transfer quenching are high energy species, they can engage in immediate back electron transfer in the successor complex within the solvent cage. The critical parameters in the establishment of η_{sc} are the rate of breakdown of the solvent cage and the rate of cage recombination. The factors that govern the events in the solvent cage have been discussed (59); micelles (46) and polyelectrolytes (60) have been used to alter the structure of the solvent cage which causes the yield of solvent separated charge carriers to be altered.

There have been a few determinations of the total quantum yields of formation of the separated charge carriers (ϕ_{cc}) from the oxidative quenching of $*Ru(bpy)_3^{2+}$. The quenchers have been Co(III) complexes which, upon reduction by $*Ru(bpy)_3^{2+}$, undergo rapid spin equilibration (61) and ligand labilization (62) in competition with cage recombination. For example, ϕ_{cc} values for the formation of Co_{aq}^{2+} from $Co(C_2O_4)_3^{3-}$ (63) and $Co(NH_3)_5Cl^{2+}$ (64) are ~ 0.85. In general, for Co(III) complexes (65), ϕ_{cc} appears to be between 0.5 and 1.0. However, because of the irreversible nature of the reduction of these and analogous Co(III) systems, their utilization in reversible photogalvanic cells is not practical (66).

What are required are experiments involving the fast generation of $*Ru(bpy)_3^{2+}$, triplet thionine, or any other potential reactive excited states by laser flash photolysis and the quantitative determination of the spectra of the redox products before any appreciable back electron transfer takes place in the bulk solution. In that way, ϕ_{cc} can be determined for the relevant photogalvanic chemical

systems and controlled by means of the alteration of
solution medium.

Behavior of the Separated Charge Carriers

The quantum yield of separated charge carriers can be
expressed in terms of the efficiencies of the various
processes that lead ultimately to their formation (equation 15). We have seen that for thionine and $Ru(bpy)_3^{2+}$,

$$\phi_{cc} = \phi_{*R} \eta_q \eta_{redox} \eta_{sc} \qquad (15)$$

$\phi_{cc} \sim 1$ for semithionine and $Ru(bpy)_3^{3+}$ or $Ru(bpy)_3^{+}$. If
these species and the complementary products from the
quenching step were capable of diffusing to the appropriate electrodes without engaging in further reaction, a
successfully efficient photogalvanic cell would have
already been created. It is, however, the necessity of
preventing back electron transfer reactions between energy-
rich species in bulk solution that challenges the
ingenuity of chemists. It is important that thermal back
electron transfer reactions be slow compared to the diffusion of the species to the electrodes and electron transfer
processes thereon even in order to achieve storage in
microtime and transduction of light into electricity.
The options for achieving practical photogalvanic storage
would be enormously broadened if the rate of thermal back
reactions were measurable in hours or days. Some variation in the rate of bulk back electron transfer can be
achieved by molecular manipulation of the redox couple and
variation of solution medium. Perhaps a more effective
way to reduce back electron transfer is to build in an
alternative pathway which diverts one or both of the
products of the redox quenching. Although contributing to
some energy loss, such a diversion can be achieved through
other bimolecular reactions or interaction at an interface.

An interesting example of successful competition with
prompt reversal of the photoredox step occurs with semithionine which undergoes bimolecular disproportionation
to thionine and leucothionine in competition with bimolecular back electron transfer with Fe_{aq}^{3+} generated in the
redox quenching step of the iron-thionine photogalvanic
cell. The rate constants of these bimolecular reactions
are somewhat dependent on solution medium so that it is
possible to employ solvent and counter ions to maximize

the disproportionation. Rate constants for disproportionation of semithionine typically fall in the range of 10^7 - 10^9 $M^{-1}s^{-1}$ and, in fact, have been correlated quantitatively with Kosower's solvent parameter, Z (67). Less is known about the variation of the second-order rate constant for the oxidation of semithionine by Fe^{3+}_{aq} with solution medium. Values in the range $2x10^4$ - $9x10^5$ $M^{-1}s^{-1}$ have been reported at pH2, depending on both solvent and counter ion (24). Conditions are, therefore, readily available for reducing the prompt back electron transfer reaction to a negligible level. Diversion via disproportionation is a promising means of obtaining charge carriers which are stable in the microtime domain because of two factors: 1) synproportionation, the reverse of disproportionation, is much slower than the latter process; the rate constants of synproportionation are of the order of 10^2-10^3 $M^{-1}s^{-1}$, depending on solution medium (54); 2) oxidation of leucothionine by Fe^{3+}_{aq} is much slower than the oxidation of semithionine by this reagent; the rate constants of the leucothionine + Fe^{3+}_{aq} reaction are in the range $3x10^2$-$2x10^3$ $M^{-1}s^{-1}$ depending on solution medium (24). Leucothionine is, accordingly, the dominant reduced dye species in photobleached iron-thionine solutions under conveniently achieved conditions (7,54). At the photostationary state under sunlight intensities, leucothionine constitutes more than 99% of reduced dye (7). The lifetime of leucothionine under these conditions, determined by measuring the rate of relaxation of the photostationary state, varies inversely with the dark concentration of dye and is dependent upon solution medium; $\tau \sim 1s$ when $[TH^+]_o$ $\sim 10^{-3}$ M in 50 v/v% aqueous CH_3CN at pH2 with HSO_4^-/SO_4^{2-} as counter ion (13). Such a lifetime is sufficient to permit diffusion of virtually all photogenerated charge carriers to the electrodes in TI-TL photogalvanic transducers (16). Extension of this classical, but nevertheless promising, system from effective storage in microtime toward storage in macrotime may evolve from research involving modification of the structure of the thionine, use of other classes of redox dyes, use of metal complexes other than those of iron, variation of the ligands around the metal center, variation of solution medium, and use of organizates.

Ru(bpy)$_3^{3+}$ and Ru(bpy)$_3^+$ are very powerful one-electron oxidizing and reducing agents with standard reduction potentials of +1.26 and -1.28V, respectively. As a result, their back electron transfer reactions are very rapid with $k \sim 10^7\text{-}10^9$ $M^{-1}s^{-1}$ (34) and their storage capability is negligible even in microtime. Furthermore, $Ru(bpy)_3^{3+}$ is kinetically unstable in aqueous solutions that are any-thing except highly acidic; the reduction, which generates O_2 over a very narrow pH range in alkaline solution, is mechanistically very complicated (68). $Ru(bpy)_3^+$ is stable in deaerated acetonitrile solution where it can be gener-ated electrochemically in the controlled-potential reduc-tion of $Ru(bpy)_3^{2+}$ (69) but it is unstable in aqueous medium, apparently undergoing slow ($k \sim 0.2$ s^{-1}) inter-action with water (70). The instabilities of $Ru(bpy)_3^{3+}$ and $Ru(bpy)_3^+$ in aqueous solution restrict their use as charge carriers in aqueous photogalvanic cells; investi-gations of their roles in non-aqueous cells should be carried out in the future. Even so, long-term storage of $Ru(bpy)_3^+$ may not be feasible because of its disproportion-ation reaction to $Ru(bpy)_3^0$ although the latter species may be kinetically more stable due to the two-electron nature of its return to the originally absorbing substrate. Back electron transfer involving $Ru(bpy)_3^{3+}$ has been shown to be effectively retarded by several orders of magnitude by the presence of polyvinylsulfate (60) and research will certainly continue in that direction. The use of hydro-phobic ligands in Ru(II) complexes, monolayer assemblies, and liquid-solid interfaces has been directed toward H_2 formation (71) but its applicability for photogalvanic conversion should be recognized.

CONCLUSIONS

It is readily seen that the value of ϕ_{cc} is not the limiting photochemical determinant of the efficiency of photogalvanic cells but that the storage of separated charge carriers in microtime and macrotime domain is the compelling challenge. There is no question that research will continue for some time to come on quenching reactions, cage recombination, and the chemistry of the charge carriers for systems closely related to thionine and $Ru(bpy)_3^{2+}$. Perhaps those investigations will produce insight into the design of more complex and more successful photogalvanic systems. Solution of the back electron transfer problem is absolutely essential and should involve basic research investigations into the construction of systems where highly exergic electron transfer reactions

can be rendered slow, even in the microtime domain. Novel
means must be sought of diverting the immediate products
of the redox quenching step via pathways, such as disproportionation, which produce separated charge carriers
stable in the microtime, and even macrotime, domain. Such
diverted species will be two-electron transfer species;
theories of two-electron transfer must be developed to the
same level of understanding as those of one-electron
transfer. Potential charge carriers can be generated by
radiolytic or electrolytic techniques so that their
chemistry can be investigated in the absence of photochemical constraints; discovery of kinetically stable
charge carriers would then lead to attempts to create the
photochemical conditions such that $\phi_{cc} \rightarrow 1$. Research
programs for the application of the techniques of interfacial chemistry to photogalvanic cells should provide
many answers to the problem.

Naturally, the quest for new photochemical systems
which can be used in photogalvanic cells must continue.
The potentiality of Ru(II)-polypyridyl complexes for solar
energy conversion has led to increased interest in
polypyridyl complexes of other transition metals. The
disadvantages of the very short lifetimes ($\tau^o \sim 20\mu s$) of
$*Fe(bpy)_3^{2+}$ (72) and $*Os(bpy)_3^{2+}$ (34) in aqueous solution
may be offset by advantages in energetics and kinetics;
$*Os(bpy)_3^{2+}$ is a better reducing agent but a weaker oxidizing agent than is $*Ru(bpy)_3^{2+}$ and $Os(bpy)_3^{2+}$ is a weaker
oxidizing agent than $Ru(bpy)_3^{2+}$ (34). Furthermore,
$Os(bpy)_3^{2+}$ absorbs strongly ($\epsilon > 10^3$ M^{-1}cm^{-1}) in the 300-
700 nm region (68) which makes it an attractive light
harvesting material. Although $Cr(bpy)_3^{3+}$ does not absorb
strongly ($\epsilon < 10^3$ M^{-1}cm^{-1}) at $\lambda > 400$nm, its lowest
excited state, which is a metal-centered doublet, has a
lifetime in aqueous solution of 63μs (73) which can be
prolonged by up to a factor of ten by altering the solution
medium or incorporating phen ligands (74). $*Cr(bpy)_3^{3+}$ is
a much better oxidizing agent than is $*Ru(bpy)_3^{3+}$ and undergoes facile reductive quenching (75) in competition with
rather complex photochemistry (76). The very low
E_{o-o}(M-*M) value of 1.71V suggests that the absorption
spectrum of molecularly modified Cr(III)-polypyridyl
complexes can be shifted significantly into the visible.

It is clear that there is a very long way to go before
an efficient and economically competitive photogalvanic

cell can be put into operation. Such a goal may never be achieved but it should not be said that this occurred through lack of effort. The studies made in understanding the nature of electron transfer reactions of excited states within this past decade augurs well for the understanding of other factors that comprise the photochemical determinants.

ACKNOWLEDGMENT

The authors are grateful to the U.S. Department of Energy for support of this work under Contract No. EY-76-S-02-2889.

REFERENCES

(1)   N. N. Lichtin in "Solar Power and Fuels", J. R. Bolton, ed., Academic Press, New York, 1977, p. 119.

(2)   R. Gomer, Electrochem. Acta, 20, 13 (1975).

(3)   W. J. Albery and M. D. Archer, J. Electrochem. Soc., 124, 688 (1977).

(4)   W. J. Albery and A. W. Foulds, J. Photochem., in press.

(5)   K. Weber, Z. Physik. Chem., B15, 18 (1931).

(6)   J. Weiss, Nature, 136, 794 (1935).

(7)   E. Rabinowitch, J. Chem. Phys., 8, 551, 560 (1940).

(8)   P. D. Wildes and N. N. Lichtin, J. Am. Chem. Soc., 100, 6568 (1978).

(9)   W. Nernst, Z. Physik. Chem., 47, 52 (1904).

(10) W. D. K. Clark and J. A. Eckert, Solar Energy, 17, 147 (1975).

(11) W. D. K. Clark, J. A. Eckert, D. E. Hall, N. N. Lichtin, and P. D. Wildes, Amer. Ceramic Soc. Bull., 56, 408 (1977).

(12) D. E. Hall, J. A. Eckert, N. N. Lichtin, and
     P. D. Wildes, J. Electrochem. Soc., 123, 1705 (1976).

(13) P. D. Wildes, M. Z. Hoffman, and N. N. Lichtin in
     "Proceedings of the International Symposium of Solar
     Energy", J. Berkowitz and I. Lesk, eds., The Elec-
     trochemical Society, Princeton, New Jersey, 1976,
     p. 128.

(14) P. D. Wildes, K. T. Brown, M. Z. Hoffman, D. E. Hall,
     and N. N. Lichtin, Solar Energy, 19, 579 (1977).

(15) D. E. Hall, P. D. Wildes, and N. N. Lichtin, J.
     Electrochem. Soc., 125, 1365 (1978).

(16) N. N. Lichtin, P. D. Wildes, T. L. Osif, and D. E.
     Hall, Adv. in Chem. Ser., in press.

(17) D. Hall, J. Electrochem. Soc., 124, 804 (1977).

(18) A. Heller, ed., "Semiconductor Liquid-Junction Solar
     Cells", Proceedings Vol. 77-3, The Electrochemical
     Society, Princeton, New Jersey, 1977.

(19) K. G. Mathai and E. Rabinowitch, J. Phys. Chem., 66,
     663 (1962).

(20) V. Srinivasan and E. Rabinowitch, J. Chem. Phys., 52,
     1165 (1970).

(21) H. Gerischer in "Physical Chemistry", H. Eyring,
     D. Henderson, and W. Jost, eds., Volume 9A, Academic
     Press, New York, 1970, p. 463.

(22) V. G. Levich, Adv. Electrochem. Electrochem. Eng.,
     4, 249 (1966).

(23) H. Gerischer in "Solar Power and Fuels", J. R.
     Bolton, ed., Academic Press, New York, 1977, p. 77.

(24) T. L. Osif, N. N. Lichtin, and M. Z. Hoffman,
     J. Phys. Chem., 82, 1778 (1978).

(25) C. T. Lin and N. Sutin, J. Phys. Chem., 80, 97 (1976).

(26) C. T. Lin, W. Böttcher, M. Chou, C. Creutz, and
      N. Sutin, J. Am. Chem. Soc., 98, 6536 (1976).

(27) P. J. Delaive, J. T. Lee, H. Abruna, H. W.
      Sprintschnik, T. J. Meyer, and D. G. Whitten,
      Adv. in Chem. Series, 168, 28 (1978).

(28) E. V. Dose and L. J. Wilson, Inorg. Chem., 17,
      2660 (1978).

(29) P. D. Wildes, D. R. Hobart, N. N. Lichtin, D. E.
      Hall, and J. A. Eckert, Solar Energy, 19, 567 (1977).

(30) J. N. Demas and G. A. Crosby, J. Am. Chem. Soc., 93,
      2841 (1971).

(31) P. D. Wildes, N. N. Lichtin, M. Z. Hoffman,
      L. Andrews, and H. Linschitz, Photochem. Photobiol.,
      25, 21 (1977).

(32) J. N. Demas and J. W. Addington, J. Am. Chem. Soc.,
      98, 5800 (1976).

(33) J. Van Houten and R. J. Watts, J. Am. Chem. Soc.,
      98, 4853 (1976).

(34) N. Sutin and C. Creutz, Adv. in Chem. Series, 168,
      1 (1978).

(35) R. C. Young, J. K. Nagle, T. J. Meyer, and D. G.
      Whitten, J. Am. Chem. Soc., 100, 4773 (1978).

(36) A. A. Lamola and N. J. Turro, "Energy Transfer and
      Organic Photochemistry", Interscience, New York,
      1969; J. N. Demas, E. W. Harris, and R. P. McBride,
      J. Am. Chem. Soc., 99, 3547 (1977).

(37) P. E. Hoggard and G. B. Porter, J. Am. Chem. Soc.,
      100, 1457 (1978).

(38) R. J. Watts, J. S. Harrington, and J. Van Houten,
      Adv. in Chem. Series, 168, 57 (1978).

(39) D. Rehm and A. Weller, Ber. Bunsenges. Phys. Chem.,
      73, 834 (1969); Isr. J. Chem., 8, 259 (1970).

(40) N. Sutin in "Inorganic Biochemistry", G. L. Eichhorn, ed., Vol. 2, Elsevier, Amsterdam, 1973, p. 611.

(41) C. Creutz and N. Sutin, J. Am. Chem. Soc., 99, 241 (1977).

(42) E. Vogelmann, S. Schreiner, W. Rauscher, and H. E. A. Kramer, Z. Phys. Chem. (Neue Folge), 101, 321 (1976).

(43) V. Balzani, F. Bolletta, M. T. Gandolfi, and M. Maestri, Topics in Current Chemistry, in press.

(44) C. R. Bock, T. J. Meyer, and D. G. Whitten, J. Am. Chem. Soc., 97, 2909 (1975).

(45) M. A. Hoselton, C. T. Lin, H. Schwarz, and N. Sutin, J. Am. Chem. Soc., 100, 2383 (1978); C. Creutz, Inorg. Chem., 17, 1036 (1978).

(46) D. Meisel, M. S. Matheson, and J. Rabani, J. Am. Chem. Soc., 100, 117 (1978).

(47) D. Meisel and M. S. Matheson, J. Am. Chem. Soc., 99, 6577 (1978).

(48) D. Meisel, J. Rabani, D. Meyerstein, and M. S. Matheson, J. Phys. Chem., 82, 985 (1978).

(49) R. Bonneau and R. D. S. Stevens, Chem. Phys. Letters, 13, 376 (1972).

(50) J. Faure, R. Bonneau and J. Joussot-Dubien, Photochem. Photobiol., 6, 331 (1967).

(51) R. Bonneau and J. Pereyre, Photochem. Photobiol., 21, 173 (1975).

(52) J. Faure, R. Bonneau and J. Joussot-Dubien, Ber. Bunsenges. Phys. Chem., 72, 263 (1968).

(53) W. M. Clark, B. Cohen and H. D. Gibbs, Public Health Rep., 23, 1131 (1925).

(54) P. D. Wildes and N. N. Lichtin, J. Phys. Chem., 82, 981 (1978).

(55) H. E. A. Kramer and M. Hafner, Ber. Bunsenges. Phys. Chem., 73, 833 (1969).

(56) C. G. Hatchard and C. A. Parker, Trans. Faraday Soc., 57, 1093 (1961).

(57) T. Osif, S. Ray, N. N. Lichtin, and M. Z. Hoffman, Abstracts of the Second International Conference on Photochemical Conversion and Storage of Solar Energy, Cambridge, England, August 9-11, 1978, p. 17.

(58) V. Balzani, L. Moggi, M. F. Manfrin, F. Bolletta, and G. L. Laurence, Coord. Chem. Rev., 15, 321 (1975).

(59) J. P. Lorand, Prog. Inorg. Chem., 17, 207 (1972).

(60) D. Meyerstein, J. Rabani, M. S. Matheson, and D. Meisel, J. Phys. Chem., 83, 1879 (1978).

(61) E. V. Dose, M. A. Hoselton, N. Sutin, M. F. Tweedle, and L. J. Wilson, J. Am. Chem. Soc., 100, 1141 (1978).

(62) J. Lilie, N. Shinohara, and M. G. Simic, J. Am. Chem. Soc., 98, 6516 (1976).

(63) J. N. Demas and A. W. Adamson, J. Am. Chem. Soc., 95, 5159 (1973).

(64) G. Navon and N. Sutin, Inorg. Chem., 13, 2159 (1974).

(65) E. Finkenberg, P. Fisher, S.-M. Y. Huang, and H. D. Gafney, J. Phys. Chem., 82, 526 (1978).

(66) B. Durham and T. J. Meyer, J. Am. Chem. Soc., 100, 6286 (1978).

(67) P. D. Wildes, N. N. Lichtin, and M. Z. Hoffman, J. Am. Chem. Soc., 97, 2288 (1975).

(68) C. Creutz and N. Sutin, Proc. Natl. Acad. Sci. U.S.A., 72, 2858 (1975).

(69) C. P. Anderson, P. J. Salmon, T. J. Meyer, and R. C. Young, J. Am. Chem. Soc., 99, 1980 (1977).

(70) Q. G. Mulazzani, S. Emmi, P. G. Fuochi, M. Z. Hoffman, and M. Venturi, J. Am. Chem. Soc., 100, 981 (1978).

(71) G. Sprintschnik, H. W. Sprintschnik, P. P. Kirsch, and D. G. Whitten, J. Am. Chem. Soc., 99, 4947 (1977).

(72) J. Phillips, J. A. Koningstein, C. H. Langford, and R. Sasseville, J. Phys. Chem., 82, 622 (1978).

(73) M. Maestri, F. Bolletta, L. Moggi, V. Balzani, M. S. Henry, and M. Z. Hoffman, J. Chem. Soc., Chem. Commun., 491 (1977).

(74) M. S. Henry and M. Z. Hoffman, Adv. in Chem. Series, 168, 91 (1978).

(75) A. Juris, M. F. Manfrin, M. Maestri, and N. Serpone, Inorg. Chem., 17, 2258 (1978).

(76) M. Maestri, F. Bolletta, L. Moggi, V. Balzani, M. S. Henry, and M. Z. Hoffman, J. Am. Chem. Soc., 100, 2694 (1978).

REVERSIBLE EXCITED-STATE ELECTRON-TRANSFER REACTIONS OF

TRANSITION METAL COMPLEXES.

B.A. DeGraff[1], J.N.Demas[2], and D.G. Taylor[2]

Departments of Chemistry, 1) James Madison
University, Harrisonburg, VA 22801 and
2) University of Virginia, Charlottesville,
VA 22901

Systems exhibiting reversible excited-state electron-transfer reactions are currently of great interest because of their utilization in photogalvanic cells and photocatalysis. These systems can be described by the following general equations:

$$D + h\nu \xrightarrow{\phi'} {}^*D \tag{1}$$

$$^*D \longrightarrow D + \Delta \text{ or } h\nu \tag{2}$$

$$^*D + Q \xrightarrow{k_q} D^{\pm} + Q^{\mp} \ (\phi_{et}) \text{ or} \tag{3a}$$

$$D + Q \ (1-\phi_{et}) \tag{3b}$$

$$D^{\pm} + Q^{\mp} \xrightarrow{k_{th}} D + Q \tag{4}$$

where D is the photosensitizer, Q is the quencher, and $D^{\pm}$ and $Q^{\mp}$ are the oxidized and reduced forms of D and Q following excited-state electron transfer. ϕ' is the efficiency of population of the sensitizing state of D, ϕ_{et} is the efficiency of detectible electron transfer per quenching encounter, k_q is the bimolecular quenching constant for deactivation of *D by Q, and k_{th} is the rate constant for the thermal back electron transfer reaction. Process (3-b) represents all quenching processes which do not lead to observable electron transfer.

Systems of practical interest should have as high a $\phi'\phi_{et}$ and as slow a k_{th} as possible. Progress in the area of reversible excited-state electron-transfer reactions

189

has, in part, been hampered by the difficulty of rapidly, inexpensively and accurately surveying large numbers of systems. We summarize here the common methods used to measure these parameters along with the advantages and dis- advantages of each method. Also presented is a flash photolytic study on the efficient photochemical generation and thermal reactions of a series of Cu(I) complexes by excited-state electron transfer from tris(2,2'-bipyridine) ruthenium(II), $[Ru(bpy)_3]^{2+}$.

AVAILABLE METHODS

Conventional Flash Photolysis. An excellent discus- sion of the flash photolysis method and instrumentation has been given by West (1). This technique finds greatest application in the determination of k_{th}'s and in providing estimates of $\phi'\phi_{et}$. The determination of k_{th} is usually quite straightforward as long as the reaction is completely reversible. Under these conditions, on flashing, the change in sensitizer concentration equals the change in the quencher concentration and that relationship is maintained throughout the back reaction. Since the sensitizer typically has the larger molar extinction coefficient, ε , its absorbance change is usually monitored. A plot of 1/Absorbance (sensitizer) vs. time provides $k_{th}/\varepsilon 1$ where 1 is the monitoring path length. If the reaction is not completely reversible but the stoichiometric relationship(s) between the main and side reaction is known, one can fit the absorbance recovery curve of the sensitizer using standard iterative techniques to obtain good estimates of the k_{th} values.

The determination of ϕ_{et} is best done by comparing the quencher of interest to a quencher whose $\phi'\phi_{et}$ is known for the particular sensitizer. For example, $[Ru(bpy)_3]^{2+}$ quenched by Fe^{3+} provides a convenient standard with an estimated $\phi'\phi_{et}$ of unity (2). This approach is based on a comparison of the relative sensitizer bleaching in two solutions, one containing the reference quencher and the other the unknown quencher. From the relative changes in optical density, the degree of excited-state quenching in each solution and a knowledge of $\phi'\phi_{et}$ for the standard quencher, the $\phi'\phi_{et}$ for the quencher of interest is readily obtained.

The instrumentation for these experiments need not be highly sophisticated though a short flash pulse, even at the expense of considerable energy, is desirable. By judicious choice of conditions and, in some cases, corrections (vide infra), $\emptyset'\emptyset_{et}$ and k_{th} for reactions with thermal back reactions near the diffusion controlled limit can be measured. The standard flash photolysis configuration is a transversely excited narrow sample tube which is longitudinally monitored. This system permits high sensitivity with up to 20 cm long sample tubes and a very uniform sample excitation. Thus, very small optical density changes can be accurately measured, and reactions with small $\emptyset'\emptyset_{et}$'s or large k_{th}'s can be studied. As an example, where $\Delta\varepsilon$ (product-reactant) at the wavelength of choice is at least $10^3 M^{-1} cm^{-1}$, we have routinely measured concentration changes in the 10^{-6}–10^{-7} $\underline{M}$ region.

The use of conventional flash photolysis is not without pitfalls however, and careful attention to certain potential problems is required. The broad continuum emission from the excitation lamp coupled with the fact that many sensitizers absorb over a broad spectral region makes this method subject to "inner filter effects" from the quencher. This problem is probably responsible for the spread of values reported in $\emptyset'\emptyset_{et}$ for Cu^{2+} quenching of $[Ru(bpy)_3]^{2+}$ (3,4). The difficulty can be minimized by use of filter solutions or materials to delimit the wavelength range of the excitation flash pulse to a clean region where only the sensitizer absorbs. If this cannot be done, an alternate method of determining $\emptyset_{et}$ should be selected. It is also important to keep the percentage of bleaching low to avoid saturation, an almost complete reaction of the sensitizer. In this case, the degree of reaction ceases to depend on pulse energy or quantum yield, and meaningful measurements of $\emptyset'\emptyset_{et}$ become impossible. As long as this error source is recognized, it is easily avoided by reducing the excitation intensity. If possible, the concentration of quenchers should be high enough to heavily quench (> 90%) the sensitizer. This minimizes possible problems due to errors in the values of k_q. Finally, when k_{th} is quite large, account must be taken of the finite width of the flash pulse.

As an example of the use of conventional flash photolysis to study reversible excited-state electron transfer

reactions, we have examined the reaction between excited
$[Ru(bpy)_3]^{2+}$ and several $Cu(II)$ complexes.

The yield of detectible electron transfer to $Cu(II)$
species was determined by a relative flash photolysis
method using the reaction

$$*[Ru(bpy)_3]^{2+} + Fe^{3+} \xrightarrow[\emptyset'\emptyset_{et}(Fe^{2+})]{} [Ru(bpy)_3]^{3+} + Fe^{2+} \quad (5)$$

as a standard. The Fe^{3+} reaction has $\emptyset'\emptyset_{et}$ (Fe^{2+}) estimated
to be unity (2). Two solutions containing the same concen-
trations of $[Ru(bpy)_3]^{2+}$ were flashed under identical experimental conditions of flash energy and cell configuration.
One solution was quenched by Fe^{3+} and the other by the
$Cu(II)$ species. The change in optical density was extra-
polated back to zero time using the measured k_{th}. Because
of the finite width of the flash, we used the terminus of
the flash (~ 30 µsec) as our zero time. Since some back
reaction occurred during the flash, this method underesti-
mates the yield for the systems with large k_{th}'s (e.g., the
chloro, formato and acetato copper complexes): we estimate
the error to be $\sim 5\%$. The flash pulse was restricted by
means of a sodium nitrite filter solution to $\lambda \geq 410$ nm
where the Fe^{3+} was transparent. Thus, the various quenchers
were not competitive with the sensitizer for the excitation
light. Use of a Plexiglass filter ($\lambda \geq 340$ nm) in place
of sodium nitrite produced $\emptyset'\emptyset_{et}$'s which were $\sim 20\%$ too
high because of the Fe^{3+} inner filter effect. The appropriate quencher concentrations were determined on the
basis of K_{sv}'s obtained from the appropriate linear Stern-
Volmer plots (Figure 1). The concentrations were chosen
to give greater than 90% quenching. The k_q's for the
various complexes were evaluated from the K_{sv}'s and γ_o's,
determined using a N_2 laser (5), via the relationship
$k_q = K_{sv}/\gamma_o$. The bleaching of the $[Ru(bpy)_3]^{2+}$ was kept
below $\sim 60\%$ to prevent saturation of the reaction. The
electron-transfer probability yield, $\emptyset'\emptyset_{et}(x)$ for the un-
knowns is then given by

$$\emptyset'\emptyset_{et}(x) = \emptyset'\emptyset_{et}(Fe^{2+}) \left[\frac{\Delta A_o(x)}{\Delta A_o(Fe)}\right]\left[\frac{f(Fe)}{f(x)}\right] \quad (6)$$

where ΔA_o's are the absorbance changes at zero time and
f's are the probability of quenching of $*[Ru(bpy)_3]^{2+}$

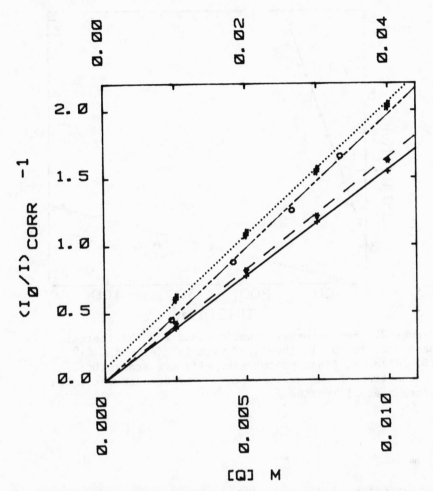

Figure 1. Stern-Volmer quenching plots for deactivation
of $[Ru(bpy)_3]^{2+}$ in deoxygenated water by —— + —— +
copper(II)-formato complexes, * —— —— * copper(II)-
acetato complexes, o—— - - ——o copper(II)-chloro
complexes, and #.....# aquo copper(II). The lower concentration axis applies to all data except the aquo
copper(II) species which uses the upper concentration
scale. The aquo copper(II) data has been displaced
vertically by 0.1 units to enhance viewing.

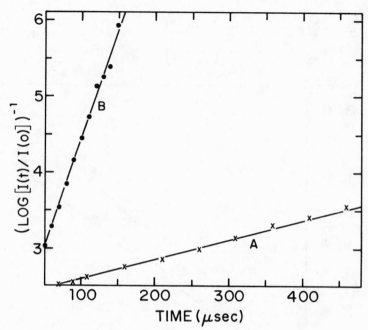

Figure 2. Second-order kinetic plots for the thermal
back reaction of $[Ru(bpy)_3]^{3+}$ with Cu^+ (A) and $CuCl^0$
(B) following flash photolysis. I's are transmitted
intensities of the monitor beam. "0" denotes the value
after reequilibration.

calculated from the Stern-Volmer quenching constants.
Figure 2 shows typical second order kinetic plots as were
used to estimate k_{th}'s and ΔA_o's. We estimate that this
method yields $\phi'\phi_{et}$'s accurate to $\sim30\%$. The $\phi'\phi_{et}$'s, K_{sv}'s,
k_q's, and k_{th}'s are given in Table I.

The $\phi'\phi_{et}$'s are high although apparently somewhat less
than unity. These subunity $\phi'\phi_{et}$'s can be attributed to
either competition of exothermic energy-transfer quenching
to the low-lying Cu^{2+} d-d state or efficient back electron
transfer in the encounter pair

$$D^+\big|M(I) \longrightarrow D\big|M(II) \longrightarrow D + M(II) \qquad (7)$$

TABLE I

Photochemical Properties of $*[Ru(bpy)_3]^{2+}$ with
Cu(II) Complexes

Quencher[a]	$K_{sv}^{\emptyset}$	$\emptyset'\emptyset_{et}^c$	$k_q \times 10^{-8b}$ $M^{-1}-s^{-1}$	$k_{th} \times 10^{-9}$ $M^{-1}-s^{-1}$
Cu^{2+}	49 ± 1	0.74	0.82	0.34 ± 0.02
$CuCl^{+d,e}$	196 ± 4	0.59	3.3	3.5 ± 0.10
Cu-formate[f]	156 ± 1	0.16	2.7	2.10 ± 0.16
Cu-acetate[g]	165 ± 4	—	2.8	2.30 ± 0.12

(a) Deoxygenated aqueous solutions at 21°C with ionic
    strength of 1.0.
(b) Accuracy $\pm$ 10 - 15%.
(c) Accuracy $\pm$ 25%. Based on media independent τ_0 of
    600 nsec.
(d) 0.10 M $HClO_4$.
(e) [NaCl] = 0.5 M. 75% $CuCl^+$.
(f) [Na(O_2CH)] = [HO_2CH] = 0.5 M. 80% $Cu(O_2CH)_4^{2-}$.
(g) [NaO_2CCH_3] = [HO_2CCH_3] = 0.5 M. 90% $Cu(O_2CCH_3)_4^{2-}$.

Since k_q's for quenching of $*[Ru(bpy)_3]^{2+}$ by energy transfer to d-d excited states can be comparable to the k_q's
for the Cu(II) species (6), energy transfer is a realistic
explanation of the subunity $\emptyset'\emptyset_{et}$'s. Back electron transfer in the encounter pair is a highly exothermic process,
however, and may well compete with the disengagement of
Cu^+ from the encounter pair. We cannot differentiate between these two pathways.

There are two possible detailed mechanisms for Cu^+
production (4) similar to those originally proposed for
Fe^{3+} (7).

$$*D + Cu^{2+} \longrightarrow *D|Cu^{2+} \tag{8}$$

$$*D|Cu^{2+} \longrightarrow D^+|Cu^+ \tag{9}$$

$$*D|Cu^{2+} \longrightarrow D|*Cu^{2+} \longrightarrow D^+|Cu^+ \tag{10}$$

$$D^+|Cu^+ \longrightarrow D^+ + Cu^+ \tag{11}$$

where $*D$ is excited $[Ru(bpy)_3]^{2+}$ and D^+ is $[Ru(bpy)_3]^{3+}$.

Equations 8 and 11 are common to both mechanisms which
differ only in the details of the electron-transfer step.
In Eq. 9, *D directly reduces Cu^{2+} while in Eq. 10, *Cu^{2+}
produced by energy transfer oxidizes D.

We cannot definitely rule out Reaction 10, but it
seems unlikely. Energy transfer to Cu^{2+} results in excitation of the d-d excited state. This state should
promptly, within no more than a few solvent vibrations,
undergo internal conversion to its excited state zero
point (8). The zero point energy of Cu^{2+} is < 0.9 ev (10)
which, when coupled with the reduction potential of 0.15 V
for the Cu^{2+}-Cu^{+} (12) yields an E^{o} for *Cu^{2+} + e $\rightarrow$ Cu^{+}
of < 1.05 V. Since E^{o} for the $[Ru(bipy)_3]^{3+}$ - $[Ru(bpy)_3]^{2+}$
couple is ∼ 1.25 V, oxidation of $[Ru(bipy)_3]^{2+}$ by *Cu^{2+} is
energy deficient by > 0.2 V. The high efficiency of
oxidation of $[Ru(bpy)_3]^{2+}$ seems inconsistent with this
weak driving force. Thus, we consider Reaction 10 unlikely.
As a final point, even if energetically favorable, the Cu^{2+}
excited-state is so close to the ground state that radiationless deactivation would probably very efficiently deactivate
the excited state before *Cu^{2+} would function as an oxidant.

The Cu(II) data in Table I show, as expected, that the
reduction of positive charge in the quencher and the reduced species enhances k_q and k_{th}, respectively. This
cannot be the only factor involved, however. The $CuCl^{+}$
system shows the fastest rates, yet it is not the most
electrostatically favored. The greater ease of reduction
(12) and a more favorable charge factor of $CuCl^{+}$ vs Cu^{2+}
may account for the faster k_q. Somewhat surprising and
anamolous are the differences in k_{th}'s for Cu^{+} vs CuCl.
Because of the lower reducing power of CuCl vs Cu^{+} (12),
we would have expected Cu^{+} to be oxidized more rapidly than
CuCl (even taking into account the somewhat more favorable
charge factors); yet CuCl is oxidized an order of magnitude
faster than Cu^{+}. Indeed since the CuCl reaction is near
the diffusion controlled limit, the differences may be even
larger. This result suggests that the ligand can play more
than just a passive role of modifying the redox potentials,
and somehow mediates the reverse thermal electron transfer
process. Thus, Cl^{-} may function as an efficient anionic
bridge which electrostatically holds the cationic $[Ru(bpy)_3]^{3+}$
and the Cu^{+} together to facilitate the outer sphere electron
transfer.

Durante and Ford have demonstrated intramolecular electron-transfer to form Cu(I) in the pentaammine(pyrazine) ruthenium(II) complex of Cu(II) (13). Our results with electron-transfer from the CT excited state of $[Ru(bpy)_3]^{2+}$ to a variety of Cu(II) complexes indicate that it is probably the CT state of the $[Ru(NH_3)_5(pyrazine)]^{2+}$ complex which is responsible for their Cu(I) production.

Laser Flash Photolysis. Laser excitation sources can improve on conventional flash photolysis in the study of reversible excited-state electron-transfer reactions. Laser systems have been described by West (1).

Lasers in principle can make ideal flash photolytic sources. They can have extremely high powers and pulse energies, short pulse durations, high monochromaticities, good collimation, and beam quality. Suitable lasers are ruby (primary or doubled), neodymium (tripled, doubled, or quadrupled) and flash lamp pumped dye lasers. Nitrogen lasers have suitable pulse energies but suffer from very inhomogeneous beams which make them particularly difficult to use in the study of bimolecular processes where the rates are proportional to the square of the reactant concentrations. The noble gas–fluorine excimer lasers offer a powerful new source of short duration high power uv radiation (14).

The least expensive and among the most versatile of these sources are flash lamp pumped dye lasers which yield 10's of mj to j's of energy in 0.1-1 µsec pulses with tunability over the 360-800 nm range. Excimer lasers are the next most expensive and offer relatively inexpensive uv pulses (50-150 mj, 10-30 nsec, and 193-351 nm lines (14)). The ruby and neodymium lasers while offering versatility are too costly and can require technological capabilities beyond that of many research groups.

Besides cost, lasers are not without their disadvantages. Their very high pulse energies can easily yield saturation. Beam inhomogeneity is also a potentially serious problem, although large beam areas permit sampling a uniform region. Goldschmidt has discussed several of these errors (15).

The final problem which is common to both laser and conventional flash photolysis is determination of the

effective quantum dose delivered to the sample. Dynamic
laser actinometers have been utilized in which a substance
with an accurately known T-T or S-T absorption spectrum is
flashed in the same configuration as is used for the electron-
transfer reactions (1,15). The difficulty of measuring
spectra to adequate accuracy has limited the usefulness of
the approach. Excited-state electron-transfer reactions
have also been used as discussed above (3,4,16).

Steady State (SS) Method. Lin and Sutin have developed
and used an extremely clever steady-state method for determination of $\phi'\phi_{et}$ (17). This method utilizes an optically
dilute solution of the photosensitizer so that the emission
intensity is directly proportional to the sensitizer concentration. The steady-state emission intensity for the
sample is then measured at a series of accurately determined light fluxes. At low light fluxes the light driven
redox reaction consumes only a small amount of the sensitizer, and the emission intensity is directly proportional
to the excitation intensity. As the flux is increased, however, an increasing fraction of the sensitizer is in the non-
luminescent $D^{\pm}$ form at steady state. The power dependent
consumption of D causes the emission flux vs excitation
flux curve to become sublinear at the higher powers. These
data permitted Lin and Sutin to calculate the steady-state
concentrations of D and $D^{\pm}$ at different fluxes. The steady-
state occurs when the forward and reverse reaction rates
are equal. If k_{th} is known from a separate experiment, the
reverse rate can be calculated. The forward photochemical
rate is determined by the flux, the extinction coefficient
of the sensitizer at the exciting wavelength, and $\phi'\phi_{et}$.
Since $\phi'\phi_{et}$ is the only parameter which cannot be directly
measured, it can be calculated by equating the forward and
reverse rates. Lin and Sutin have described the method in
detail including a linearized method of data evaluation
(17). The authors have used this approach to calibrate
the $[Ru(bpy)_3]^{2+}$-Fe^{3+} system and find $\phi'\phi_{et} = 0.81 \pm 0.16$.
Similar measurements have also been carried out on a series
of polypyridine-ruthenium(II) photosensitizers with Fe^{3+}(16).

The steady state method is especially valuable because
it is an absolute one. Because it is done under static
conditions, geometries and light intensities are all fully
controllable and measurable with high accuracy. Very high
accuracy is thus, in principle, possible. Excluding the

need to measure k_{th}, the equipment requirements are minimal. A spectrofluorimeter and an actinometer are all that
is required. The SS method is immune to saturation effects.
Therefore, it is particularly useful for generating standard
electron-transfer reactions which can be used as actinometers in dynamic flash lamp and laser flash photolysis
experiments. By use of a strong enough light source such
as a cw ionized Ar laser, the Lin-Sutin method should be
adaptable for k_{th}'s which approach the diffusion-controlled
limit.

The steady state method has several disadvantages.
It is time-consuming and tedious. k_{th} must be derived
from an independent measurement technique such as stopped-
flow kinetics or flash photolysis. Also, if the reaction
products are not very stable (i.e. τ <10s), then this
method fails completely because reliable steady-state values
are unavailable. At the very highest k_{th}'s and with
quenchers which absorb appreciably at the exciting wavelength, thermal transients may cause improper steady-state
emission intensities. For example, we have observed that
the $[Ru(bpy)_3]^{2+}-Cu^{2+}$ system does not yield accurate steady-
state values at high fluxes and with high Cu^{2+} concentrations because of the absorption of the Cu^{2+}. The sensitizer must also emit, but this is rarely a problem
especially since the emitting and sensitizing states do
not have to be the same (e.g. triplet sensitization and
singlet emission).

Step-Excitation (SE) Method. Wight, Turley, and Demas
(18) have developed a transient modification of the Lin-
Sutin steady-state method. The SE method appears to overcome all of the disadvantages of the SS method. Wight et al.
reasoned that more information would be provided in a
dynamic rather than steady-state experiment. This method
utilizes a step excitation function which is initially zero
and then rises instantly to a constant value. As in the
Lin-Sutin approach, the solutions are optically dilute
and relative concentrations are monitored fluorometrically.
When the source is turned on, the emission rises to a
maximum and then decays to a steady-state as luminescent D
is converted to $D^{\pm}$. From this single experiment, and with
the values of the flux, the sensitizer extinction coefficient
at the exciting wavelength, and the degree of quenching of
the excited state, both $\phi'\phi_{et}$ and k_{th} can be determined.

Interested readers are referred to the original theoretical paper in which suitable linearized equations are derived as well as an analysis of the influence of errors in the various measurement parameters.

This SE method has enormous advantages. It is simple and can be implemented using readily available equipment. For example, for systems with slow k_{th}'s, a photographic shutter in the excitation beam makes an excellent source of the step function; the opening time is < 1 msec. Both $\phi'\phi_{et}$ and k_{th} are obtained from data taken in a single experiment lasting < 1s. Using a laser source such as the ionized Ar laser (available in many Raman systems), k_{th}'s at the diffusion controlled limit should be measurable. For example with a $\phi'\phi_{et}$ of 0.5, a sensitizer concentration of $\sim 10^{-7}M$, a quenching efficiency of *D near unity, ε = 4000, and a flux of $1.3 \times 10^{-4} ein/cm^{-2}\text{-}s^{-1}$ (0.5 W in 1.4 mm diameter beam at 488 nm), the steady-state emission intensity will be 50% of the peak value for $k_{th} = 10^{10}M^{-1}s^{-1}$. Also, as with the Lin-Sutin method, the step-excitation method is an absolute one and is therefore suitable for calibration of dynamic actinometer reference systems. Unlike the SS method, the SE method lends itself to computerized data acquisition and reduction and is capable of extremely high accuracy and precision. Further it does not require that $D^{\pm}$ and $Q^{\mp}$ be stable for more than about 1 s. It is insensitive to saturation effects.

An example demonstrates the utility of the SE method. Taylor and Demas have redetermined $\phi'\phi_{et}$ for $[Ru(bpy)_3]^{2+}$-Fe^{3+} system. Using a microcomputerized data-acquisition system, they were able to collect and reduce the data in a complete experiment to obtain both $\phi'\phi_{et}$ and k_{th} in 5 min. They obtained $\phi'\phi_{et}$=0.94 ± 0.05 (2) which is more consistent with the spectroscopic and kinetic information on this system than the 0.81 estimate of Lin and Sutin (17) or the $\phi' \sim 0.5$ flash photolysis results (19).

The principal disadvantage of the SE method is that transient information must be obtained. This would usually necessitate using an oscilloscope or transient recorder, although for systems with very low k_{th}'s at low light intensities a recorder would be adequate. The sensitizer must emit as in the SS method. $D^{\pm}$ and $Q^{\mp}$ must be stable for at least several tens of milliseconds. Also, at high fluxes the necessary condition of a uniform flux across the excitation beam may be difficult to achieve.

ACKNOWLEDGMENTS

We gratefully acknowledge assistance of James Madison University, the National Science Foundation (CHE 77-20379) and the Air Force Office of Scientific Research (AFOSR 78-3590).

REFERENCES AND NOTES

(1) M.A. West in "Creation and Detection of the Excited State", W. R. Ware, editor, Marcel Dekker, New York, 1976, p. 217.

(2) D. G. Taylor and J. N. Demas, submitted

(3) D. Meisel, M. S. Matheson, W. A. Mulac, and J. Rabani, J. Phys. Chem., 81, 1449 (1977).

(4) M. A. Hoselton, C-T. Lin, H. A. Schwarz, and N. Sutin, J. Am. Chem. Soc., 100, 2383 (1978).

(5) D. G. Taylor, T. J. Turley, M. Rodgers, S. H. Peterson, and J. N. Demas, submitted.

(6) J. N. Demas and J. W. Addington, J. Am. Chem. Soc., 88, 5800 (1976).

(7) C. R. Bock, T. J. Meyer, and D. G. Whitten, J. Am. Chem. Soc., 96, 4710 (1974).

(8) For example, Cr(III) complexes undergo both internal conversion within the quartet manifold and inter-system crossing to the lowest doublet state in ≤ 5 psec.

(9) A. D. Kirk, P. E. Hoggard, G. B. Porter, M. G. Rockley, and M. W. Windsor, Chem. Phys. Lett., 37, 199 (1976).

(10) The zero point excited state energy was estimated for the absorption spectra using the Fleischauer criterion (11).

(11) P. D. Fleischauer, A. W. Adamson, and G. Sartori, in "Inorganic Reaction Mechanisms", Part II, J. O. Edwards, Ed., Wiley, New York, N.Y., 1972, p. 1.

(12) W. M. Latimer, "The Oxidation States of the Elements and Their Potentials in Aqueous Solution", 2nd

Edition, Prentice Hall, Inc., New York, N.Y. (1952).

(13) V. A. Durante and P. C. Ford, J. Am. Chem. Soc., 97, 6898 (1975).

(14) See for example literature on the Tachisto TAC II laser system, Tachisto Inc., 13 Highland Circle, Needham, Mass., 02194.

(15) C. R. Goldschmidt, Proc. Soc. Photo-Opt. Instrum. Eng., 99, 252 (1977)).

(16) C-T. Lin, W. Büttcher, M. Chou, C. Creutz and N. Sutin, J. Am. Chem. Soc., 98, 6536 (1976).

(17) C-T. Lin and N. Sutin, J. Phys. Chem., 80, 97 (1976).

(18) C. A. Wight III, T. J. Turley, and J. N. Demas, J. Chem. Phys., 68, 5486 (1978).

(19) R. Bensasson, C. Salet, and V. Balzani, J. Am. Chem. Soc., 98, 3722 (1976).

PHOTOGALVANOVOLTAIC CELLS AND PHOTOVOLTAIC CELLS USING GLASSY CARBON ELECTRODES

H. Ti Tien, John Higgins, and James Mountz

Biophysics Department, Michigan State

University, East Lansing, Michigan 48824

ABSTRACT

Numerous electrochemical photocells have been proposed for the direct conversion of solar energy to electricity or chemical fuel. These cells, on the basis of
two well established photoelectric effects, can be clas-
sified as either photovoltaic (PV) or photogalvanic (PG).
The operation of a PV cell depends on the generation of
an EMF as a result of the absorption of light, whereas
the operation of a PG cell relies on the excitation by
light of photoactive species in solution which induces a
Faradaic process at the electrode. This paper describes
a new type of electrochemical photocell, the operation
of which is based on the combined principles of the PV and
PG phenomena. This system, having the advantages of both
the PV and PG cell, is therefore called the photogalvanovoltaic (PGV) cell. The key element of the cell responsible for the PV effect is a porphyrin-coated glassy carbon electrode. Either Pt or glassy carbon can serve as
the counter electrode. The results of an electrochemical
PV cell using porphyrin-coated glassy carbon electrodes
are also described.

INTRODUCTION

     Recent increase of interest and research in photo-
electrical energy conversion is partly due to the general
awareness of an impending shortage of easily recovered
and processed fossil fuels, with a concomitant need to
discover and develop new energy sources. Solar powered
cells are a promising source of energy, and the practical
motivations for their study is having an influence on the
type of research undertaken recently. Diverse scientific
and technical fields are brought to bear upon the problems
of the commercial refinement of the various solar energy
converters. Incidental problems, such as minute side reactions and deterioration, must be addressed and solved
if the practical work is to attain its goals. Even in
the random search for new materials or phenomena which may
be applied to photoelectric energy conversion, the questions of utility and costs ultimately surface. In general,
methods of photoelectric conversion seek to be more efficient and to deliver the energy in a more versatile form
than does thermal conversion. Two major problems in the
development of photoelectric conversion are the high cost
of the most efficient solar cells (pn-junction Si cells)
and the low efficiency of cheaper designs, coupled with
fragility and deterioration in the form of corrosion,
oxide formation, and anodic dissolution (1). Before describing our studies on two approaches that show promise
in alleviating both these problems, a photovoltaic cell
constructed of glassy carbon coated with tetraphenylporphyrin pigments, and a concept for combining two types of
solar cells to produce a photogalvanovoltaic cell more
efficient than either type alone, we shall first of all
give a classification of photoelectric energy converters
and then summarize their mechanisms of operation. Per-
tinent background materials and references can be found
in recent publications (2-5).

Types of Photoelectric Energy Converters

     Photoelectric energy converters, or solar cells, are
generally divided into three overlapping groups: photovoltaic, photogalvanic, and photoelectrochemical, with the
common property of partially converting the energy of some
of the absorbed photons into useful work. A number of
criteria are generally used to classify the type of cell.
If the absorption occurs in the region of a solid-state

junction (semiconductor-semiconductor or semiconductor-
metal) and charge transport is exclusively electronic,
then the cell is termed photovoltaic. If the electric
current driven by absorption of photons is partly ionic,
the junction of the electronic and ionic regions implies
chemical reaction, so the cell is called photoelectrochemical. The chemical reaction may be cyclic, with no
net consumption or production of chemical species in the
electrolyte, the only net change being transport of charge
so as to create direct current; or charge transport may
occur concomitantly with net chemical reaction in the
electrolyte. If the reaction is endergonic and the products do not suffer substantial back or side reactions
(or can be separated before reacting), part of the photon's
energy is stored as chemical energy. Any photoelectrochemical cell could also be considered a photogalvanic
cell, but this term is normally reserved for those photoelectrochemical cells in which the photons are absorbed
by components of the electrolyte, and for which the useful
work produced is primarily electrical (6). This brings us
to a second approach for distinguishing photoelectric
energy converters, and that is by the way in which photoconversion occurs. In photovoltaic cells a semiconductor
band gap assists the transfer of free energy to electrons.
In photogalvanic cells absorption occurs in the liquid
(electrolyte) phase, and in photoelectrochemical cells
absorption may occur in either manner, depending upon the
design of the cell. For the purpose of our discussion,
we consider any solar cell for which the electronic band
structure of the electrode determines conversion and whose
primary work is electrical to be a photovoltaic cell.
Accordingly, one may distinguish three classes of photovoltaic cells: solid state cells, semiconductor-liquid
junction cells (7), and sensitized semiconductor cells (8),
which we also refer to as pigmented-semiconductor cells,
whenever the pigment is not in solution. In the above
classification, there are varieties of each; for example,
solid state cells may be homotype pn-junction, heterotype
pn-junction, or Schottky barrier. Light directly strikes
the active (energy converting) electrode of the semiconductor-liquid junction cell, which is in contact with an
electrolyte transparent to the wavelength of interest.
The pigmented electrode is coated with a chemisorbed pigment or dye that does not enter into solution. Both metals
and semiconductors may be pigmented, but the most practical
for solar energy conversion is a pigmented-semiconductor

type. The semiconductor-liquid junction and pigmented-semiconductor cells may also be considered as photoelectrochemical cells, and they are sometimes referred to as "wet" solar cells (1) because of their electrolytic solution. Liquid junction cells are sometimes called Schottky barrier type. Hybrids of photoelectrochemical cells may be conceived, and a following section describes our photogalvanovoltaic cell, a hybrid between a modified iron-thionine photogalvanic cell and a pigmented-semiconductor photovoltaic cell (9).

Mechanisms of Energy Transduction

For the energy of the photon to be converted into useful work, it must be absorbed in a region of strong local internal field. In bare semiconductors,   the field separates the high-energy charge carriers (after first dissociating any excitons into electron and hole), and carries one type into the bulk of the semiconductor and the other type to the surface of the semiconductor, where it reacts with the electrolyte. In coated semiconductors, the electric field either pulls an electron from the excited pigment molecule into the bulk of the semiconductor or transports an electron from the semiconductor to the vacant lower energy level of the pigment (a process which is conveniently represented as injection of holes into the semiconductor). The field may also assist the transfer of electrons between the absorbed molecules and other molecules or ions in solution in the electrolyte. For both types of cells, the photogeneration of charge carriers and their injection into the semiconductor puts the electron and hole constituents of the semiconductor and electrolyte out of chemical equilibrium. Thus, an electrochemical potential difference is established, and a current arises upon connecting these two phases via a circuit that permits charge transport, such as semiconductor/metal wire/counter electrode/electrolyte/pigment.

The initial internal field, maintained at equilibrium in the dark, is established by the exchange of electrons as the various phases (semiconductor, pigment, and electrolyte) come into equilibrium. The initially unequal exchange of charge across a phase boundary naturally charges the two phases and creates internal, electrostatic fields. High charge density phases, such as metals and electro-

lytic solutions with concentrations of electrolytes greater than about 10^{-3} M, cannot maintain internal fields; the
charge exchange manifests itself as a surface charge or a
dipole layer at the phase boundary. The low free charge
density and charge traps (impurity ions and defects) of
semiconductors means that the internal electrostatic field
cannot be shielded immediately below the surface, so that
it extends for hundreds to thousands of Angstroms, and
that region is called the space charge layer. In effect,
the free charge density of the majority carriers of the
semiconductor is reduced as the charges adjust to screen
the internal field in the bulk of the semiconductor. The
exchange of charge required to establish equilibrium can
proceed to the point that substantially all the free
charge is swept from the space charge layer, in which case
it is called a depletion layer. Further charge exchange
can lead to the dominance of minority charge carriers in
the space charge layer, which is then termed the inversion
layer. Optimal operating conditions for a semiconductor
photoelectrochemical cell obtain when the semiconductor
and electrolyte are matched to put the semiconductor in
the depletion mode, so that separation of photogenerated
charges is enhanced. The layer of ionized pigment molecules may be modeled by the specific and diffuse electric
charge layers of Gouy-Chapman theory, with modifications
for the fact that the molecular layers are strongly adsorbed.

Charge exchange continues in equilibrium, but at
equal forward and reverse reaction velocities, the magnitude of current in one direction being called the exchange
current. The rate constants are governed by a number of
factors. Two important ones are the density of states
available to each phase and the distribution of electrons
in these states. The density of states functions are represented by conduction and valence bands and trapping
levels in semiconductors; conduction bands in metals; and
molecular energy levels of finite width, normally assumed
Gaussian, in the molecules of the pigment and electrolyte.
The distribution of electrons determines the Fermi levels
of semiconductors, metals, and redox solutions; and, for
the pigments, which molecular orbitals are occupied. The
Fermi level, E_F, of a redox solution is related to its
oxidation potential (electrode potential), U_{redox}, by

$$\frac{1}{e} E_{F, redox} \ (eV) = constant - U_{redox}(V), \tag{1}$$

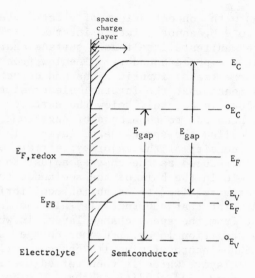

Figure 1.   Band bending in p-type semiconductor at liquid
             junction interface. $^{O}E_C$ and $^{O}E_V$ are initial band
             edges, before contact with electrolyte. Equi-
             librium between semiconductor and electrolyte
             results in a transfer of electrons to the semi-
             conductor, and is represented by raising of
             Fermi level $^{O}E_F$ to $E_F = E_{F,redox}$, and band
             bending to interior levels E_C and E_V.   The
             electrolyte Fermi level $E_{F,redox}$ changes little,
             and is related to the redox potential. If the
             redox level is E_{FB}, the equilibrium bands are
             flat.

where the constant is −4.5 V if $E_{F,redox}$ is referred to
the standard hydrogen electrode (10) and the Fermi level
of each phase is measured with respect to the vacuum level.
The generation of a space charge layer and the modifaction
of energy levels in a semiconductor in contact with a
redox electrolyte solution (the liquid junction photocell)
is illustrated in Figure 1. Should the initial redox
potential have the particular value U_{FB} so that the corresponding redox Fermi level E_{FB} equals the initial semiconductor Fermi level $^{O}E_F$, then the two phases are initially in equilibrium, there is no net charge exchange,
and the semiconductor energy band remains flat. Such a
redox electrolyte is said to have the flat band potential
(8). In general, the initial redox potential U_O gives a

Fermi level $E_{F,redox}$ that is not equal to ^OE_F, with their
difference representing a driving force for net charge
exchange. For $E_{F,redox}$ greater than ^OE_F electrons enter the semiconductor, creating a space charge layer that
gives an internal electrostatic field retarding further
entry of electrons. Net charge exchange falls to
zero at equilibrium, and the initial ^OE_F is raised to its
equilibrium value, E_F. The requirement of uniform E_F at
equilibrium and an interfacial surface charge density
different from the interior charge density is equivalent
to band curvature at the interface. The charge density
in the interior of the semiconductor is unchanged, so that
E_V, E_F, and E_C have their initial _relative_ values there.
The charge layers formed at the interface cause an electrostatic field, resulting in a potential energy difference
between an electron at rest _in vacuo_ in the region of the
interface and one at rest _in vacuo_ in a region far from
the interface. This is equivalent to E_V and E_C in the
bulk of the semiconductor being higher, with respect to
the _vacuo_ state at the interface, than before the charge
layers were formed. The high density of ionic charges
in any electrolyte used for practical cells means that
$E_{F,redox}$ is changed imperceptibly by the charge exchange.

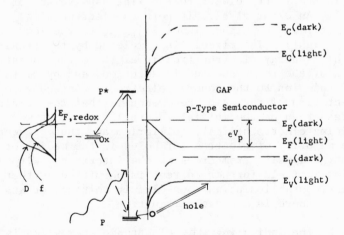

Figure 2.   Sensitized semiconductor photoexcitation. Electron from excited pigment P^* reduces Ox to Ox$^-$
of redox couple. Vacant orbital of P accepts an
electron from the semiconductor, thus injecting
a hole, reducing space charge and band bend, and
lowering E_F, whose shift gives photo-potential

V_p. Some Ox^- reduces P+, but most carries cell
current. Molecular levels have Gaussian density of states function D and distribution
function f such that $f = \frac{1}{2}D$ at $E_{F,redox}$.

Both the pigmented and bare semiconductor-liquid
junction cells have been studied in detail (2-4) and the
physical and chemical basis of their operations are fairly well understood. The highest reported power conversion
efficiency of a dye sensitized photoelectrochemical cell
is 1.5% for sintered ZnO at 653 nm sensitized by rose
bengal and under anodic bias (1). In constrast, the Si
pn-junction photovoltaic cell has an efficiency of 12-18%.
Figure 2 illustrates photoexcitation of a pigmented semiconductor electrode. An electron is excited from energy
level P of the pigment molecule to level P^*, an excited
state. Under the influence of the space charge layer's
electric field, an electron leaves the semiconductor to
occupy level P, creating a hole. The electric field tends
to inject the hole into the semiconductor (double arrow),
with the result that the space charge is reduced, and the
energy bands and Fermi level shift from the equilibrium
dark values (dotted lines) to steady state values (solid
lines) in the light, closer to the flat band values of
Figure 1. An external circuit permits electron flow
through it from the higher level $E_{F,redox}$ to the lower
level E_F(light). The circuit is completed by the transfer
of the high-energy electron from level P^* to the vacant
level Ox of the redox species in solution, and by migra-
tion of this ion to the counter electrode. An equivalent
representation of the same phenomenon is that the injection
of charges reduces the initial internal field that was
maintaining electrochemical equilibrium and thus preventing the majority of electrons or holes in the medium of
high charge density from crossing the interface. With a
reduction in field further current is permitted, either
transient, establishing an open-circuit voltage, or sustained, if there is a closed circuit.

Each line indicating the electronic energy levels
of a molecule represents a Gaussian density of states,
as shown at the left. The fraction of a molecular species
with an occupied orbital is given by the distribution
function f, which depends upon the Fermi level. Energy
of hydration is ignored here. One desires to adjust the
redox potentials of the redox electrolyte and of the ground

state and excited pigments, and to choose a semiconductor
with a work function and energy band gap so that the amount
of electrical work derived from the absorbed photon is
maximized. As suggested by the above discussion, quantum
efficiencies for pigmented-semiconductor cell reactions
are very much dependent on the electrode surface, the pigment, and the redox agents. Favorable conditions give
quantum efficiencies of the order 10^{-1} (12). The energy
difference between the electronic energy levels of P^* and
the redox couple represents an energy loss, as does the
amount of band bending (electric field) seen by the in-
jected hole (some band bending is necessary for charge
injection at a useful rate).

Photogalvanic cells are based on reversible photoredox reactions. The photosensitive electrolyte near one
electrode is illuminated. The electrolyte contains a
photo-sensitive species that is excited from the ground
singlet state S_0 to the first excited singlet state S_1.
There are several modes of decay, but the sequence for the
photoreduction half reaction, which gives electrical work,
is shown in Equation 2, where D represents the dye (with
its state in parentheses), Re and Ox are the reduced and
oxidized forms of the other species, and W is electrical
work.

$$D(S_0) \xrightarrow{h\nu} D(S_1) \qquad \text{(2a)}$$

$$D(S_1) \longrightarrow D(T_1) \qquad \text{(2b)}$$

$$D(T_1) + Re \longrightarrow D^-(S_0) + Ox \qquad \text{(2c)}$$

$$D^-(S_0) \longrightarrow D(S_0) + e^- + W \qquad \text{(2d)}$$

State S_1 decays to the triplet state T_1 with energy between the singlet states. The relatively long lived triplet state may be reduced by another species in solution
(including the ground state molecules of the photosensitive species) with only a small loss of free energy and a
small increase (or even a decrease) in entropy (6). The
species $D^-(S_0)$ may adsorb onto the inert metal electrode.
Because the photoexcitation of D and subsequent reduction
to D^- has upset the previously established equilibrium,
there will be a net transfer of electrons to the electrode.
In effect, the light has lowered the oxidation potential
of the half-cell, generating a cell potential, if the
potential of the other electrode/electrolyte is not changed

the same amount. Normally the counter electrode is kept
dark.

Other steps, producing no electrical work, compete
with these desired reactions. By far the most deleterious
to the cell's useful operation is the back reaction

$$D^-(S_o) + Ox \longrightarrow D(S_o) + Re + Q, \qquad (2e)$$

where Q is heat of reaction. A number of different cells
have been designed to eliminate or compensate for this de-
fect (6,13). Other modes of decay also waste energy. These
are   fluorescence, internal conversion, and transformation to a different chemical species (a photo-product)
in place of Equation 2b, with the result that the initial
photo energy is transformed primarily into a lower energy
photon, heat, or chemical energy, respectively (14).
Phosphorescence or a photoproduct may occur in place of
step 2c. Minimization of these effects requires consideration of the dye or pigment and the redox couple to be used.

PHOTOGALVANOVOLTAIC CELLS

As the name implies, the photogalvano-voltaic (PGV)
cell is a novel type of photoelectrochemical cell based
on a combined principle of photogalvanic and photovoltaic
effects. The principle element of the cell consists of
a pigmented semiconductor electrode and a counter electrode reversible to a photoactive dye in solution. The
photo-emf generated across the cell is equal to the sum
of the voltages derived from the photogalvanic and photovoltaic processes occurring at the respective electrodes.

Background

In the course of our investigation of photoelectric
effects in chlorophyll- containing bilayer lipid membranes,
which have been used as a model for the photosynthetic
thylakoid membrane (15), we came upon the idea that a new
type of photoelectrochemical cell might be developed. In
the pigmented bilayer lipid membrane (BLM) system, the
photoactive material (e.g., chlorophyll) is localized in
an ultrathin hydrophobic membrane less than 100 Å thick,
which is interposed between two aqueous solutions containing redox compounds. Thus, light-induced charge separa-

tions occur only at the solution/membrane interfaces leading eventually to a voltage across the membrane. Energy
conversion is assured if the consequence of light-initiated
coupled redox reactions results in a flow of electrons
through the membrane in a direction opposite from that
which is thermodynamically favorable in the dark. In
principle, the photoelectric BLM system could be developed
for the conversion of light to electrical energy. However, there were at least two major shortcomings which
arise from the properties of the membrane. First, it
was very difficult to fabricate large areas of unsup-
ported membranes less than 100Å thick of sufficient mechanical strength, and second, most of the light used to
excite the membrane passed through it without causing any
useful effect owing to the minute quantity (10^{-10} mole/
cm^2) of pigment in the membrane. This results in a very
inefficient use of light energy.

To overcome the first problem, we have formed BLM
in the pores of polycarbonate films (commercially known
as Nuclepore Membranes). Pigmented BLM on this kind of
support not only exhibited superior photoelectrical properties compared with those of conventional BLM, but also
possessed far greater stability and manipulatability. In
order to overcome the second shortcoming of poor light
utilization, we reasoned that the light, passing through
the membrane, could be made to interact with photoactive
species such as thionine and Fe^{2+} ions, if they were present in the bulk of the solution. Such interaction should
result in the usual photogalvanic action and thus contribute further to light conversion efficiency. The first
version of our PGV cell consisted essentially of a photogalvanic cell in series with a photovoltaic cell (9). The
key element responsible for the photovoltaic effect of the
test cell was made from a Nuclepore membrane which was
coated with a lipid solution of phosphatidyl choline, Mg-
meso-tetraphenyl porphyrin, and oxidized cholesterol in
a choloform:n-octane (1:1) solvent mixture. The coated
membrane mounted on the opening of the right chamber was then
placed in position as shown in Figure 3. The PGV test cell
had three electrodes. Two of them were calomel electrodes
with saturated KCl bridges which were used to detect the
membrane photovoltaic effect. The photogalvanic effect of
the iron/thionine system was measured between the platinum
electrode and the shunted calomel electrodes, whereas the
combined photoeffects of the cell were monitored between

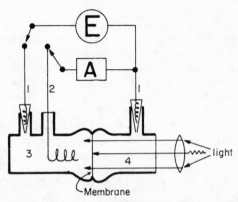

Figure 3.   Schematic diagram of the photogalvanovoltaic
(PGV) test cell. E-electrometer; 1, calomel
electrode with saturated KCl bridge; 2, platinum wire electrode; 3, Fe^{2+}/Fe^{3+}-thionine solution; 4, transparent electrolyte solution containing electron donors; membrane, porous polycarbonate membrane coated with a layer of eosin
and lipid solution made of tetraphenyl porphyrin, oxidized cholesterol, phosphatidyl choline
and n-octane.

the Pt electrode and the calomel electrode positioned in
the right chamber (marked 4). The photopotential of this
PGV cell was equal to the sum of the voltages derived from
the PG and PV effects. Our explanation for the operation
of the PGV cell is straight-forward. Exciting light after
passing through transparent redox solution (right side),
impinged on the membrane, a tiny portion of it being absorbed by the photoactive material (eosin and tetraphenyl
porphyrin) as is illustrated in Figure 3. This absorbed
light resulted in electronic excitation of molecules and
in charge separations followed by coupled interfacial
electron-transfer reactions in that ascorbate was oxidized
to dehydroascorbate in the right chamber. Concurrently,
Fe^{3+} in the left chamber was reduced to Fe^{2+}. Most of the
unabsorbed light was then absorbed by thionine present in
the left chamber and this caused the usual photogalvanic
effect. In brief, light-initiated redox reactions, both
in the membrane and in the bulk solution, moved the electrons from the right half of the cell to the left as indicated by the observed polarity (left side negative).
Although the measurement of photo-induced voltages relative to the dark voltage has provided a critical test of
the concept of the PGV cell under development, this PGV

cell had at least two drawbacks: the high internal resistance and longterm instability of the membrane. To
overcome these shortcomings, we have developed a PGV cell
without the membrane (28). Before describing the details
of the second version of PGV cell, a discussion of some
of the properties of different types of PG and PV cells
using glassy carbon electrodes, which were investigated
during the course of this study, is in order.

Photogalvanic Cells

The usual photogalvanic effect is observed by im-
mersing two identical platinum electrodes into the iron
thionine solution and shielding one in the dark while the
other is illuminated. The usual solution consists of
10^{-4}M thionine, 10^{-2}M $FeSO_4$, 4 x 10^{-4}M $Fe_2(SO_4)_3$, and
10^{-2}M H_2SO_4. We found that by changing the electrolyte
solution to 10^{-4}M thionine, 10^{-2}M $FeCl_2$, 10^{-4}M $FeCl_3$, and
0.1M sodium acetate at pH 4 a photoresponse of more than
200 mV could be observed in the identical experimental
setup. The magnitude of the response had an interesting
dependence on the pH of the solution and thus was further
investigated.

The results of our study of the effect of pH on the
maximum light-induced potential are shown in Figure 4.
Below pH 3, the photopotential behaves in a complex manner. Above pH 3.5 the photopotential varies linearly with
pH, with a slope of 60 mV following the Nernst equation.
The maximum change is obtained at a pH equal to 3.4.

Following the suggestions by Rabinowitch (29) and Gomer
(6), we can write the following for the potential-deter-
mining reaction of the illuminated electrode

$$\cdot TH_2^+ \rightarrow TH^+ + H^+ + e \tag{3}$$

where $\cdot TH_2^+$ and TH^+ are the semiquinoid form and oxidized
form of the dye thionine, respectively. The photopotential
E^* can then be given by

$$E^* = E^o + \frac{RT}{F}\ln \frac{(TH^+)(H^+)}{(\cdot TH_2^+)} \tag{4}$$

where E^o denotes the standard redox potential for $\cdot TH_2^+$
to TH^+. Equation 4 predicts a decreasing photopotential
for the cell with increasing pH. This has been found to be

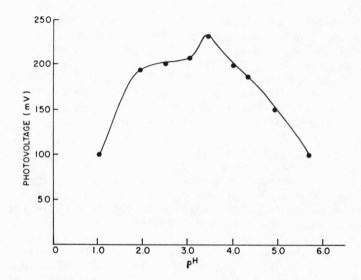

Figure 4.   Photopotential of the PG cell versus pH of the
            modified Fe-thionine electrolyte solution (see
            text).   Irradiance was 8.4×10^2 W m^{-2}.

the case at pH higher than 3.5.   As to why the maximum
photopotential should occur around pH 3.4, we have no ready
answer, for this phenomenon has not been studied.   However,
it appears that, since the species $\cdot TH_2^+$ is generated by
light, it is evident from Eq. (3) that the observed
photopotential should depend on the stability of the entity at the electrode surface.   Conceivably the acetate
ion acting as a ligand can somehow either alter the redox potential of the system or increase the rate of
interfacial electron transfer.   Additional experiments
using a variety of ligands are of obvious interest (11).
We also found that the PG effect can be made to occur on
much cheaper material such as glassy carbon, which has
displayed a PG response nearly equal to that obtained
using platinum.   Glassy carbon is more desirable for use
since it is relatively inexpensive.   A description is
given below.

Glassy Carbon

The material of the phctoelectrode of the second version of our photoelectrochemical cell is glassy carbon,
also called vitreous, amorphous, or turbostratic carbon.

The counter electrode was made of either glassy carbon
or platinum. This subsection places glassy carbon in the
context of other types of electrode materials and summarizes some of its physical and electrical properties perti-
nent to its use in a photovoltaic cell.

Glasses have been tried as photoelectrodes. Their
electronic theory is less developed, but it is now clear
that the two requirements for a photoelectrochemical cell,
an energy gap for mobile charge carriers and a space charge
layer, can be met by glasses. All glasses have enough
local order to generate conduction and valence bands and
a band gap (16,17). Probably the major electronic dis-
tinction between crystalline and glassy media is that
many (but not all) glasses have a mobility edge (18,19).
This is an energy level in the electronic conduction band
separating localized and extended quantum states. An
electron whose energy lies below the mobility edge may
contribute to the electrical conductivity of the glass by
hopping from one localized state to another (17). The
temperature dependence of conductivity and other electrical transport properties depends upon the presence or absence of a mobility edge. Furthermore, it may affect the
rate constant for exchange of electrons with an absorbed
pigment, and so have a bearing on what redox potential
should be used. Recent experiments (20) show glassy carbon has a mobility edge after appropriate heat treatment.

In the last ten years the synthesis of organic high
polymers resulted in very hard, glass-like carbons that
were recognized as being distinctly different from soft
carbons in mechanical, physical, and electrical properties. These are produced under trade names such as Cellulose, Glassy, and Vitreous Carbon. A summary of their
properties is given in reference (21). Measurements of
electrical properties have been reported (20). Studies
on the structure (bond lengths and angles, and coordination numbers and distances) have been reported (22).

Glassy carbon is distinguished physically from soft
and graphitized carbons by its concoidal fracture and
glassy appearance, impermeability to most gases, resistance to chemical attack and by its inability to transform
into graphite under the usual graphitizing process (21).
The origin of these differences may lie in its manufacturing process, which is the carbonization of high polymers.

(High polymer solids are heated to temperatures up to
3000° C in reducing atmospheres.) In contrast, soft carbons are manufactured from carbonaceous gases and crystallize at temperatures around 500° C. The polymers used to
make glassy carbon are highly cross-linked, and experiments suggest that enough linkages remain to prevent graphitization and to impart the other properties.

The nature of these linkages is not clear, but they
are probably carbon atoms (21). Radial distribution
curves obtained from x-ray diffraction on different types
of carbons give nearest neighbor bond lengths about 1.50
Å (compared with 1.41 Å for graphite and 1.54 Å for diamond) and the coordination number for the first coordination sphere is between 3.2 and 3.4 (compared with 3 for
graphite and 4 for diamond). The particular value depends
upon the heat treatment to form the glassy carbon, and its
physical and chemical properties are strongly dependent
upon these two parameters, with a fairly clear division
into (1) vacuum deposited (soft) carbons and (2) glassy
carbons from the carbonization of certain cross-linked
polymers (22). Both sp^2 bonding (trigonal, as in the plane
of graphite crystals) and sp^3 bonding (diamond structure)
are present. A model in which the coordination numbers of
both graphite and diamond were given equal weight was able
to fit x-ray data closely. The data appear to support a
model with small islands of graphite-like structure in a
diamond-like matrix (23). Truly amorphous material has
no unsatisfied bonds, but the models so far proposed make
probable a higher fraction of unsatisfied surface bonds
(21). This is an important consideration with respect to
the chemisorption of pigments.

The electrical resistivity of glassy carbon (10 - 200
x 10^{-4} Ohm-cm) is greater than that of graphitized carbon
(6 - 12 x 10^{-4} Ohm-cm), but is big enough for electrodes.
A typical electronic charge density is 10^{19} cm^{-3} although
this can increase or decrease orders of magnitude with
different heat treatment. Saxena and Bragg (20) conclude
from measurements of conductivity and Hall effect that the
conduction mechanism depends upon the heat treatment. When
the glassy carbon was formed at temperatures greater than
2000° C the conductivity was found to be the sum of a tem-
perature-independent contribution due to diffuse boundary
scattering and a hopping contribution. The mean free path

of the first carriers was of the order of the diameter of
the crystallites. The hopping contribution came from
localized states in the conduction band but below the mobility edge, just as in other glasses. Characterization of
the band structure is complicated by a metal-insulator
transition, as shown in Figure 5. High polymers heat
treated below 700° C have a semiconductor band structure,
with a wide energy gap, a full valence band V.B. (no hole
carriers), and an empty conduction band, C.B.   As the heat
treating temperature increased, the low-lying conduction
states become occupied but they also turn into localized
states, and there was a band of localized states (middle
diagram of Figure 5). With complete graphitization, extended states are occupied. These results concur with
other findings of both localized and extended electron
states of comparable density in glassy carbon (24).

The foregoing properties of glassy carbon may explain
the absence of a significant photoelectric effect in our
unsensitized glassy carbon samples and electrolyte solutions
(acidic, aqueous solutions of iron and sometimes thionine
and acetate). Voltages were usually less than 2 mV and
currents less than 5 μA for a photoelectrode and counter
electrode of glassy carbon. Such a low response would be
consistent with both an energy gap too wide for the visible spectrum to raise an electron across and with a Fermi
level pinned down by localized, glass-structure conduction
band states.

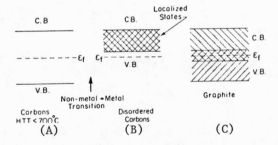

Figure 5. Band structure of carbons varies with heat treating temperature. Three samples carbonized at successively
higher T: (A) Wide, insulator or semiconductor band gap.
(B) Localized states lie between valence and conduction
bands, with E_F near top of valence band permitting creation
of hopping charge carriers. Glassy state. (C) Graphite has
overlapping bands with extended states occupied, giving metal-

lic properties. High polymers do not reach this state.
(From ref. 20.)

In summary, glassy carbon has a number of features
that make it attractive as a pigmented or sensitized photoelectrode. First, under proper heat treatment its electronic band structure has a gap between extended, conducting states. Second, its density of charge carriers is of
the order to permit a space charge layer. Third, it is an
inexpensive material to produce, not requiring high toler-
ances, at least not at this stage of development. Fourth,
both the band gap and the density of charges may be varied
under heat treatment. Fifth, glassy carbon is much more
inert to corrosive conditions than most materials, and is
nearly impermeable to gases. Graphite is easily attacked
by acids, it undergoes anodic dissolution, and in extreme
environments many types of small molecules and ions may
penetrate between the layers to form ordered compounds
(intercalates) or to decompose the graphite. Finally,
glassy carbon is mechanically strong and can be manufactured in intricate shapes.

Photovoltaic Cells

The photovoltaic (PV) cell was constructed as follows:
Glassy carbon electrodes were coated with meso-tetraphenyl-
porphyrin (TPP) both metal-free and chelated to the metals
Cu(II), Fe(II), Mg(II), Mn(II), and Zn(II). The dry,

Table 1. Some Oxidation Potentials of Tetraphenylporphyrin
(25).

Metal	$E_{\frac{1}{2}}(1)$	$E_{\frac{1}{2}}(2)$
none	.97	1.12
	-1.08	-1.52
Mg(II)	.54	.86

crystalline TPP was dissolved in chloroform of 1.3% (w/v
in mg/100 ml) concentration. The carbon was cleaned with
chloroform in a reflux still, thoroughly dried in a stream
of hot air (about 80° C), and while still hot was dipped
half way into the TPP/CHCl$_3$ for a few seconds, removed,
and again dried in hot air. Figure 6 shows the design of
the PV cell. Electrical connections to the electrodes were
made with alligator clips and measurements were made with
a Keithley 610 digital electrometer and ammeter. The

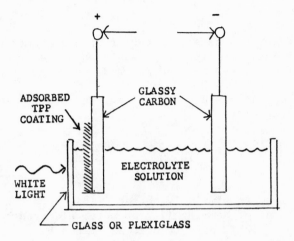

Figure 6.   Diagram of pigmented photovoltaic (PV) cell.
             Area of each electrode face that is immersed is
             2.2 cm^2.

counter electrode was either uncoated glassy carbon (GC)
or platinum. All data reported are for the GC counter
electrode. Both of the GC electrodes were 1.5 cm wide,
about 5 cm long, and 0.2 cm thick. They were dipped 1.5
cm into the electrolyte solutions so that the area was 2.25
cm^2. The electrolyte solution was contained in a plexiglass or pyrex glass cell open to room air and at ambient
temperature. The photoelectrode (PE) was 0.1 to 0.2 cm
from the face of the cell through which it was illuminated
with white light from a projector lamp (see Fig. 6). A
heat shield screened infrared waves. The light intensity
was 4.0 x 10^2 Watt m^{-2} for all the data of this section.
This equals a typical summer noon time intensity of the sun's
light at the earth's surface in our area, as measured by
a silicon solar cell. The compositions of the two electrolyte solutions used are given in Table 2. For some experiments these were diluted with water or acid, as explained
in the corresponding figures. The compositions of the
electrolyte solutions that we used were chosen to give the
biggest photoresponses after experimenting with different
components and concentrations. Concentrations of the TPP
dipping solution from 0.25% (w/v) to 3% were tested for
photovoltage and photocurrent responses. For TPP and Mg-TPP,
solutions between 1.0% and 1.5% yielded the biggest photoresponse, but other concentrations between 0.5% and 3% also

work well.

In contrast to the strong photoresponse (photopotential
and photocurrent) from TPP on glassy carbon electrodes, the
metals Ca, Ti, Al, Fe, Pb, and Zn produced either a small

Table 2.   Compositions and Stoichiometric Molarities of
           Electrolyte Solutions

Electrolyte	$FeSO_4$	$Fe_2(SO_4)_3$	H_2SO_4
A	.02 M	.01 M	0.1 M
B	.01	.001	.01

or no response when similarly coated. This agrees with
the observations of Memming (12) on metal electrodes
sensitized by soluble dyes, and is explained by the lack
of an energy gap in the metal conduction band: an excited
dye or pigment molecule simultaneously injects a high
energy electron into the metal and receives a low energy
electron from the metal.

Figure 7 shows the photopotential (photovoltage across
the open-circuit cell) from six different TPP. The polarity
of the photoelectrode was positive for all six TPP coat-

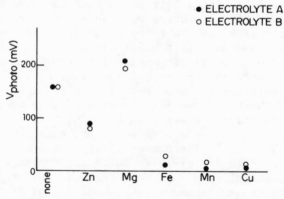

Figure 7.   Photopotential for glassy carbon electrode coat-
           ed with TPP containing different metal ions, or
           none (i.e. free TPP) in electrolyte A.   Ir-
           radiance was 4.0×10^2 W m^{-2}.

ings for both a glassy carbon counter electrode and a
platinum counter electrode (data not shown). The photopotentials were essentially the same (within 5 mV) for
both glassy carbon and platinum counter electrodes against
all the pigments. This is strong evidence that the glassy
carbon is close to being an inert electrode in these electrolytes, in agreement with its resistance to chemical
attack as referenced in the section on glassy carbon. We
did not observe any indications of significant chemical
reaction or deterioration on a carbon counter electrode
that was in use for several months.

There was apparently little or no contribution to
the photoresponse from the carbon electrode itself; when
two bare carbon electrodes were used, one illuminated and
the other shielded from the light, the response was less
than 2 mV and $5 \mu A$. Furthermore, the action spectrum re-
sembles that of a metalloporphyrin (Figure 14C). The dark
TPP pigmented-semiconductor cell did not act as a galvanic
cell; i.e. when the cell was dark its open-circuit voltage
and short-circuit current were zero. Several tests determined that the source of the electrical power was true
photoelectric energy conversion and not some irreversible,
photocatalyzed, galvanic redox reaction with the TPP. From
constant short-circuit currents over about a day, it was
determined that if the current were due to the irreversible
reaction of only that portion of the coat that was illuminated and in contact with the electrolyte solution, each
metal-free TPP would have to lose or gain more than 225
electrons. If every molecule contributed charge, the
change would be 38 electrons per molecule. Another test,
although performed on the photogalvanovoltaic cell to be
described in the next section, was also a check that the
redox reaction of the TPP was reversible. The amount of
TPP was determined by weighing the coat on a piece of mylar,
to which TPP similarly absorbed. A coating of 3.2×10^{-7}
moles delivered a current of $30 \mu A$ over 24 hours without
diminution. This is a charge transport of 2.7×10^{-5} Faradays, about 100 times the number of moles of TPP. A third
test attempted to rule out irreversible redox reactions
among the constituents of the electrolyte, conceivably
catalyzed by the simultaneous presence of TPP and light.
From the Faradays passed by a current of $250 \mu A$ over three
hours in a cell of very small volume of electrolyte, the
ratios of moles of electrons to moles of other species were
determined to be .373 for $Fe(II)$, .0373 for $Fe(III)$, and

.00848 for SO_4^{2-}. While these ratios are certainly not conclusive, in combination with the fact that the voltage had not dropped they suggest that electrolyte was not being consumed.

A typical quantum efficiency for a metal-free or Mg-TPP was 0.4%. This low value is probably due to only the first absorbed monolayer of pigment participating in the conversion of photo energy into electrical energy through the transfer of charge to the carbon. The spectral irradiance F_λ (energy/area·time·wavelength) is related to the number of incident photons $N\lambda$ within the interval $d\lambda$ by

$$F_\lambda = \frac{N_\lambda hc}{At} \tag{5}$$

where h is Planck's constant, c is the speed of light, A the area of the electrode, and t the time of irradiation. Integration gives the total number of incident photons, N, in terms of the irradiance F (energy/area·time). For white light N_λ is independent of wavelength, so the integration of Equation (5) gives

$$N = \int N_\lambda d\lambda = \frac{FtA}{hc} \frac{(\lambda_2 - \lambda_1)}{\ln(\frac{\lambda_2}{\lambda_1})} \tag{6}$$

Assuming a white light band from 400 nm to 700 nm gives

$$N = \frac{FtA}{hc} \cdot (5.36 \times 10^{-7}) \quad \text{(mks units)} \tag{7}$$

The total number, n, of electrons transferred at the electrodes is given by n = It/e, where e is the electronic charge. The quantum efficiency is then

$$\frac{n}{N} = (1.86 \times 10^6) I \frac{hc}{FA} \quad \text{(mks units)} \tag{8}$$

At $F = 400$ W m^{-2} and $I = 1.5 \times 10^{-4}$ A, the quantum efficiency is 0.39%.

Under irradiance of 400 W m^{-2} a typical free or Mg-TPP PV cell with electrolyte A could deliver 150 μA through a short circuit over a 24 hour test period with a current decrease of less than 5%. Cell resistance was around 10^3 Ohms. The resistance of the bare carbon electrode was about 0.7 Ω.

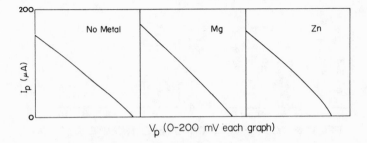

Figure 8. Photocurrent-photovoltage characteristic curves
for three different TPP coatings on PV electrode. Electrolyte solution A was used. The load resistance varied from
1 Ω at maximal I to $10^{14}\Omega$ at zero I. The nearly straight
lines suggest the photocell may be modeled by an ideal
potential source in series with a constant total internal
resistance, as in Fig. 9. Irradiance was 4.0 x 10^2 W m^{-2}.

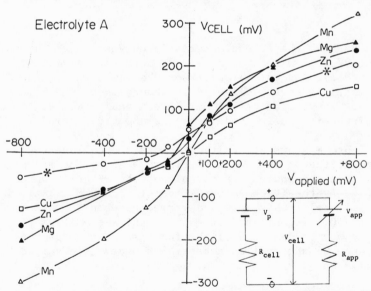

Figure 9. Cell voltage V_{cell} versus applied voltages V_{app}
for PV cell. (*) marks metal-free TPP. The polarities
are identified on the equivalent circuit, where V_p is the
photo potential. R_{cell} includes all cell resistance.

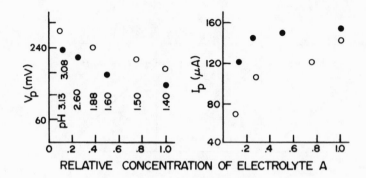

Figure 10.   The photopotential and photocurrent of PV cell
             versus the concentration of solution A, defined
             as (Vol. of Soln. A)/(Vol. of Soln. A + Vol.
             of Water Added).   The photoelectrodes were
             coated with (•) metal-free TPP and (o) Mg-TPP.
             Irradiance was 4.0 x 10^2 W m^{-2}.

Figure 8 shows the IV characteristic curves (power curves)
for the PV cell with the three most sensitive pigments,
free, Mg(II), and Zn(II), with somewhat less photoresponse.
The electrical power delivered is about 7.4 x 10^{-6} W, and
the power conversion efficiency (electrical power delivered/
light intensity x electrode area) is about 8.2 x 10^{-5}.

     Figure 9 shows the voltage across the cell versus
applied voltage.   The insert of Figure 9 shows the measuring
circuit and the equivalent circuit of the PV cell, which is
justified by the IV characteristics of Figure 8. R_{cell} includes all cell resistance and varied between 800 and 1000
at V_{app} = 0.   The slight tendency of the extremities to
saturate may be due to a slight increase in the internal
resistance of the applied potential source.   (It varied between 1.3 and 2 kΩ.)

     Figure 10 shows the variation of photopotential and
photocurrent of PV with metal-free and Mg-TPP as the concentration of electrolyte is reduced.   The reduction in
current is expected from the increased electrolytic ionic
resistance as the ionic strength decreases.   The voltage
change may have to do with the changing double charge
layers as the ionic strength is reduced.

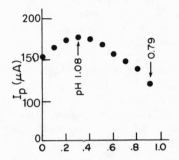

Figure 11.   Photocurrent versus acidity of PV cell with
metal-free TPP coating. Abscissa shows ml of sulfuric acid
(95-98%) added to initial 40 ml of electrolyte solution A.
The resultant pH values are (1. to r.):   1.37 (at zero ml
of acid), 1.24, 1.15, 1.08, 1.01, 0.96, 0.91, 0.86, 0.83,
and 0.79.   Irradiance was 4×10^2 W m^{-2}.

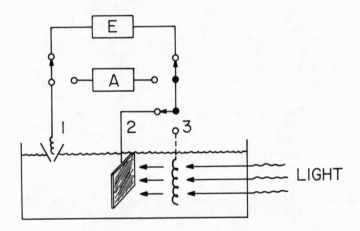

Figure 12.   The PGV cell. Electrometer and ammeter are represented by E and A. Numbers 1, 2, and 3 correspond to a
saturated calomel (reference), TPP coated carbon (PV), and
bare platinum (PG) electrode, respectively.   The PV, PG,
and PGV measurements were made across electrodes 2 and 1, 1
and 3, and 2 and 3, with the first electrode of each couple
being the cathode and connected to each meter's high input
terminal.   The PV glassy carbon electrode was coated with
meso-Mg(II)-tetraphenylporphyrin.

Figure 11 shows the photocurrent versus the acidity
of the electrolyte. The initial rise in current is probably due to the rise in Fe^{3+} concentration as increasing
acidity reduces hydrolysis. The maximum may be due to
reaching the diffusion limiting current due to the presence
of a high concentration of indifferent ions (H^+ and SO_4^{2-})
carrying the current in the diffusion layers at each electrode. The cell was apparently operating in the rate-
limited regime rather than diffusion limited because stirring the solution made no difference in current regardless
of the position of the electrodes.

A Mg-TPP PV cell with solution B had a photoresponse
of 300 mV and 64 μA cm^{-2} with an internal resistance of 600
Ohms. The irradiance was 8.4 x 10^2 W m^{-2}.

Photogalvanovoltaic Cells Using
Glassy Carbon Electrode

The photovoltaic (PV) electrode of our PGV cell is a
glassy carbon electrode coated with meso-Mg(II)-tetraphenyl-
porphyrin which does not enter into solution in the electrolyte. The photogalvanic (PG) electrode is the platinum/
iron-thionine electrode. This combination was suggested by
our observation that the carbon-TPP electrode was the
cathodic electrode in the PV cell, and that the Pt electrode is anodic in the PG cell, permitting a series connec-
tion to add the electrode potentials and increase the fraction of the visible band that would be absorbed. It also
permits the possibility having a single cell both store
photo energy chemically (assuming the electroactive species
produced by the light can be separated or otherwise prevented from reacting immediately) and delivering the energy
directly in the form of electricity when needed.

Results for PGV Cells

Experimental methods. Figure 12 shows the design of
the PGV cell. The electrolyte was the modified iron-thionine solution described above, but at pH 4.5. The presence
of the thionine and acetate had no effect on the photoresponse of the PV electrode. In this study Pt was used
instead of carbon for the PG electrode. The PV electrode
was a glassy carbon rod 0.3 cm in diameter and coated with
2% (w/v) TPP in CHCl$_3$. The electrode area was 0.57 cm^2
(80% apportioned to the PV carbon electrode). Light was

from a tungsten halogen lamp with a heat shield.

Photoresponse. Figure 13 shows the photoresponses
for the PG and PV electrodes to be 168 mV and 214 mV,
respectively. The additive effect of the cell was con-
firmed, as shown in Figure 13, by measuring the PGV response between the platinum (PG electrode) and the TPP
coated carbon (PV electrode). The observed 377 mV agrees
almost exactly, within experimental error, with the expected result of the sum of the PG and PV cells. The total
short circuit current for the PGV cell is also shown in
Figure 13. Table 3 summarizes the characteristics of this
test cell. The current limiting electrode in this cell
was the PG electrode.

Spectral characteristics of the PGV cell are shown
in Figure 14. The peaks of both the PGV and the PV photopotentials at 434 nm correspond to the Soret peak of TPP
at 424 nm (30) red shifted at pH 4.5 (27). Both photoelectrodes appear to be absorbing light independently.

Figure 15 shows the equivalent circuit for the PGV
cell. There is a similarity to the circuit representing a
cell that has a photoactive membrane as an electrode (28).
The TPP layer can be visualized as the membrane, with the

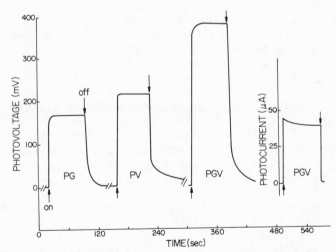

Figure 13. The photoresponse of the PGV cell and its component PV and PG electrodes. An additive effect is observed.
The response in the far right is the PGV photocurrent.

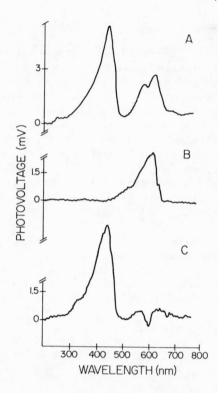

Figure 14. Action spectrum of (A) the PGV cell, and its
component (B) PG and (C) PV electrodes. PG and PV are versus SCE. Irradiance was 8.4×10^2 W m^{-2}.

carbon electrode replacing one of the aqueous solutions of
the membrane. C_m arises because a boundary layer of un-
neutralized charge can presumably exist across the TPP
layer. This particular PV electrode can also exhibit a
dark potential, E_d, and resistance, R_d, in addition to the
photopotential, E_v, and photoresistance, R_v. The photogalvanic cell can be represented by a light generated
battery, E_g, and a resistance, R_{sg}, between the solution

Table 3. Summary of PG, PV, and PGV Test Cells. Irradiance 10^3 W m^{-2}. (Ref. 28).

Cell	Dark Resistance	Light Resistance	I (µA)	V (mV)
PG	1.5×10^4	3.5×10^3	7.	168.
PV	3.4×10^4	5.0×10^3	42.	214.
PGV	5.0×10^4	7.7×10^3	43.	377.

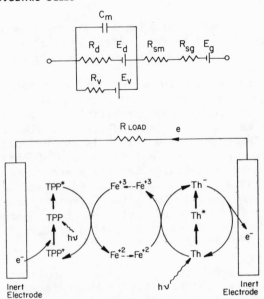

Figure 15. Upper: An equivalent circuit of the photo-galvanovoltaic (PGV) cell. R_{sg}, solution/Pt electrode resistance; E_g photogalvanic potential; C_m pigment coat capacitance; R_d pigment coat ionic resistance; E_d–PV electrode dark potential; R_v pigment electronic resistance, E_v photovoltaic potential; R_{sm} solution calomel electrode resistance. Lower diagram shows a mechanism explaining the operation of the PGV cell (see text).

and the PG electrode. This is equivalent to series con-nection of the circuits representing the PV and PG cells separately.

Charge transport studies of the type described in the previous section showed that the dominant reactions are cyclic, with net conversion of energy. Upon excitation of TPP, either one of these sequences could follow: photo-excited TPP* gives an electron to the ferric ion to become TPP+ and then accepts an electron from the electrode; or, TPP* may be reduced by the electrode to TPP⁻, which then reduced ferric ion to become TPP again.

CONCLUSIONS

Our investigation of a solar cell designed around some very simplified analogies with the complex and elusive

photoelectric mechanisms driving the photosynthetic conversion of energy has led to some discoveries of interesting
phenomena that show promise for photoelectric applications.

First, with respect to our objective of fashioning
the beginnings of a practical solar cell, we conclude that
the glassy carbon-tetraphenylporphyrin electrode is worthy
of further investigation. Although the overall power conversion efficiency of 7 to 8 x 10^{-5} is low by comparison
with other solar cells with a long history of optimization,
the comparatively inexpensive materials and manufacturing
process and the chemical stability of the glassy carbon are
very desirable features. The rather straightforward construction design of our first cells offers much opportunity
to investigate refinements that could dramatically increase
power output.

The photogalvanovoltaic cell has the possibility of
simultaneous generation of electricity and chemical storage
of photoenergy for later electrical power from the same
cell. Our studies have proven the validity of the concept
of adding the voltage from each mechanism with no decrease
in current. Again, our cells were not extensively opti-
mized in design, and some of the new designs improving the
efficiency of the iron-thionine cell could be applied here,
perhaps resulting in orders of magnitude increase in power
conversion efficiency. The photogalvanovoltaic concept is,
of course, applicable to other, presently more efficient,
photoelectrochemical cells. This area has not received
much attention in the literature.

Glassy carbon may be a useful inert electrode, as suggested by the absence of dark or photopotentials from the
bare electrodes, and the substitution of it for platinum
in the iron-thionine cell to obtain nearly equivalent photo-
voltages.

We speculate whether the special properties of the
glass localized states (Figure 5) could be a tool for the
investigation of the photosensitization by electrochemical
techniques. For example, this band acts like a zone of
high density traps (20). One of the problems of both
sensitized and unsensitized semiconductor-liquid junction
cells is reduction of the photocurrent by trapping of electrons at defects and grain boundaries in both the semiconductor and any polycrystalline pigment phase present.

It may be possible to devise experiments using this band to gain insight into the trapping mechanism and thus improve the power conversion efficiencies of the various semiconductor-electrolyte combinations.

Finally, a PGV battery may be constructed, which could be even more efficient if the electrodes could be fabricated in such a manner as to minimize the shadowing effect (e.g., using transparent SnO_2 electrodes) by connecting a number of PGV cells in series. This PGV solar battery can then generate a photovoltage of sufficient magnitude for the electrolysis of water.

ACKNOWLEDGEMENTS

This work was supported in part by a NSF energy-related postdoctoral fellowship to J. M. M. and by funds from the College of Natural Sciences, Michigan State University.

REFERENCES

1. H. Tsubomura, M. Matsumura, K. Nakatani, K. Yamamoto, and K. Maeda, Solar Energy, 21, 93-98 (1978).

2. G. Porter and M.D. Archer, Interdisc. Sci. Rev.,1, 119-143 (1976).

3. M. Almgren, in Project Results: Solar Energy/Photochemical Conversion and Storage, S. Claessen and L. Engstrom, Ed. (National Swedish Board for Energy Source Development, Stockholm, (1977), Chapter VII.

4. J.R. Bolton, (Ed.) Solar Power and Fuels, Proc. First Int. Conf. on the Photochemical Conversion and Storage of Solar Energy, New York, Academic Press. (1977).

5. V. Balzani, F. Bolletta, M.T. Ganolfic, and M. Maestri, T. Curr. Chem., 75, 1-64 (1978).

6. R. Gomer, Electrochim. Acta, 20, 13-20, (1975).

7. A. Heller, K.-C. Chang, and B. Miller, J. Am. Chem. Soc., 100, 684-688, (1978).

8. H. Gerischer, J. Electroanal. Chem., $\underline{58}$, 263-274
    (1975).

9. H. T. Tien and J. M. Mountz, Int. J. Energy Res., $\underline{2}$,
    197-200 (1978).

10. F. Lohmann, Z. Naturforsch, $\underline{22a}$, 843 (1967).

11. H. T. Tien and J. M. Mountz, J. Electrochem. Soc.,
    $\underline{125}$, 885-886 (1978).

12. R. Memming and H. Tributsch, J. Phys. Chem., $\underline{75}$,
    562-568 (1971).

13. D. E. Hall, J. A. Eckert, N. N. Lichtin and P. D.
    Wildes, J. Electrochem. Soc., $\underline{123}$, 1705-1707 (1976).

14. J. B. Birks, Photochem. Photobiol., $\underline{24}$, 287-289 (1976).

15. H. T. Tien, Bilayer Lipid Membranes (BLM): Theory
    and Practice, Dekker, Inc. New York (1974).

16. B. T. Kolomiets, Phys. Status Solidi, $\underline{7}$, 359 (1964).

17. N. F. Mott, Rev. Modern Phys., $\underline{50}$, 203-208 (1978).

18. N. F. Mott, Adv. Phys. $\underline{16}$, 49 (1967).

19. M. H. Cohen, H. Fritzsche, and S. R. Ovshinsky, Phys.
    Rev. Lett., $\underline{22}$, 1069 (1969).

20. R. R. Saxena and R. H. Bragg, J. of Non-Cryst. Solids,
    $\underline{28}$, 45-60 (1978).

21. T. Noda, M. Inagaki, and S. Yamada, J. Non-Cryst.
    Solids, $\underline{1}$, 285-302 (1969).

22. R. Grigorvici, J. Non-Cryst. Solids, $\underline{1}$, 303-325 (1969).

23. K. Furukawa, J. Cryst. Japan, $\underline{6}$, 101 (1964).

24. S. Orszesko and K. T. Yang, Carbon, $\underline{12}$, 493 (1974).

25. J.-H. Fuhrhop, in Structure and Bonding (Springer-
    Verlag, N.Y., 1973).

26. J. Kiwi, and M. Gratzel, J. Am. Chem. Soc., <u>100</u>,
     6314-6320 (1978).

27. J. E. Falk, <u>Porphyrins and Metalloporphyrins</u>, (B.B.A.
     Library, Vol. 2; Elsevier, N.Y., 1964), pp. 22, 73.

28. J. M. Mountz and H. T. Tien, Solar Energy, <u>21</u>, 291-
     295 (1978).

29. E. J. Rabinowitch, J.Chem Phys., <u>8</u>, 551 (1940).

30. D. Dolphin, Z. Muljiani, K. Rousseau, D. C. Borg,
     T. Fajer, and R. H. Felton, in <u>The Chemical and</u>
     <u>Physical Behavior of Porphyrins and Related Structures,</u>
     Alan Adler, ed. (N.Y. Acad. Sci., <u>206</u> (1973)), pp.
     177ff.

GROWTH AND CHARACTERIZATION OF THIN FILM III-V COMPOUND
SEMICONDUCTOR MATERIAL FOR SOLAR CELL APPLICATIONS

W. D. Johnston, Jr.

Bell Telephone Laboratories

Holmdel, New Jersey   07733

I. GENERAL CONSIDERATIONS

The United States National Photovoltaic Program Plan
(1) has as a specific goal the achievement of 10 percent
efficient photovoltaic arrays at a cost (1975 dollars) of
$0.50 per peak watt and an annual production rate of
5×10^8 peak watts by the end of 1986. The primary effort
toward meeting this intermediate term goal is directed to
reduction in cost of single-crystal silicon solar cells
of essentially traditional design. For the longer term,
for which costs below $.30 a peak watt and a hundred-fold
greater production rate are envisioned, other materials
and processes may be required.

The family of III-V compound semiconductors include
several members which are attractive in this context.
Single crystal heterojunction cells based on GaAs or InP
have been fabricated with efficiencies which substantially
exceed the 10 percent requirement (see Table I). These
cells also present the advantage that the efficiencies decrease less rapidly at elevated temperatures than is the
case with Si cells. (CdS/Cu_2S cells, also commercially
available, are not useful at elevated temperatures because
of the instability of the Cu_2S).

237

TABLE I

EFFICIENCIES OF GaAs AND InP BASED SOLAR CELLS

Cell Type	Epi-Growth	Single Crystal	Thin Film
$Al_xGa_{1-x}As/GaAs$	LPE	24.6% (a)	N.A.
$Al_xGa_{1-x}As/GaAs$	MOG	19.0% (b)	?
AlAs/GaAs	VPE	20.0% (c)	?
GaAs/GaAs	VPE	20.0% (d)	?
MOS/GaAs	VPE	18.0% (e)	6.3% (f)
CdS/InP	VPE	15.0% (g)	5.7% (g)
ITO/InP	–	14.0% (h)	?

a) H. A. Vander Plas, L. W. James and R. L. Moon, Proc. 13
    Photovoltaic Spec. Conf., p. 934 (IEEE,
    New York, N.Y., 1978).
b) N. J. Nelson, K. K. Johnson, R. L. Moon,
    H. A. Vander Plas and L. W. James, Appl. Phys.
    Lett. 33, 26 (1978).
c) W. D. Johnston, Jr., J. Cryst Gr. 39, 117 (1978).
d) J. C. C. Fan, C. O. Bozler and R. L. Chapman, Appl.
    Phys. Lett. 32, 390 (1978).
e) R. J. Stirn and Y. M. C. Yeh, Appl. Phys. Lett. 27,
    95 (1975), also R. J. Stirn, Priv. Comm., (1978).
f) Ref. 5
g) Ref. 8
h) K. Sreeharsa, K. J. Bachman, P. H. Schmidt,
    E. G. Spencer and F. A. Thiel, Appl. Phys. Lett. 30,
    645 (1977).

The direct band gap of these materials gives rise to high values of optical absorbance and hence relatively thin layers of active semiconductor may be used. This amounts to 2-3 micrometers for GaAs or InP as compared to the ∿100 micrometers needed to obtain equivalent optical absorption in Si, for instance. The value of the energy gap in these compounds also provides a more nearly optimum match to the solar spectrum. Another III-V compound, AlAs, has a larger, indirect band gap and is nearly transparent to most of the solar spectrum in layers of ∿10 μm thickness (see Fig. 1). This compound has virtually the same lattice constant as GaAs and can be used efficaciously as a transparent top contact to reduce series resistance. The AlAs

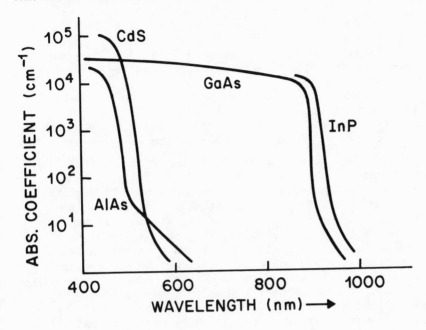

Figure 1   Absorption coefficient vs. wavelength for CdS,
AlAs, GaAs and InP.

layer also serves to reduce the recombination of photogenerated minority carriers which occurs at the bare GaAs
surface.   CdS or ITO (indium-tin-oxide, typically 92 percent indium) can be used for a similar purpose in
conjunction with InP.

Solar cells of GaAs or InP prepared on single crystal
GaAs or InP substrates are very expensive, and it is unlikely that they will prove cost-effective without high-
ratio optical (lens or mirrors) concentrator systems.   Accordingly, practical use of these materials for solar
photovoltaic conversion requires the development of techniques to fabricate them in thin film form on inexpensive
substrate material such as glass or steel.   The substrate
chosen should, ideally, provide a physically adherent bond
to the semiconductor with low-resistance, nonrectifying
electrical contact.   It should be stable against the formation of undesired chemical species in the film growth
environment, and should not dope the semiconductor film so

as to yield unwanted electrical properties. A close match
in thermal expansion coefficient to the semiconductor film
is necessary. Electrically conductive (i.e., metallic)
substrates simplify the contact problem and are desirable.

With regard to the cost of the substrate in question,
it is important to consider that it be quoted for the appropriate thin-sheet, large area form needed. The ultimate
target of $.20 - .30 per peak watt implies an _array_ cost of
$20 - 30/m^2 and substrate costs exceeding $5.0/m^2($.50/ft^2)
seem excessive. This would rule out specialty glass or
stainless steel sheet, for instance, but soda-lime glass or
cold-rolled steel would be acceptable. In most of the
current research on properties of III-V thin film semiconductors, the layers studied are grown on graphite
ceramic. This material is low cost on a _weight_ basis but
is not presently available at a low cost on the required
area basis. It is manufactured by high pressure pyrolysis
of oil and it seems doubtful that thin sheets (rather than
'bricks') of the material can be formed economically by
this approach. Nevertheless, graphite has many desirable
properties as a substrate and is clearly less expensive
than single or polycrystal semiconductor alternatives.
The thermal expansion coefficient (300 K) may be varied
over the 2 - 10 x 10^{-6}/K range by varying the temperature
and pressure during formation, so that a good match to the
thermal expansion of semiconductors (typically
4 - 6 x 10^{-6}/K) can be formed. The electrical conductivity
is high, it is chemically inert, and it is less brittle and
more easily handled than single crystal wafers or soda-lime
glass of equivalent thickness. Thin-film on graphite cells
would be cost effective at the more moderate optical concentration ratios which might be achieved with Fresnel lens
or parabolic trough systems, even if direct thin-sheet
graphite production does not eventuate.

The characteristics of the thin film semiconductor of
importance to solar cell operation are for the most part
the characteristics of the low-gap component or the
'absorber' layer of the cells (an exception is the
AlAs/GaAs heterojunction cell, where properties of the AlAs
also pose important performance limitations as will be discussed below). Accordingly the problem is primarily one of
growing a thin layer of GaAs or InP on a cheap, foreign

substrate in such a way as to obtain suitable grain size, adhesion, electrical contact, and layer continuity. In practice these characteristics are largely determined by choice of substrate and growth process and are not subject to much control by varying growth parameters. Clearly only growth processes for which there is hope of high areal production rates are of interest. Fortunately the vapor-phase techniques which have given the best polycrystal semiconductor layers seem to fulfill this requirement.

There are several approaches to characterization of the polycrystalline thin films after they are grown. The physical structure is best examined at low to moderate magnification (200-1000X) in the scanning electron microscope (SEM) which provides the necessary depth of field. Grain size may be 'guess-timated', with good reproducibility if not with arguable precision, from SEM photographs. Pinhole and/or microcrack density may also be determined; this relates directly to leakage current paths. For Schottky barrier cells particularly, but for heterojunctions as well, pinholes lead to 'shorted' cells with low fill factor (ratio of maximum power to short circuit current - open circuit voltage product) and little or no useful output.

The electrical properties of the films may be determined by assumed equivalence to measured properties of single or polycrystal films grown on other substrates at the same time, or by direct measurement of the actual film on the substrate of interest, either in a test structure or as part of the finished device. Typically the conductivity type, carrier concentration, and resistivity in the plane and normal to the plane of the substrate are measured. Because current flow normal to the substrate plane may take place primarily within rather than across grains, there may be very anisotropic transport properties for polycrystalline films. Ultimately the power conversion efficiency of finished solar cells is of primary interest. Evaluation of solar cell structures is based on measurement of open-circuit voltage (V_{oc}), short circuit current (I_{sc}), and fill factor (ratio of maximum power to $V_{oc} \cdot I_{sc}$ product). Dark log current density - applied voltage plots are useful in relating performance deficiencies to particular material aspects.

Basic studies of the properties of polycrystalline semiconductors are at a very rudimentary state. The effect of grain boundaries is the issue, but little is known either empirically or theoretically about the electronic structure of grain boundaries in GaAs, and essentially nothing is known about detailed properties of grain boundaries in InP. Unlike the II-VI compounds, polycrystalline films of III-V compounds show high lateral (cross-grain) resistivity $(\gtrsim 10^5 - 10^6 \ \Omega \cdot cm)$ even when the resistivity along the grain is apparently $\sim 10^{-2} \ \Omega \cdot cm$. This is the result of a potential barrier at grain boundaries which reflects majority carriers, evidently associated with a boundary depletion layer. The grain boundaries must then clearly be electrically active and can be expected to affect junction current-voltage behavior; the simplest model for which must include a parallel combination of 'grain-face' and 'grain-boundary' diodes, with the latter having a lower blocking voltage. This can easily arise if an excess density of states or 'exponential band tails' are present at the grain boundaries where bonds are distorted and stretched, as shown schematically in Fig. 2 (2).

For several reasons work on basic properties of polycrystal layers as well as on thin film solar cell structures has tended to concentrate on GaAs rather than InP. Single crystal GaAs is well-understood, and is readily grown in reproducible, well-characterized form. The growth of single-crystal InP is still something of an art in comparison. Single crystal solar cells fabricated from GaAs have been made with efficiency over 24 percent while the best InP cell reported is only 14 percent. Theoretically GaAs provides a better material match to the terrestrial solar spectrum. The slightly lower band gap of InP (1.35 vs 1.45 ev) is not offset by an increase in current collected since a strong absorption notch due to atmospheric water vapor occurs in the terrestrial insolation spectrum at 1.3 ev (see Fig. 3). Thin films of GaAs also seem easier to grow than InP and there seems to be more doubt as to the long-range supply of In. For these reasons the discussion in this paper will concentrate on the preparation and properties of GaAs thin films. The reader interested in other thin films approaches involving II-VI compounds, amorphous Si, etc., is referred to the review by Bachmann which has just appeared (3).

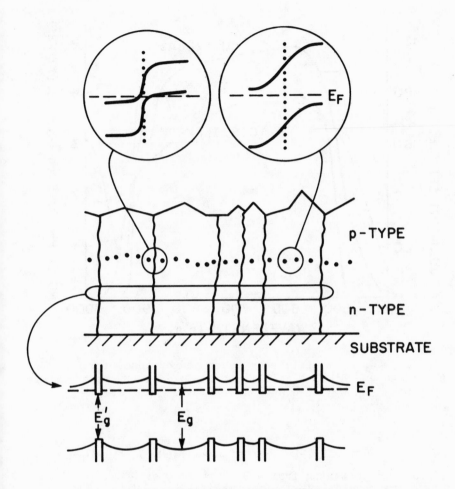

Figure 2   Schematic cross-section of a polycrystalline film
containing a p-n junction, showing the effect of an
enhanced grain boundary density of states in producing
local energy gap reduction and barriers to intergrain
majority carrier flow.

II.   GROWTH TECHNIQUES AND PHYSICAL CHARACTERIZATION

        Thin films of GaAs may be grown on foreign substrates
by close-spaced vapor transport, by chemical vapor deposition (VPE) from a mixture of GaCl and As vapor, by pyrolytic

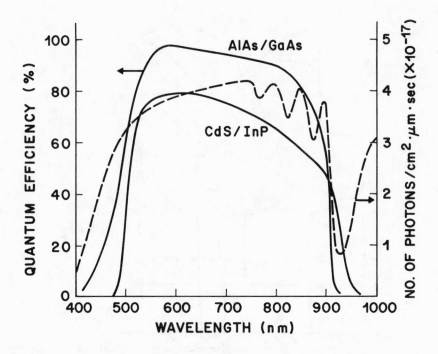

Figure 3 Quantum efficiency (current response) for
AlAs/GaAs and CdS/InP cells superimposed on AM2 spectrum.

reaction of trialkyl-gallium with AsH_3 (MOG), by glow-
discharge sputtering from a GaAs target, or by evaporation
of Ga and As as in 'molecular-beam' epitaxy (MBE). The
first thin-film GaAs solar cells were prepared as metal
Schottky-barrier devices on films grown by close-spaced
vapor transport (4). The best reported efficiencies for
thin film GaAs cells is 6.3 percent, again Schottky-barrier
MOS cells, but made from films grown by the chloride VPE
process (5). The properties of AlAs/GaAs heterojunctions
with both layers prepared by chloride VPE have also been
studied extensively (6). Organometallic pyrolysis growth
(7) appears to offer the potential of good film quality at
lowest mass production cost. Sputter-deposited films
suffer from small grain size and poor stoichiometry. Low
deposition rates and high equipment costs make molecular

beam epitaxy unsuitable for inexpensive thin-film fabrication.

The chemical reactions and the quality of the films produced in the close-spaced and chloride VPE processes are essentially the same. In the former a substrate is placed above and close (with respect to lateral dimensions) to a source of polycrystalline GaAs (see Fig. 4). The source and substrate are heated in a H_2 - HCl atmosphere with source maintained 10-20K hotter than the substrate. GaAs is grown on the substrate by the reversible reaction

(1) $2GaAs_{(solid)} + 2HCl \leftrightarrow As_2 + 2GaCl + H_2$

The disadvantages of this process relate to the geometrical constraints against scaling to large area production.

In the open-tube chloride VPE process (Fig. 5), a boat of Ga metal is used as the source of Ga and As is introduced as AsH_3 in H_2 carrier. The reactions

(2) $2AsH_3 \rightarrow As_2 + 3H_2$

(3) $2Ga + 2HCl \rightarrow 2GaCl + H_2$

(4) $2GaCl + As_2 + H_2 \leftrightarrow GaAs + 2HCl$

describe the growth. AlAs may be grown analogously with the complication that an Al_2O_3 tube must be used as AlCl rapidly attacks fused quartz components. The equivalent reactions with In and PH_3 may be employed but InCl is relatively stable and reacts inefficiently with the P_2 and P_4 species. Direct reaction with PH_3 is preferable but even in that case nearly all the indium chloride and most of the phosphorus deposit at the exhaust end of the reactor. The close-spaced process is more satisfactory in this regard as the reactor may be closed and unreacted materials 'recycled', but the process is still slow and is not likely to be practical.

The efficiency of AlAs formation on the other hand is very high since the equivalent reaction to (4) must be run well below equilibrium temperature ($\sim$1600K) if deposition

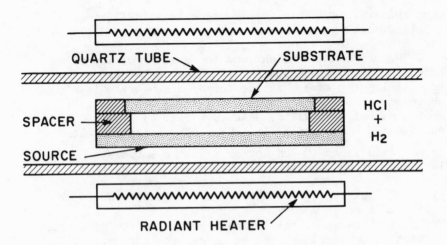

Figure 4   Schematic of close-spaced growth apparatus.

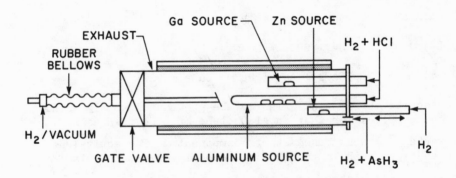

Figure 5   Schematic of chloride-hydride VPE growth apparatus for AlAs or GaAs.

upon GaAs is desired, since GaAs decomposes above 1400K. Growth at 1270-1300K for AlAs and 900-950K for GaAs is optimum for graphite substrates.   The metal source temperatures are not critical so long as the chloride conversion is nearly complete.   GaAs layers grown in this way may be doped p-type with Zn vapor introduced from a heated boat

into the H_2 stream, or n-type either by variation of the
Ga/As ratio (HCl/AsH_3 inlet ratio) or introduction of
H_2S or H_2Se. The AlAs is n-type as grown, and variation of
carrier concentration from 3×10^{17} to 3×10^{18} cm^{-3} is
possible by changing substrate temperature. Little control
over the AlAs carrier concentration by use of external dopants is possible, but appropriate material for hetero-
junctions to GaAs ($n \sim 8 \times 10^{17}$, $\rho \sim .02$ ohm·cm) is readily
produced. $Al_xGa_{1-x}As$ alloys cannot be grown by this process since the minimum temperature for good AlAs growth is
about 1240K and the maximum temperature for GaAs growth is
about 1170K. At temperatures in between, polycrystalline
AlAs and dense AlAs inclusions in poor quality GaAs and
$Al_xGa_{1-x}As$ are produced.

Some control over grain size is possible in the VPE
process, however. Since GaAs is a cubic material, growth
rates tend to be isotropic in three dimensions, i.e., the
width and thickness of grains tend to be similar and grain
shapes resemble regular polyhedrons. Growth in the hexagonal form is possible but needle rather than plate growth
results, which is highly undesirable since a continuous
film is then never obtained. Evidently the grain size cannot exceed some fraction of the average film thickness if
coalescence of grains and filling of cracks, voids, and
'pinholes' is to be obtained. One would want to provide
conditions favoring a high nucleation probability but a low
post-nucleation growth rate to ensure continuity. Direct
manipulation of the growth variables has little effect.
Substrate temperature, thermal gradient at the growth position, Ga and As mole fractions, and total flow velocity
affect growth rate on single crystal substrates but growth
will occur elsewhere in the tube on walls rather than on a
foreign substrate if the latter is not optimally located.

For p-type films, Zn doping to produce $p \sim 5 \times 10^{17}$ or
above also enhances nucleation and dramatically reduces
pinholes, while at the same time effecting an ohmic contact
to graphite or molybdenum substrates and maintaining a
large grain size (see Fig. 6).

For n-type layers, gas phase dopants have no such
effect. A thin (30 nm nominal thickness) layer of Sn or Ge
promotes dense nucleation but grain size is reduced somewhat. It is interesting to note that the close lattice
match between Ge and GaAs is not the cause of this growth

Figure 6   SEM views (45° perspective) of VPE GaAs films on
a) bare graphite, undoped; b) bare graphite, Zn doped
$(p \sim 1 \times 10^{18} \text{ cm}^{-3})$; c) undoped (n-type) on bare/Ge coated
graphite and d) undoped on bare/Sn coated graphite.

enhancement since the Ge layer is amorphous and the Sn
layer, which gives essentially the same effect, is above
its nominal melting point at the growth temperature.
Either interlayer produces ohmic contact without excessive
doping of the bulk of the GaAs layer. The principal drawback of the chloride VPE technique is the low efficiency
for conversion of Ga to GaAs on the desired substrate. Undesired growth on the reactor walls not only competes with

the substrate but necessitates frequent cleaning for reproducible results. The situation for InP growth is, of
course, even worse in this regard. Another difficulty encountered is that InP films do not make ohmic contact to
graphite substrates and an interfacial layer of GaAs must
be grown for this purpose (8).

Metallorganic growth offers the immediate advantage
that growth occurs only on the heated substrate and the
cold reactor walls do not interfere. In an appropriate
vertical reactor geometry, efficiencies of conversion of
$Ga(CH_3)_3$ to GaAs on the desired substrates can exceed 90
percent. Growth usually takes place with As/Ga mole fraction ratios of $\sim$ 10:1 so that the efficiency of AsH_3
utilization is always less than 10 percent, however. In
the usual arrangement, induction heating of a graphite
susceptor on which the substrates are placed is employed
(Fig. 7). The fused silica reactor tube may be water-cooled
or allowed to rise somewhat above room temperature.
$Ga(CH_3)_3$ vapor is picked up by a H_2 stream in a bubbler,
typically held at -10°C. This is mixed with an AsH_3/H_2
stream at room temperature and introduced into the reactor.
The detailed chemical reaction at the heated substrate is
not known, but presumably the overall reaction

(5) $(CH_3)_3Ga + AsH_3 \xrightarrow[\text{(600-750°C)}]{\Delta} GaAs_{(solid)} + CH_4$

is arrived at stepwise:

(6) $(CH_3)_3Ga + AsH_3 \longrightarrow (CH_3)_2Ga^*AsH_2 + CH_4$ ⎤

    $GaAs + CH_4 \longleftarrow CH_3Ga^*AsH + CH_4 \longleftarrow$

etc. where the asterisk denotes a surface complex.

At excessive temperatures CH bonds may be broken
giving rise to contamination with carbon, while too low
temperatures permit inclusion of CH_3 groups or addition
complexes of the $(CH_3)_3Ga: AsH_3$ type. The last problem is
more severe with the In-P analog of (5), since $(CH_3)_3In:PH_3$
is a relatively stable black powder with fairly low vapor
pressure. In this system there is no particular difficulty

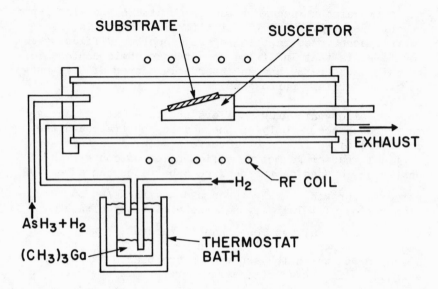

Figure 7 Schematic of apparatus for growth of GaAs by
metallorganic pyrolysis.

to add $(CH_3)_3Al$ to the reaction (5) and $Al_xGa_{1-x}As$ alloy
may be grown across the full range of x from 0 to 1.0, with
x in the solid corresponding closely to the Al/(Al+Ga) mole
fraction in the gas phase. Good growth may be obtained at
700-750°C for all values of x with growth rates of 10-20
micrometers/hr. In the absence of deliberately added dopants, the net carrier concentration and conductivity type
of metallorganic-grown GaAs depends on the As/Ga mole
fraction ratio with minimum values for N_D and N_A of
$\sim 5 \times 10^{15} cm^{-3}$ at a As/Ga mole ratio of ~ 10. P-type films
are obtained as the metal content in the gas phase is in-
creased, or $(C_2H_5)_2Zn$ vapor may be added with an additional
bubbler. N-type doping can be accomplished with H_2Se or
H_2S as for chloride VPE.

     No simple method of nucleation control for grain size
modification or continuity improvement has been developed
for metallorganic growth. In general films on graphite
tend to have larger grain size and require a greater

thickness ($\sim$ 10-15μm) to insure freedom from pinholes as
compared to films on molybdenum sheet (see Fig. 8). The
substrate temperature and the flow velocity have limited
effect, but some enhancement has been obtained by starting
growth at a lower temperature for nucleation and raising
the temperature to finish the layer (9). The surface
treatments with thin films of Zn, Ge, or Sn have much less
effect on nucleation or growth than in the chloride VPE
process, but just as in that case permit ohmic contacts to
be realized. Rather thick (2000 nm) Ge coatings may be
applied to substrates within the reactor by pyrolysis of
GeH_4, but GaAs films grown on such coatings appear to be

electrically inferior to those grown on graphite or molybdenum substrates coated thinly with evaporated, amorphous
Ge. A major problem with large-area metallorganic growth
is obtaining pinhole and crack-free films. The open-pore
structure of graphite often results in the deeper pores
not filling with GaAs and acting as a source of pinholes.
Pyrolytic graphite can be predeposited over the substrate
to yield a substantially pore-free surface but this layer
does not match the expansion coefficient of GaAs and the
semiconductor layers flake off when cooled down from the
growth temperature.

A greater variety of substrates are compatible with
MOG than VPE film formation as compatibility with an HCl
containing ambient is not required. Ferrous alloys are
unfortunately not suitable as iron reacts with AsH_3 to form

Fe_3As_2 which grows preferentially to the desired GaAs. Ni
behaves similarly. Ni-Fe alloy of 50-50 composition would
otherwise be a candidate substrate since this provides an
excellent thermal expansion match. It may be possible to
coat ferrous alloys with a barrier metal, but Cr, Mo, and
W have proved disappointing in initial experiments at 100
nm thickness. Pd, Pt, and Ir are too expensive to be
suitable. Further work in this direction is needed.

The final difficulty to be mentioned is the present
high cost of $(CH_3)_3Ga$ - about \$80.00/gm. This works out
to about \$.50 per cm^2 under optimistic assumptions about
conversion efficiency. By comparison the cost of elemental
Ga in six 9s purity is only $\sim$ \$.80/gm in kilogram quan-
tities. Use with optical concentrators would ease the

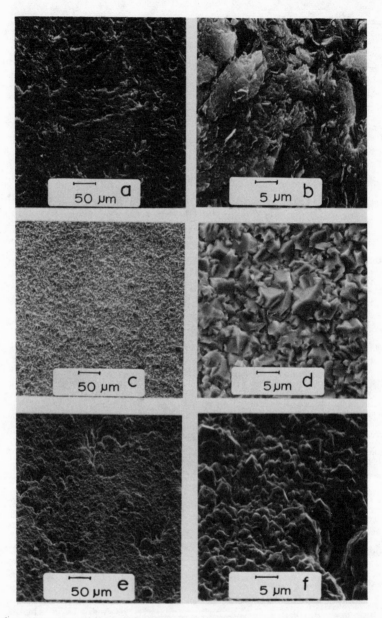

Figure 8   SEM views of graphite substrate pore structure
(a and b, 45° perspective); structure of GaAs grains on
Ge coated Mo (c and d, 0° perspective) and on Ge coated
graphite (e and f, 45° perspective).

requirement for cost reduction, the potential for which certainly exists since $(CH_3)_3Ga$ is now sold only as a specialty research item in very limited quantity.

III. ELECTRICAL CHARACTERISTICS OF THIN GaAs AND InP FILMS

The parameters of interest to be determined for the characterization of a thin-film GaAs/substrate composite are the carrier concentration and resistivity of the grain 'bulk', the effective minority carrier diffusion length, the lateral sheet resistivity for majority carrier flow across grain boundaries, the effective carrier concentration and vertical resistivity of the grain boundaries, and the nature of the substrate - semiconductor contact. In general these quantities must be determined (in so far as possible) directly on the sample in question since films grown under somewhat different conditions or on a different substrate will not have similar properties. This necessitates measurements on a conducting substrate, for example. This principally affects measurement of lateral current transport, which then requires that a p-n junction be used to isolate the layer for which sheet resistance is being determined. Since this quantity is really of interest only for heterojunction cells, this is not a major problem.

Otherwise the procedures are similar in principle whether heterojunction (AlAs/GaAs or CdS/InP), Schottky barrier or liquid junction cells are intended as the final structure. Capacitance-voltage measurement samples the large-area, lower carrier concentration regions represented by the grain 'faces' and can be used to determine the carrier concentration representative of the grain interior. The reverse breakdown voltage reflects the influence of grain boundaries, on the other hand. A semilog plot of the dark I-V characteristic (Fig. 9) permits separation of linear series or shunt resistance effects from factors which relate to junction dark current or tunneling. Measurement of quantum efficiency versus wavelength of exciting light (Fig. 10) permits values of the effective diffusion length ℓ to be determined from the relation

$$\eta(\lambda) = \frac{\alpha(\lambda)\ell}{1+\alpha(\lambda)\ell}$$

where $\alpha(\lambda)$, the optical absorption coefficient, may be assumed the same as for single crystal material in the above-band gap range.

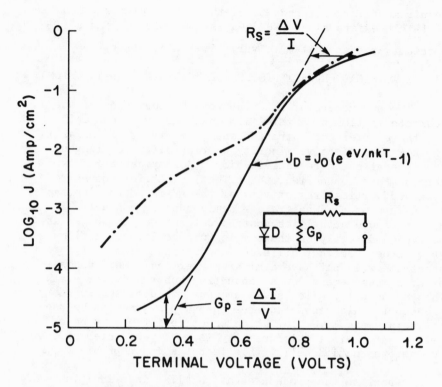

Figure 9 Typical dependence of log current density on
voltage for a diode with series resistance and shunt conductance (solid curve). Poor diode (dash-dot) characteristic is typical of junction formed by post-growth diffusion.
J_o, n, G_p and R_s may be determined from the value and slope
in the linear region and the current and voltage 'offsets'
shown, respectively, for diodes which are sufficiently good
to show a well-defined linear region.

     The model which emerges from the studies which have
been done is one in which the film is envisioned as consisting of columnar grains within which essentially single
crystal behavior is observed. These are connected by grain
boundary regions which pose reflective barriers to the
transfer of majority carriers from grain-to-grain, but
which, rather surprisingly, do not act as minority carrier
sinks. High internal quantum efficiency indicating the

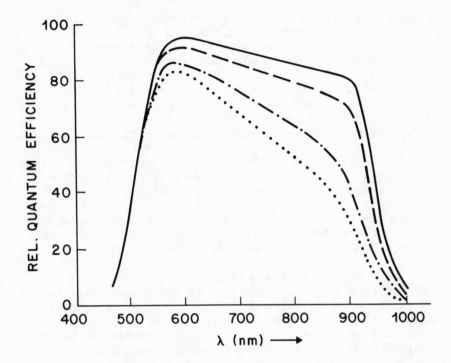

Figure 10 Dependence of quantum efficiency on wavelength
for single crystal (solid) and polycrystal AlAs/GaAs
heterojunction diodes as a function of wavelength. The
doping level in the GaAs layer increases from the dashed
to dash-dot to dotted curves and the diffusion length
falls from approximately 4 to 3.5 to 2 to 1 μm.

absence of recombination centers at the grain boundaries
is commonly observed provided the doping level is kept
below about 10^{18} cm^{-3}. This places no restriction on cell
construction since doping levels of $\sim 10^{17}$ cm^{-3} are adequate for p-n junction cells and even lower levels are
desirable for Schottky barrier cells. Capacitance-voltage
measurements indicate the carrier concentration in the
grain bulk is essentially the same as that determined from
Hall effect measurements on layers grown on semi-
insulating single crystal substrates under 'identical'
conditions, e.g., in the same growth run. The reverse

breakdown for Schottky diodes on thin films occurs at a
much lower voltage and is soft compared to the breakdown of similar diodes prepared on epilayers grown on
single crystal substrates with the same C-V characteristics, consistent with the assumed presence of high
charge density at the grain boundaries/surface inter-
section. One is then led to postulate that a Schottky
barrier thin film diode acts as a parallel combination of
grain-face and grain-boundary diodes as in Fig. 11.

The major effect of the soft grain boundary diodes is
to increase the dark current leakage (reverse saturation
current) which is reflected for solar cells in a reduced
open circuit voltage, and to reduce the fill-factor. The
fill-factor is further degraded by shunt resistance due to
pinholes and microcracks, and series resistance due to an
imperfect semiconductor-substrate contact.

The performance of p-n junctions or heterojunctions
need not suffer from the presence of grain boundaries to
the extent Schottky barriers suffer. The 'grain boundary'
diodes soften the reverse breakdown characteristic but have
little effect on the forward current in contrast to the
Schottky barrier case. This is because heavy doping
(i.e., $\sim 10^{18}$ cm^{-3}) gives rise to much enhanced tunneling
current in an M-S structure but not in a p-n junction unless a density of states within the band gap is introduced.
This does not occur as evidenced by the lack of recombination centers associated with the grains. Thus high open-
circuit voltages, comparable to single-crystal cells, have
been observed for polycrystal AlAs/GaAs heterojunction
material (6).

The InP/CdS interface appears to act as a p-n junction
rather than a Schottky barrier, even though the CdS is
heavily n-type. The fact that high open circuit voltages
(comparable to p-n homojunctions in InP) are not observed
presumably reflects the imperfect match between CdS and InP.
This match is good only between the (0001) plane of the CdS
wurzite structure and the (111) plane of the InP
(zincblende) structure. Obviously this match cannot be
obtained on the polyfaceted grains and one must expect a
high density of interface states. Since the depletion
layer lies almost entirely within the InP, these may not
have a large effect on the current collection (quantum)

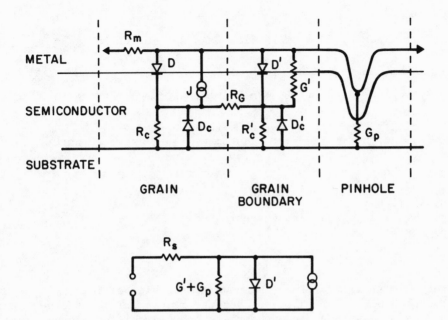

Figure 11 Model of a Schottky barrier polycrystalline
cell. If the contact resistances to the substrate (R_c and
R_c') are negligible, D_c and D_c' may be neglected and the
lower equivalent circuit used. G' and G_p form a shunting
conductance due to grain boundaries and pinholes, and the
open circuit voltage is limited by turn-on of the grain
boundary diodes D'.

efficiency. They do however serve to enhance the dark
current and hence reduce the open circuit voltage, and as
a result, the polycrystal InP/CdS cell does not quite match
the polycrystal GaAs MOS cells in performance so far (see
Table II).

The promise for highest potential solar cell
performance from these materials seems to be held by the
AlAs/GaAs cell, which presently suffers from a
'nonintrinsic' but very serious series resistance problem.
Of the several cell configurations discussed, only this one
is affected by the high lateral resistance of the

TABLE II

PROPERTIES OF THIN FILM GaAs AND InP CELLS

Growth Proc.	VPE	VPE	MOG	CVD
Cell Type	GaAs/AlAs (a)	GaAs-MOS	GaAs-MOS	InP/CdS (d)
Gr. Size	3-5	3-5	1-3	1-3
Efficiency	?	6.3% (b)	4.0% (c)	5.7%
Diff. length	1.5 μm	1.5 μm	1.3 μm	1.3 μm
V_{oc}	.95	.65	.63	.65

a) Ref. 6
b) Ref. 5
c) W. D. Johnston, Jr. (unpublished)
d) Ref. 8

polycrystal III-V films. No power conversion efficiency
has been cited for this cell because meaningful tests cannot be made without an appropriate transparent top contact
which will serve to 'short' over the grain boundary resistance. Finely gridded cells have shown open circuit
voltages and fill factors comparable to single crystal
cells of 17 percent AM1 efficiency, with spectral response
curves indicative of high internal quantum efficiency. The
use of conductive oxides as transparent top contacts is
made difficult by solid state reaction with the AlAs to
yield an Al_2O_3 insulating interphase, but work to alleviate
this problem is continuing.

IV. SUMMARY AND PROGNOSIS

The long-range prospects for practical thin film III-V
solar cells remain uncertain, but clearly continued work
on a research scale is warranted. Growth by pyrolytic
reaction of metal-alkyls with the group V hydrides seems
most promising for eventual production, but $(CH_3)_3Ga$ is
presently very expensive. The performance of films grown
this way is not yet quite so good as that of the chloride-
VPE films, presumably because of the somewhat finer grain
structure. Graphite and molybdenum are the most
satisfactory substrates overall and are likely to be

acceptable if thin-film cells can be used with hundred-fold optical concentration.

Cells for flat-plate arrays will require development of a sheet-graphite production process or development of a cheap, sheet-form, compatible substrate. Given the problems of availability of Ga, difficulty of making large areas of semiconductor free from pinholes, and the current high technology required for satisfactory contacts, and on the other hand the advantage GaAs represents for hybrid photovoltaic/thermal systems; the concentrator scenario seems more plausible.

The prospect for surpassing the 10 percent efficiency goal with a III-V thin film solar cell seems best for a p-n junction configuration. For GaAs this requires the achievement of a transparent top contact to the AlAs or $Al_xGa_{1-x}As$ top polycrystal layer. The other configurations (InP/CdS and GaAs - MOS) presently suffer from junction performance limitations directly associated with the polycrystalline nature of the material, which may be of a more fundamental nature.

Present experience in environmental tests suggest that any photovoltaic array for practical terrestrial use (silicon cells included) will have to be provided with a complete hermetic seal (10). Thus the relative stability of one material combination over another at moderately elevated temperatures or humidities in unencapsulated condition is largely irrelevant. There is no clear cost advantage for one of these approaches and ultimate performance will be the deciding factor - indeed, it is doubtful that serious development efforts will be directed toward III-V thin-film solar cells until laboratory efficiencies of at least 10 percent are reported. If that goal is met under conditions compatible with at least moderate optical concentration, attainment of the cost goal of $.50 per peak watt seems possible so far as the semiconductor cell element is concerned.

REFERENCES

1. United States Dept. of Energy, DOE/ET-0035 (78), (March 1978).

2. Similar conclusions have independently been drawn by
    L. M. Fraas and K. Zanio, Hughes Research Report #521,
    August 1978 (unpublished).

3. K. J. Bachmann in Current Topics in Material Science,
    Vol. 3 E. Kaldis, Ed. North Holland Press, New York,
    N. Y. (1978).

4. P. Vohl, D. M. Perkins, S. G. Ellis, R. R. Addis,
    W. Huri, and G. Noel, IEEE Trans ED 14, 26 (1966).

5. S. S. Chu, T. L. Chu, and H. T. Yang, Appl. Phys.
    Lett. 32, 557 (1978), also Proc. 13th Photovoltaic
    Specialist Conference IEEE, New York, N.Y. (1978)
    (p. 956).

6. W. D. Johnston, Jr. and W. M. Callahan, J. Electrochem.
    Soc. 125, 977 (1978).

7. H. M. Manasevit, J. Electrochem. Soc. 118, 647 (1971).

8. J. L. Shay, M. Bettini, S. Wagner, K. J. Bachmann, and
    E. Buehler, Proc. 12th Photovoltaic Specialist
    Conference, IEEE, New York, N.Y. 1977 (p. 540).

9. A. E. Blakeslee and S. M. Vernon, Paper A-4, 20th
    Electronic Materials Conference, Santa Barbara,
    California, June 1978.

10. This conclusion on the author's part arises from a
    conference on "Stability of (Thin-Film) Solar Cells
    and Materials" held at the National Bureau of Standards
    Washington, D.C. in May 1978, particularly from
    presentations by P. Iles and R. Ross (Proceedings
    unpublished).

NITROGEN REDUCING SOLAR CELLS

G.N.SCHRAUZER, T.D.GUTH, M.R.PALMER
and J.SALEHI

Department of Chemistry, University
of California at San Diego, Revelle
College, La Jolla, California   92093

The increasing awareness of the limits of our supplies
of natural gas has created some interest in recent years in
replacing existing methods of industrial ammonia synthesis
either with less energy-demanding variants, or by altogether
new methods of synthesis that would be independent of conventional energy sources. While it is fully recognized that the
HABER-BOSCH Process of ammonia synthesis will probably never
be fully replaced by other methods, it is nevertheless worth-
while to examine the feasibility of alternative approaches to
the problem. Nature's way of fixing nitrogen is one such alternative which obviously works and which in fact exhibits all
desirable features of a process which, if it could be carried
out on a large scale technically, could replace the HABER-
BOSCH Process.

It is for this reason that we have been investigating the
chemical basis of biological nitrogen fixation in our laboratory during the past 10 years. It was our aim not only to establish the fundamental chemical aspects of the mechanism of
this intriguing process but also to develop functional models
of the nitrogen fixing enzyme that could perhaps become the
basis of new catalytic processes. Our work has been successful academically and also provided us with several functional
model systems that duplicate the reactions of nitrogenase, but
it also revealed that even our simplest models were still too
complicated to be used anywhere else but in the academic laboratory. This not only applies to our models. It may in fact be
argued that any homogeneous catalyst of nitrogen reduction would
have to be inexpensive enough to be "thrown away". The chances
of finding such a catalyst are slim.

Another weakness of all existing homogeneous nitrogen reducing systems is their dependence on an external reductant. Of all possible agents, hydrogen is obviously the best but must be generated with expenditure of energy. Hence, it seems that the more obvious alternatives of reducing nitrogen at low temperatures and pressures must be discarded.

Faced with this state of affairs it became clear that only one theoretical possibility exists to accomplish our task. It would be necessary to *combine the features of plant photosynthesis with those of biological nitrogen fixation* in a simple solid state device, constructed from inexpensive materials. Prototypes of solar cells which reduce molecular nitrogen according to this principle were designed in 1975-1977(1). In the following, some of the features of the reactions of these cells will be described.

In the nitrogen reducing solar cells, chemisorbed nitrogen is reduced with photochemically generated " reducing equivalents" under simultaneous evolution of oxygen. The photolysis of water is well known to occur on certain semiconductor surfaces, notably TiO_2(2) and is related to nitrogen photoreduction inasmuch as in this case photochemically generated reducing equivalents are reacted with protons to yield hydrogen rather than with nitrogen. The formation of *ammonia* in nitrogen reducing solar cells can accordingly be formulated in terms of the reaction in eq [1]:

$$N_2^g + 3H_2O^l + mh\nu \longrightarrow 2NH_3^g + 1.50_2^g$$
$$\Delta G^{298} = 766 \text{ kJ/mol} \qquad [1]$$

A related reaction is the formation of *hydrazine*, eq. [2]:

$$N_2^g + 2H_2O^l + xh\nu \longrightarrow N_2H_4^l + O_2^g$$
$$\Delta G^{298} = 626 \text{ kJ/mol} \qquad [2]$$

Titanium dioxide appeared to be a logical *substratum* for experiments aimed at verifying eqs [1] and [2] not only in the light of its ability to induce the photolysis of water, but also in view of other published work on the photoreduction of organic and inorganic compounds. Titanium dioxide has been recommended as the base-material in photographic devices where a latent metallic image is generated after exposure to light.

In this process metal ions such as Pd^{+2} at the TiO_2 surface are reduced on illumination to form metal nuclei. Using chemical developers visible images can be generated in a second step(3). Other authors have demonstrated the photoreduction of reducible dyes on TiO_2, notably of methylene blue(4). Of particular interest is the observed photoreduction of *acetylene* and of *ethylene* on incompletely outgassed TiO_2 powders. Boonstra and Mutsaers(5) demonstrated that acetylene is reduced to a mixture of methane, ethylene, ethane and C_3-hydrocarbons if illuminated in the presence of TiO_2. Ethylene is photoreduced to methane, ethane and *n*-butane. These reactions are dependent on the presence of surface –OH groups. Completely outgassed TiO_2, obtained by heating the sample to 500° *in vacuo* no longer photoreduces acetylene or ethylene. Acetylene is instead cyclotrimerized to benzene (both thermally and photochemically), while ethylene is dimerized to 1-butene. These reactions are typical of reactions of unsaturated organic compounds with low-valent transition metal complexes and are thus indicative of the presence of reducing centers at the TiO_2 surface.

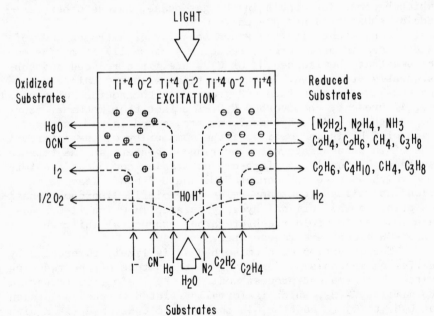

SCHEME I. Schematic representation of photooxidation and –reduction reactions on incompletely outgassed rutile. From Ref. (1), reproduced with permission.

Photooxidations of substrates on illuminated titania powders have also been reported. We only mention the oxidation
of mercury to mercuric oxide(6), the oxidation of cyanide to
cyanate(7) and of iodide to iodine(8). Most recently(9) acetic acid was shown to undergo photodecomposition into CH_4 and
CO_2 on illuminated TiO_2. Some of these reactions are represented graphically in SCHEME I.

The primary steps in the illumination of TiO_2 may be pic-
tured as the formation of pairs of negative and positive centers. The negative centers will ultimately become localized
in the valence shell of the titanium ions. The formation of
Ti(III)-ions on the surface of illuminated TiO_2 has been de-
monstrated(10). However, it is likely that titanium centers
of even lower oxidation states are generated through charge
transfer processes. In addition, the reduced centers may form
clusters and need not be mononuclear, giving rise to the actual reducing sites. The positive centers or holes may react
with surface -OH groups to form •OH radicals which initially remain chemisorbed but ultimately terminate to yield O_2,
which becomes detectable in the gas-phase, unless other reducing substrates are present.

Consistent with the known behavior of nitrogen and of
acetylene in other protic reducing systems(11),it may be concluded that chemisorbed N_2 or C_2H_2 are first reduced and subsequently protonated. In the reactions with incompletely outgassed TiO_2, the surface -OH groups are the sources of protons,
or, if present, chemisorbed H_2O. The fact that methane, ethylene, ethane and C_3-hydrocarbons are formed in the photoreduction of C_2H_2 is indicative of the complexity of the interactions of this substrate with the reduced centers at the titania
surface. Obviously, surface-organotitanium compounds are formed which undergo subsequent C-C and C-H bond cleavage reactions as well as reactions related to olefin disproportionation prior to the terminal hydrolysis step in which Ti-C bonds
are severed to yield the hydrocarbon product(s) while the surface -OH groups are regenerated.

The reduction of chemisorbed N_2 is likely to proceed by
different mechanisms, depending on the nature of the reducing
sites. The simplest process would be the 2-electron reduction
to diimide, N_2H_2, which is formally related to the conversion
of C_2H_2 to C_2H_4. However, in contrast to C_2H_4, N_2H_2 is a highly
reactive, unstable species which may either disproportionate
to yield N_2H_4 and N_2 or decompose into the elements N_2 and H_2.
We presently do not know how N_2H_2 reacts on the surface of TiO_2,
but it may be suspected that its decomposition into the ele-

ments is a main, if not *the* main reaction. Thus, under certain experimental conditions with specially prepared, metal-doped samples of TiO_2, we have consistently observed a *nitrogen dependent* evolution of H_2 (12) which we attribute to the formation and preferred decomposition of chemisorbed diimide. A similar N_2-dependent H_2 evolution is not seen with the best N_2-*reducing* TiO_2 samples. With these we usually observe an evolution of H_2 and O_2 on irradiation under argon. In the presence of nitrogen the H_2- but not the O_2 evolution is significantly inhibited. These results indicate that the photocatalytic properties of TiO_2 depend on details of sample pretreatment, as will be in discussed in greater detail below. Just as C_2H_2 is reduced to several products (CH_4, C_2H_4, C_2H_6 and C_3-hydrocarbons) and thus by more than one mechanism, it must also be supposed that the reduction of N_2 occurs by several mechanisms, in part by way of diimide, but possibly also via nitride intermediates which on hydrolysis yield ammonia or hydrazine directly.

Pure TiO_2 is a relatively inefficient substratum for water photolysis and is also only slightly active in nitrogen photoreduction. A substantial improvement of the photocatalytic properties was observed with iron-doped titania powders prepared by heating $Fe_2(SO_4)_3$-impregnated anatase powder to $1,000°$. Fe_2O_3 greatly accelerates the rate of the anatase → rutile conversion, which is thermodynamically favored [$\Delta H° = -5.30$ kJ/mol(13)] and irreversible. The transformation occurs in the temperature range between $750°$ and $1,000°$ at rates which are greatly dependent on the type and concentration of impurities and various other experimental details, including the composition of the gas-phase during the heat-treatment(14). The most active photocatalysts for nitrogen photoreduction were obtained from samples of anatase of diameter of about 2 microns. These were impregnated using $Fe_2(SO_4)_3$ and heated to $1,000°$ for 1 hr. This did not cause complete transformation of the anatase into rutile, but afforded aggregates of small crystals of about 0.2 micron diameter with a large active surface. On prolonged heat-treatment, the crystals grew appreciably, causing a diminution of active surface and of photocatalytic activity.

The optimal concentration of Fe_2O_3 was found to be 0.2 wt.-%. Higher concentrations of dopant produced samples with low activity. Under these conditions the anatase → rutile transformation occurred rapidly, and grain-growth was also accelerated, giving rise to rutile crystals of 3 micron length and low photocatalytic activity due to diminished active surface (Fig.1).

It should also be mentioned that not all commercially available batches of anatase are suitable for the preparation of active photocatalysts. Highly aggregated samples tend to "clump" together during the heat treatment, giving rise to inactive specimens.

Interesting results were obtained in experiments with dopants other than iron. It was noted that metal oxides which accelerate the anatase → rutile conversion as a rule also produce active photocatalysts(1). These metals included Co, Mo and Ni, which are known to promote the anatase → rutile conversion as well as the grain-growth of TiO_2 at elevated temperatures(15). A number of other metals, $i.e.$, Pd, Pt, Ag, Au, V, Cr, Pb, Cu, did not produce significantly active photocatalysts. Most of them also do not accelerate the anatase → rutile conversion under our reaction conditions. On the other hand, MoO_3 promotes grain-growth but apparently has only a slight effect on the phase-transformation(15). The fact that it nevertheless produced ac-

ANATASE, 9000x
RUTILE CONTENT: CA 5%

RUTILE/ANATASE, DOPED WITH
.2% FE, HEATED 1 HOUR AT 1000°
RUTILE:ANATASE = 23:77   9000x

RUTILE/ANATASE, DOPED WITH
.5% FE, HEATED 1 HOUR AT
1000°. RUTILE:ANATASE =50:50.

9500x

RUTILE CRYSTALS FROM ANATASE
DOPED WITH 1% FE, HEATED 1 HOUR
AT 1000° (RUTILE:ANATASE=99:1)

9500x.

Fig.1: Scanning electron microphotographs of anatase, and of iron-doped rutile-anatase phases generated after heat treatment.

tive photocatalysts suggests that the effects of the metal dopants are not necessarily specific.

The reduction of N_2 takes place optimally at temperatures
of about 30-50°; radiation in the near uv-region (390–420 nm)
is photochemically active, just as in other TiO_2 sensitized
reactions. The photoreduction of N_2 was observed also on exposure of doped TiO_2 samples to normal sunlight. It also occurs
in *air*, albeit to a somewhat lesser extent than in atmospheres
of pure nitrogen(1). Whether ammonia can be produced in this
manner under practical field conditions will depend on the outcome of further studies presently in progress. Theoretically,
the ammonia output of a N_2-photofixing field of the size of 1
acre (4047 m^2) is 50-100 kg of NH_3 *per annum*, assuming current
activities of doped TiO_2 and the absence of catalyst-inactiva-
ting effects on prolonged operation. The lifetime of the photocatalysts under various conditions is currently under investigation, as are attempts to increase their efficiency.

Nitrogen Photoreduction in Nature.
The wide distribution of titanium on the earth's crust and
the possibility that minerals other than those containing titanium could possess nitrogen photoreducing activity prompted us
to conduct a number of experiments to be described in the following.

Since N_2-photoreduction reactions were expected to take
place preferentially in arid- or semiarid regions with an abundance of solar radiation, samples of common rocks, clays and of
sand were collected in desert regions of Arizona and S.Califor-
nia. Rocks and clay were powdered prior to the experiments designed to demonstrate nitrogen photoreduction under simulated
natural conditions. Samples of the powdered minerals or of heat-
sterilized desert sand (10 g) were placed into Pyrex test tubes
of 10 cm length and 1 cm width. After removal of the normal
air through evacuation, the test tubes were filled with a mixture of $^{30}N_2$ and O_2 (78:22), sealed, and exposed to sunlight.
Simultaneous controls were run with the same mineral- or sand
samples in foil-covered test tubes. After exposure, the test
tubes were opened and the solids extracted with 2 M HCl.

The extracts were made alkaline and distilled in a Micro-
Kjeldahl Apparatus. The distillates were oxidized with sodium hypobromite and the nitrogen released was collected   and
analyzed mass-spectrographically (see Fig.2).

Clays and common rocks (granite, dolomite etc.) exhibited
little if any nitrogen reducing activity. However, sand obtained from the *Imperial Sand Dunes*, a desert area 100 M east of

San Diego, showed definite N_2 reducing activity on exposure
to normal sunlight. Typically, 0.14 μmoles of NH_3 were formed
by exposing 10 g of sand to La Jolla sunlight (July) for 5 days.
N_2 was also photoreduced in an atmosphere of pure $^{30}N_2$, using
a 360 W Hanovia mercury arc lamp at a distance of about 20 cm
as the light-source. The irradiation temperature was maintained at 28° by placing the lamp into a water-cooled special
container. Under these conditions, 0.6 μmoles of NH_3 were obtained from 10 g of sand after 12 hrs of irradiation.

The simultaneous controls did not reduce $^{30}N_2$ if protec-
ted against light [the N_2 released from the Kjeldahl distillates on hypobromite oxidation contained $^{29}N_2$ in the expected ratio to $^{28}N_2$ (0.74 %)].

The sand sample from the *Imperial Sand Dunes* contained 86
% of SiO_2, 0.5 % of H_2O (estd. from weight loss after drying
at 200°), a small percentage of common minerals such as iron
oxides, aluminum silicates, $CaCO_3$, etc. .Its titanium content
was 0.08 %. It was established that neither SiO_2 nor lighter
mineral fractions ($CaCO_3$, aluminum silicates, etc.) were photoactive. The nitrogen photoreducing mineral(s) were found in
a heavy residual fraction obtained by suspending the crude sand
in C_2Br_4 (d = 3). Although this residue was present in only 1 %
of the untreated sand, it exhibited 60-80 % of the N_2 photore-

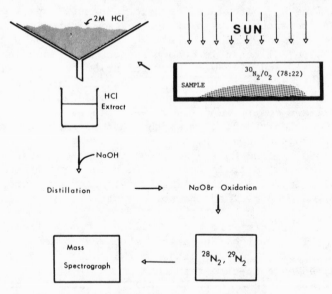

Fig.2: Experimental design of nitrogen
photoreduction experiments with desert
sand under simulated natural conditions.

ducing activity. According to analysis, the heavy residue contains 17 % SiO_2, 25 % Fe, and 9.2% Ti, as well as a number of other heavy elements in lower concentrations. From x-ray powder patterns obtained it is concluded that most of the titanium is present in the form of rutile

Sand of this composition may be estimated from these results to produce from 1 to 10 kg of NH_3 per year and acre, considering variations in sunlight intensity and temperature (lower yields of NH_3 are generally observed at higher irradiation temperatures). Our experiments establish furthermore that this process of nitrogen photoreduction takes place in other arid or semiarid regions on earth, since titanium is widely distributed on the earth's crust. Conservatively, the amount of NH_3 formed by photoreduction is estimated to $10x10^6$ tons per year, assuming a total desert area of $19x10^6$ square miles. The amount of N_2 fixed is equal to about 1/3 of the N_2 which is normally oxidized by lightning discharges, and to about 10% of the ammonia produced by biological nitrogen fixation. The possibility that N_2 photoreduction processes occur on the surface of certain planets(e.g. on Mars) has been suggested(1) and should be explored.

Acknowledgements:

This work was initially supported by a grant from the University of California Board of Patents and subsequently by NSF-Grant CHE76-10890.

(1) G.N.Schrauzer and T.D.Guth, J.Am.Chem.Soc. 99,7189 (1977).

(2) A.Fujishima and K.Honda, Nature(London), 238, 37 (1972).

(3) H.Jonker, C.J.G.F.Janssen, C.J.Dippel, Th.P.G.W.Thijssens and L.Postma, Phot.Sci.Eng. 13, 45 (1969).

(4) H.Yoneyama, Y.Toyoguchi and H.Tamura, J.Phys.Chem. 76, 3460 (1972).

(5) A.H.Boonstra and C.A.H.A.Mutsaers, J.Phys.Chem. 79, 2025 (1975).

(6) U.Kaluza and H.P.Boehm, J.Catal. 22, 347 (1971).

(7) S.N.Frank and A.J.Bard, J.Am.Chem.Soc. 99, 303 (1977).

(8) A.Fujishima and K.Honda, J.Chem.Soc.(Japan) 74, 355(1971).

(9) B.Kraeutler and A.J.Bard, J.Am.Chem.Soc. 99, 7729 (1977).

(10) R.I.Bickley and R.K.M.Jayanty, Disc.Farad. Soc. 58, 194 (1974).

(11) G.N.Schrauzer, Angew. Chem.Internatl.Ed.Engl. 14, 514 (1975).

(12) Unpublished observations.

(13) A.Navrotsky and O.J.Kleppa, J.Amer.Ceram.Soc. 50, 626 (1967).

(14) R.D.Shannon and J.A.Pask, J.Amer.Ceram.Soc. 48, 391 (1967).

(15) Y.Iida and S.Ozaki, J.Amer.Ceram. Soc. 44, 120 (1961).

PHOTOSENSITIZATION MECHANISMS FOR ENERGY STORING ISOMERIZATIONS

Guilford Jones, II, Phan Thanh Xuan, and
Sheau Hwa Chiang

Department of Chemistry, Boston University

Boston, Massachusetts 02215

ABSTRACT

The prospects for driving endoergic reactions of simple, relatively abundant, organic chemicals by photochemical means have been examined. Photoisomerization reactions which have some potential for storage of solar energy as latent heat in kinetically stable products are surveyed. Emphasis is placed on methods for the photosensitization of storage chemicals to visible light. Mechanisms include excited state complexation of isomerizable substrates through electron donor-acceptor attraction and conventional energy transfer photosensitization. Efficient isomerization of a norbornadiene derivative using the latter technique and photosensitizers absorbing past 500 nm is described. Factors controlling the efficiency of endothermic triplet energy transfer are outlined. The concept of "thermal upconversion" of the excitation energies of visible absorbing sensitizers is introduced, including a demonstration of improved quantum efficiency as a function of temperature for an isomerization which utilizes a very low energy sensitizer.

INTRODUCTION

Interest in solar energy conversion has expanded
rapidly in recent years due in part to the recognition
that use of solar power represents a sensible match between energy supply and demand. The wide distribution
and low energy density of sunlight are especially well
suited for a host of applications with low power requirements and little economy of scale (1). It is widely
appreciated, nonetheless, that the success of solar
devices will depend on the development of energy storage
capability. Much research effort is now directed to
methods of storage of solar energy as latent heat (2).
Use of phase change materials (e.g., eutectic salts) (3)
has received most attention and is most advanced; thermochemical methods have been proposed (4).

Photochemical latent heat storage has been the subject of several recent reviews (5-8). The proposed energy
storage scheme involves a closed cycle of reactants

$$A \underset{\Delta}{\overset{h\nu}{\rightleftharpoons}} B$$

which is light driven in one direction and releases energy
in the reverse. Selection of A and B depends on the
following general criteria. (1) A must absorb or be
sensitized to atmospherically filtered sunlight. The most
important, high energy portion of the solar insolation
spectrum (fig. 1) is 300 - 700 nm, including most of the
visible region. (2) To insure that photochemistry proceeds in one direction, B should not absorb solar intensity or be involved in energy transfer with a sensitizer.
(3) The quantum efficiency of photoreaction A → B must
be near unity. (4) Thermal reaction (B → A) should have
a large negative enthalpy. (5) Photoproduct B should be
kinetically stable (i.e., revert to A negligibly at
ambient temperatures). Ideally a catalyst can induce back
reaction at a controllable rate and temperature. (5)
Chemical components must survive a large number of energy
storage and reversion cycles. Chemicals, A - B, must be

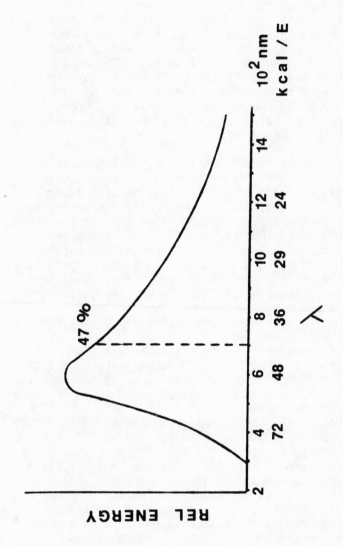

Figure 1. The Solar Insolation Spectrum.

inexpensive, available in large quantity, readily handled
and relatively non-toxic.

Advantages of a photochemical heat storage system
are readily identified. Many light driven reactions have
large energy storage capacities. Storage of energy is
possible at low collector temperatures and on cloudy days
(efficient use of shorter wavelengths). Compared to other
solar-thermal transducers, the photochemical system will
have lower requirements for collector size, storage area,
and insulation (9). Heat can be retrieved (through back
reaction of B) at relatively high temperatures (even above
that of the collector).

Selected organic photochemical reactions meet a
number of the criteria for latent heat storage. Simple
unimolecular rearrangement reactions, or photoisomerizations, are most attractive. Many of these rearrangements, particularly valence isomerizations, have large
storage enthalpies and involve relatively low molecular
weight materials. A number of the potential photoisomerization substrates are liquids of moderate volatility
which can be readily circulated and stored. The simple
inexpensive engineering of flat plate solar collectors
(1,2) and storage heat exchangers could be readily
adapted for use with these photochemical working fluids.

Four isomerization pairs are shown with structures
1 - *8*, part of a large family of isomerizations related
to the interconversion of norbornadiene (*1*) and quadricyclane (*2*). Data for these examples are shown in Table I
including quantum efficiencies for the photochemical step,
storage enthalpies, and thermal stabilities. The ability
of these systems to store impressive amounts of chemical
potential energy in relatively "stable" molecules is well
known. The thermodynamic instability of quadricyclane
photoproducts is understood in terms of bond angle strain
introduced when small (three or four membered) rings are
formed (15). The kinetic stability of photoisomers is
possible (back reaction is slowed) due to a high potential
barrier which separates the isomers which is imposed by
orbital topology factors (16). The rates of back reaction
can be modified using a variety of transition metal catalysts which provide low energy pathways for thermal
rearrangement, a general result for small ring organic
systems (17).

1	R = H , H	2
3	R = H , CO$_2$Me	4
5	R = CO$_2$Me , CO$_2$Me	6
7	R = CO$_2$Me , C$_6$H$_5$	8

Table I. Photochemical and Thermal Reaction Parameters for Norbornadiene-quadricyclane Isomerizations

Isomerization	Quantum efficiency (λ, nm) (ref)	ΔH_{AB} (kcal/mol) (J/g) (ref)	Back Reaction Temperature (°C)[b]
1,2	0.7[a](313)(10)	26.7(1190)(11)	135–200
3,4	0.9[a](313)(12)		
5,6	0.5(300)(13)	18.5(372)(14)	115–180
7,8	0.6(334)(15)	23.0(433)(5)	100–160

[a]Photosensitized. [b]Temperature range for back reaction at moderate rate (halflife = minutes).

As shown in Figure 2 for a hypothetical reversible photoisomerization, the combination of thermodynamic instability (large ΔH) and kinetic stability (large $\Delta H\ddagger$) presents something of a dilemma. As longer wavelengths of light are used and increasing amounts of chemical potential energy are stored, excited states with reduced driving force are required to mount a barrier of increasing size in the ground state surface in order to reach the photoisomer. Although the severity of this complication is not yet known (vide infra) it seems prudent to project that 700 nm will be a boundary for wavelengths of solar emission useful for photoisomerization, if storage capacity and energy retention are to remain high. The organic isomerization systems provide an interesting contrast with inorganic electron transfer reactions (18) which in principle allow use of longer wavelengths but with severely reduced storage capability (low barrier to back electron transfer). The molecular inorganic system, $NOCl \rightarrow NO + \frac{1}{2}Cl_2$, is another photochemical latent heat storage system for which the back reaction is spontaneous at ambient temperatures (19).

Progress made in the last several years in the assessment of photon energy storage properties of organic isomerizations can be summarized as follows.

(1) A number of reactions or reaction types which have some promise for photon energy storage have been identified and quantitatively studied (5, 20-24).

(2) The norbornadiene system (1 → 2) has been studied in detail, including extension of the threshold wavelength for efficient isomerization past 400 nm (using organic photosensitizers) (10).

(3) Metal ions or transition metal complexes have been used to drive reaction of *1* and the mechanisms of this new mode of photosensitization have been explored (25,26).

(4) The secondary chemical components of a solar photo-thermal system involving *1* → *2* have been developed. A novel feature of this demonstration unit is the use of polymer supports to anchor organic photosensitizers (10) and transition metal catalysts (for back reaction) (27,28). The deployment of these components in a photochemical solar

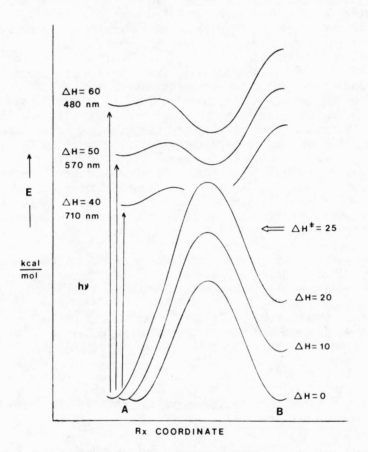

Figure 2.   Potential surface diagram for a hypothetical photochemical reaction which is
thermally reversible and which stores
variable amounts of excitation energy in a
kinetically stable product.

conversion device is shown in Scheme 1.

(5) The penultimate intermediates in energy storing photoisomerizations have been proposed in several mechanistic studies (21,29,30). Thus formation and partitioning of biradicals are expected to ultimately control the quantum efficiencies of most potentially useful organic photoreactions (vide infra).

(6) The comparative economics of a photochemical latent heat storage system has been assessed with particular reference to the storage capacity, service life and cost of the primary storage chemical (9).

(7) New developments in photocalorimetry techniques have provided a ready means for measuring latent heat storage capacities (31).

Photochemical latent heat storage of solar energy is far from demonstrated utility. It is important to note what performance features of an eventual solar photothermal unit remain in doubt. Practicability does not appear to rest with the storage capacity of (organic) photoreactions or with other intrinsic chemical limitations. Systems with unidirectional, high yield photochemistry, large storage enthalpies, and kinetic stability in photoproducts are readily found. The more elusive performance objectives are the extension of the wavelength threshold for photochemical use of solar radiation, demonstration of the durability of working chemical components, and development of systems within rather strict limitations on the cost of chemicals and hardware.

The following passages include recent photochemical results directed to meeting the first of the following objectives which we consider of highest priority: (1) the development of photosensitization mechanisms which allow use of very low energy photons in driving photoisomerizations (extension of the solar threshold to at least 600 nm); (2) the identification of inexpensive primary chemicals (isomerization substrates) which have the required thermochemical properties and are compatible with developing photosensitization techniques; (3) continued matching or tailoring of photosensitizers and catalysts with the storage medium to insure the efficiency of light harvesting and recycling; (4) demonstration of the

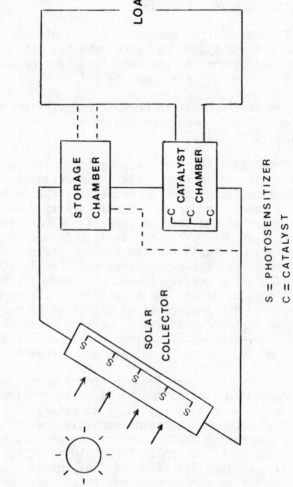

Scheme 1

LOAD

STORAGE CHAMBER

CATALYST CHAMBER

SOLAR COLLECTOR

S = PHOTOSENSITIZER
C = CATALYST

resistance of a system to fatigue (a proper order of
magnitude = 100 photo-thermal cycles).

PHOTOSENSITIZATION INVOLVING EXCIPLEXES

The unit of merit for storage of solar energy as
latent heat is 10^6 kJ. This figure somewhat more than
accounts for the daily heating load in winter for an
average dwelling in a moderately cold climate in the
U.S. (2). Assuming a capacity for a storage chemical of
100 kJ (about 20 kcal)/mol and 10 mol/kg, a material requirement for reversible storage of 10^6 kJ of 1000 kg is
readily calculated. If the cost of a storage raw material
is to stay below a recommended (3) $2,000, (a three-year
lifetime for the storage medium, probably 2-300 latent
heating cycles), then the cost of the primary chemical
should be no more than $2/kg or something less than $1/lb.

These cost figures place rather severe limitations
on the eventual choice of an economically competitive
storage chemical. If storage capacity per unit weight
is to remain high and the cost of synthetic modification
of basic industrial raw materials held to a minimum, then
the choice of chemicals for use on a large scale rests
with a relatively small number of abundant photochemically
active materials. Since it is not in general possible to
incorporate elaborate light harvesting chromophores in
isomerization substrates, photosensitization must be
employed.

Several inexpensive industrial chemicals have been
suggested (6,7) as candidates for relatively large scale
photochemical latent heat storage including dicyclopenta-
diene (9) and 1,5-cyclooctadiene (11) whose isomerizations
are shown below. The list may include a number of
norbornadienes (especially 1 and 5) whose large scale
manufacture should be straightforward.

The classical mechanisms of sensitization in solution
photochemistry involve bimolecular encounters of light-
capturing agent and reactant leading to either energy
transfer or electron transfer. An important rendition of
the latter is exciplex formation, the binding of sensitizer and substrate as an electron donor-acceptor pair.
Excited complexes, which depend for their existence on

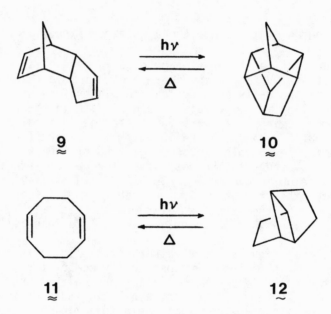

the powerful redox properties of excited states, have been
implicated in a host of photoreactions in a variety of
spectroscopic and photokinetic studies (32,33).

A number of rearrangements that qualify as exciplex
isomerizations have been identified (34-39). Importantly
these photoisomerizations are induced by electron donor-
acceptor complexation of a light absorbing agent with a
substrate for which low lying excited states do not exist
or are not deployed. In a recent publication (7) we describe the application of this principle to the activation
of a number of photoisomerizable non-conjugated dienes
which have energy storage potential. The mechanism proposed for exciplex isomerization is shown in Scheme 2.
The induction of isomerization may be viewed as a template
effect resulting from electronic polarization by the
excited sensitizer. Exciplex decay leading to photoisomer
is thought possible since either electron loss or electron
gain in the diene portion of the exciplex leads to trans-
annular bonding (40).

Scheme 2

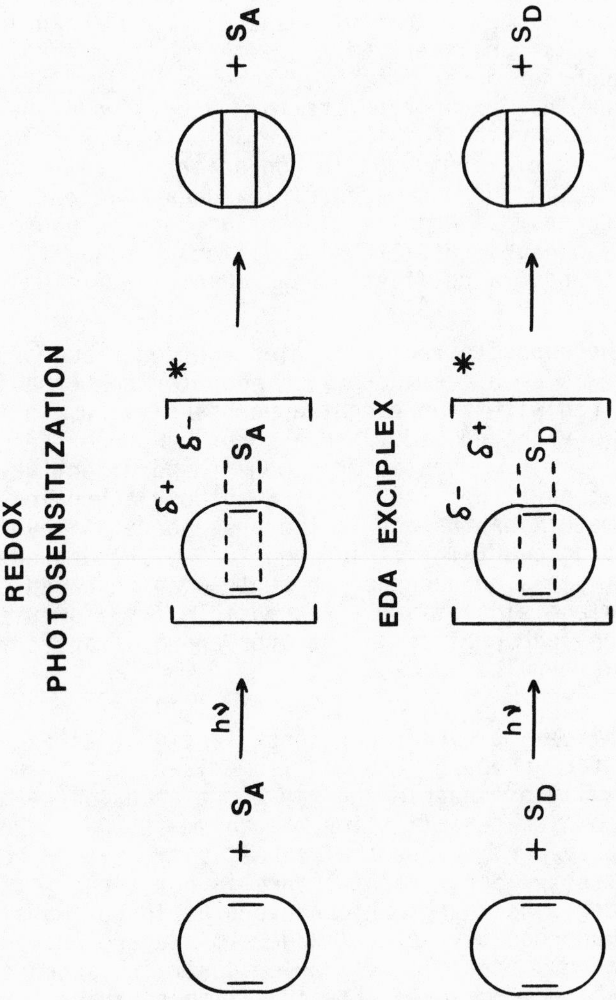
REDOX PHOTOSENSITIZATION
+ S_A
hν
δ+
δ−
S_A
*
+ S_A
EDA EXCIPLEX
+ S_D
hν
δ−
δ+
S_D
*
+ S_D

In our study a number of hydrocarbon dienes with
relatively low ionization potentials were paired with
aromatic photosensitizers. The electron donor dienes
were shown to quench the fluorescence of the electron
acceptor aromatics at variable rates which depended on
sensitizer and substrate ability to accept and donate
electrons. The falloff of quenching rate shown in
Figure 3 is consistent with the thermodynamics of electron transfer in an encounter complex (41). Independent
evidence that exciplexes are formed between dienes and
sensitizers came from the observation of weak, long wavelength exciplex emission in non-polar solvents. Unfortunately, in this series exciplex decay did not lead to
photoisomers. A competing reaction between photosensitizers and substrates (cycloaddition) was identified and
shown to have a relatively high quantum efficiency.

The opposite regime of electron donor-acceptor interaction was also examined (7). Acceptor norbornadiene 5
was paired with fluorescent donor sensitizers which displayed a range of redox and absorption properties. In
this series, two sensitizers were found effective.
Irradiation (290 - 340 nm) of 2-methoxynaphthalene or
2,6-dimethoxynapthalene in the presence of 5 gave 6 with
a limiting quantum efficiency of 0.3. The dependence of
quantum yield on quencher concentration revealed that
naphthalene singlets were responsible for isomerization.
Emission-reabsorption and Forster energy transfer mechanisms were ruled out.

This example of redox sensitization (42) via exciplexes for an energy storing isomerization (5 → 6) is
potentially extendable and may be related to the metal or
metal complex sensitization mechanism (25,26). However,
the improvement in spectral sensitivity of 5 in combination with the donor naphthalenes is not significant and
the redox sensitizers which absorb at longer wavelengths
led to unproductive exciplex decay. We are encouraged
nevertheless that all desirable substrates appeared to
form excited complexes with sensitizers having a range of
absorption properties. Two features of the exciplex
mechanism for diene isomerization should be kept in mind
in designing new systems. (1) Reactive positions in the
sensitizer may promote exciplex cycloaddition at the expense of exciplex isomerization. (2) Appreciable overlap

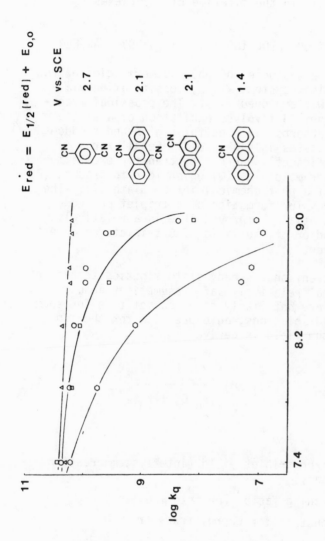

Figure 3. Stern-Volmer constants for the quenching of aromatic nitrile fluorescence in acetonitrile as a function of ionization potential of non-conjugated dienes (including 2, IP = 7.4 eV). Nitriles are shown opposite appropriate curves along with excited state reduction potentials (V vs. SCE).

of orbitals in non-conjugated dienes (more pronounced in
5 than in 9 or 11) may be required for nascent trans-
annular bonding within the lifetime of exciplexes.

PHOTOSENSITIZATION INVOLVING TRIPLET ENERGY TRANSFER

In the earliest studies of photoisomerization of the
parent norbornadiene system, 1 → 2, sensitization via
energy transfer was employed (43). The classical mechanism shown in Scheme 3 involves excitation of a sensitizer (S) which absorbs in a useful region and produces
triplet excited states in high yield. Sensitizer
triplets, if sufficiently energetic, transfer excitation
energy to quencher molecules (Q) which lead to product (P)
usually by way of a photochemical intermediate (I). The
quantum efficiency for formation of a triplet derived
product is substantially improved if Q is a relatively
poor absorber and/or if the yield of Q triplets on direct
irradiation is low.

Using a conventional steady state kinetics technique
and making several relatively safe assumptions about
processes which are not likely to be competitive for room
temperature fluid solutions, an expression for the isomerization quantum yield is derived.

$$\phi_{isom} = \phi_{isc} P_i \frac{k_q[Q]}{k_q[Q] + k_d}$$

where

ϕ_{isc} = sensitizer triplet yield (intersystem crossing efficiency)

P_i = partitioning factor for intermediate I

k_q = rate constant for energy transfer

k_d = rate constant for sensitizer triplet decay

Scheme 3

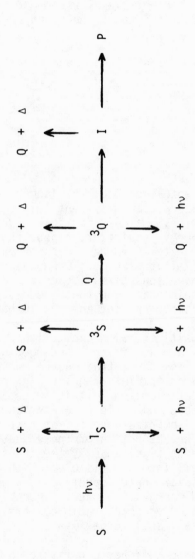

PHOTOSENSITIZATION - TRIPLET ENERGY TRANSFER

Inversion of this expression results in a function linear in reciprocal quantum yield and reciprocal quencher concentration, with an intercept to slope ratio relating bimolecular and unimolecular components of triplet decay (and triplet lifetime in the absence of quencher, τ_o).

$$\frac{1}{\phi_{isom}} = \frac{1}{P_i \phi_{isc}} + \frac{k_d}{P_i \phi_{isc} k_q [Q]}$$

$$\frac{i}{s} = \frac{k_q}{k_d} = k_q \tau_o$$

A concentration plot for isomerization $5 \rightarrow 6$ using two sensitizers with different triplet energies is shown in Figure 4. The almost negligible dependence on [5] for the high energy sensitizer, benzophenone, is consistent with a high rate of energy transfer quenching (near the diffusion limit). The more pronounced sensitivity for camphorquinone sensitized isomerization reflects a reduced quenching rate constant, no doubt the result of a lower sensitizer triplet energy. Other data shown in Table 3 confirm the dependence on sensitizer triplet level. Sensitizers with excitation energies 53 kcal/mol and greater are sufficiently energetic to drive the reaction at low concentrations of quencher whereas less robust excited triplets require high concentrations of 5 for efficient reaction. It is important to note that camphorquinone and acridine orange are effective sensitizers for isomerization of 5 at high substrate concentration. The absorption of these sensitizers just past 500 nm extends significantly the threshold of photoactivity well into the visible. With this system, 17% of the total solar energy spectrum is absorbed and about 20% of absorbed energy is stored. Indeed the low energy sensitizers bring about rapid conversion of 5 to 6 in sunlight in small scale experiments.

The norbornadienes constitute an interesting structure - reactivity series for photoisomerization utilizing photons of minimal energy. Previous studies have shown that efficient reaction of 1 and 3 requires sensitizer

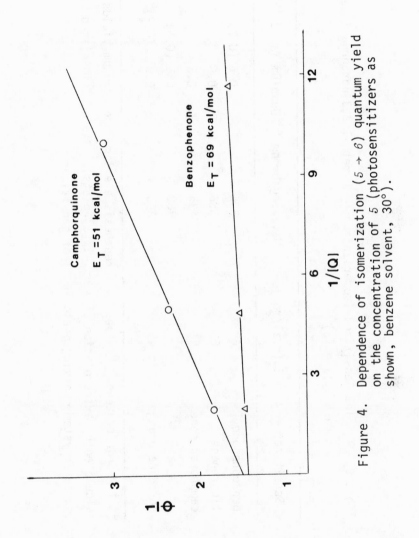

Figure 4. Dependence of isomerization ($5 \rightarrow 6$) quantum yield on the concentration of 5 (photosensitizers as shown, benzene solvent, 30°).

Table 3. Triplet Sensitized Isomerization of Dimethyl 2,3-norbornadiene-dicarboxylate (*5*)

Sensitizer	E_T(kcal/mol)	Wavelength, nm	Quantum Yield [a,b]
benzophenone	68	366	0.6 (0.7)
fluorenone	53	366	0.6
camphorquinone	51	435	0.3
		480	0.3 (0.6)
acridine orange	49	366	0.01 (0.2)[c]

[a]0.1 M *5* in benzene. [b]Values in parenthesis are limiting quantum yields calculated by extrapolation (1/φ vs. 1/[*5*] plots). [c]Values diminished by low intersystem crossing yield (0.4) of sensitizer.

triplet energies of 65 (10) and 58 (12) kcal/mol, respectively. From the data above, a minimum triplet energy of about 50 kcal/mol for isomerization of diester derivative *5* is indicated. Corresponding minimum singlet energies and absorption thresholds for *1*, *3*, and *5* would be about 70 (400), 63 (450), and 55 kcal/mol (520 nm).

Are these triplet energy minima insurmountable barriers to the use of longer wavelength light? We think not based on our analysis of the important parameters controlling energy transfer efficiency. Where energy transfer from sensitizer triplets to substrate is endoergic, the quantum yield of sensitized isomerization is sharply dependent not only on sensitizer triplet lifetime and quencher concentration as shown above but also on the temperature of the medium. Large temperature dependences of the rate of endoergic triplet energy transfer have been observed for several systems (44-45). We have used relationships developed in the pioneering energy transfer study of Sandros (44) to express the quantum yield of sensitized isomerization as a function of temperature as follows.

$$\phi_{isom} = \phi_{isc}P_i \frac{A_{df}e^{-E_{df}/RT}e^{-\Delta E_T/RT}[Q]}{A_{df}e^{-E_{df}/RT}e^{-\Delta E_T/RT}[Q] + A_d e^{-E_d/RT}}$$

where

A_{df}, E_{df} = Arrhenius parameters for diffusion rate constant

A_d, E_d = Arrhenius parameters for sensitizer triplet decay

ΔE_T = Sensitizer - quencher triplet energy difference

The results of a model study clearly reveal the salutary effect of higher temperatures on photoisomerization efficiency. Parameters were chosen as shown below defining a system with a rate constant for diffusive encounter of sensitizer triplets and substrate of

10^{10}M^{-1}sec^{-1}, a triplet lifetime of 10^{-5}sec (both at
25°), a quencher concentration of 1 M, a sensitizer intersystem crossing efficiency of 1.0, and an intermediate
(biradical) partitioning constant (limiting quantum yield)
of 0.7 (that found for 5).

$$A_{df} = 10^{12}M^{-1}sec^{-1} \qquad\qquad E_{df} = 3 \text{ kcal/mol}$$

$$A_{d} = 10^{7}sec^{-1} \qquad\qquad E_{d} = 3 \text{ kcal/mol}$$

The parameters are similar to those found for rates of
diffusion   and triplet decay (47) for relevant systems.   The temperature independence of ϕ_{isc} and P_i is
assumed, the latter substantiated by recent findings (29)
concerning the internal cycloaddition of an enone triplet.
Calculated isomerization quantum yields for three sensitizer triplet deficiencies (7-9 kcal/mol) for a 150° temperature range are shown in Figure 5.

The enhancements in quantum efficiency at elevated
temperatures (up to 25X) for sensitized isomerization of
the model system are impressive. We have sought to confirm this improvement in performance with reference to
the system $5 \rightarrow 6$. A sensitizer, benzanthrone, was chosen
whose triplet energy(E_T = 46 kcal/mol from emission (48)
and energy transfer (49) measurements) falls well below
that of 5 ($E_T \sim 53$ kcal/mol according to the camphorquinone and acridine orange sensitization results).
Quantum yield data for four concentrations and three
temperatures are shown in Figure 6. The combined effects
of temperature and concentration amount to enhancement
of quantum efficiency of ~ 50X (.005 $\rightarrow$ .21). We are presently extending the study to higher temperatures in order
to show that this reaction which stores an enormous fraction of excitation energy (19/46 = .42) can proceed with
high quantum efficiency.

The possibility of making up excited state energy
deficiency with thermal energy (as much as 10 kcal/mol)
is most attractive. If the "thermal upconversion" of
triplets is translated into absorption thresholds (singlet energies), the use of, for example, 610 nm light
(47 kcal/Einstein) to drive a reaction normally considered
to require 510 nm light (57 kcal/Einstein) is readily

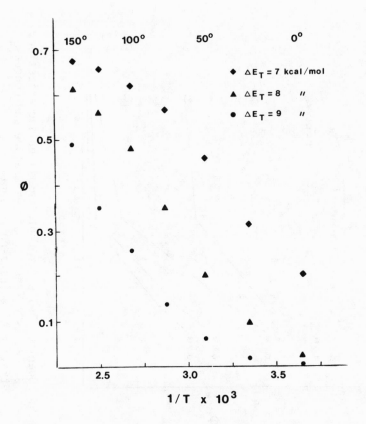

Figure 5. Dependence on temperature of the quantum yield of a model photosensitized reaction involving endoergic triplet energy transfer (for parameters, see text).

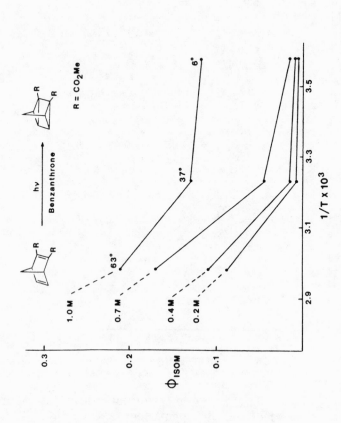

Figure 6.   Quantum yields for photosensitized isomerization,
$5 \rightarrow 6$, as a function of temperature and concentration of 5 (benzene solvent).

envisioned. This shift in wavelength to the red corresponds to an increase in fraction of total solar energy absorbed from 17 to 33% (50). The potential benefits in exploiting the elevated temperatures of solar collectors by using solar energy which is not absorbed by photochemical components but which heats the photoactive medium are largely unexplored. The sensitized organic isomerizations will provide an important test of performance and solar spectral response due to a fortunate coincidence of properties which include the suitability of long lived dye triplets for photosensitization, the desirability of high concentrations of substrate (for high volumetric storage capacity), and the persistence of photoproducts at well above ambient temperatures.

ACKNOWLEDGEMENT

We thank the Department of Energy, Division of Basic Energy Sciences, and the Office of Naval Research for support of our work.

REFERENCES

(1) F. Daniels, "Direct Use of the Sun's Energy," Yale University Press, New Haven (1964).

(2) J.A. Duffie and W.A. Beckman, "Solar Energy Thermal Processes," John Wiley and Sons, New York, 1974.

(3) M. Telkes, ASHRAE J., 38 (1974).

(4) T.A. Chubb, Solar Energy, 17, 129 (1975).

(5) W.H.F. Sasse, in "Solar Power and Fuels," J.R. Bolton, Ed., New York, 1977.

(6) G. Jones, II, T.E. Reinhardt, and W.R. Bergmark, Solar Energy, 20, 241 (1978).

(7) G. Jones, II, S.-H. Chiang, and P.T. Xuan, J. Photochem., in press.

(8) T. Laird, Chem. and Ind., 186 (1978).

(9)   S.G. Talbert, D.H. Frieling, J.A. Eibling and
      R.A. Nathan, Solar Energy, 17, 367 (1975).

(10) R.R. Hautala, J. Little, and E.M. Sweet, Solar
      Energy, 19, 503 (1977).

(11) K.B. Wiberg and H.A. Connon, J. Am. Chem. Soc., 98,
      5411 (1976).

(12) A.J.G. Barwise, A.A. Gorman, R.L. Leyland,
      P.G. Smith, and M.A.J. Rogers, J. Am. Chem. Soc.,
      100, 1814 (1978).

(13) G. Kaupp and H. Prinzbach, Helv. Chim. Acta, 52,
      956 (1969).

(14) D.S. Kabakoff, J.-C. G. Bunzli, J.R.M. Oth,
      W.B. Hammond, and J.A. Berson, J. Am. Chem. Soc.,
      97, 1510 (1975).

(15) J.F. Liebman and A. Greenburg, Chem. Rev., 76, 311
      (1976).

(16) H.E. Zimmerman, Accounts Chem. Res., 4, 272 (1971).

(17) K.C. Bishop, III, Chem. Rev., 76, 461 (1976).

(18) V. Balzani, L. Moggi, M.F. Manfrin, F. Bolletta, and
      G.S. Laurence, Coord. Chem. Rev., 15, 321 (1975).

(19) B. Carlsson and G. Wettermark, Solar Energy, 21, 87
      (1978).

(20) G. Jones, II and B.R. Ramachandran, J. Org. Chem.,
      41, 798 (1976).

(21) G. Jones, II and L.J. Turbini, J. Org. Chem., 41,
      2362 (1976).

(22) R.E. Schwerzel, R.J. Bartlett, J.R. Kelly,
      N.E. Losterman, R.A. Nathan, B.E. Sherwood,
      G.H. Stickford, and R.E. Wyant, 2nd International
      Conference on the Photochemical Conversion and Storage
      of Solar Energy, Cambridge, England, August 10-12,
      1978, abstract 1.2.

(23) T. Mukai and Y. Yamashita, Tetrahedron Lett., 357
      (1978).

(24) D.P. Fisher, V. Piermattie, and J.C. Dabrowiak,
      J. Am. Chem. Soc., 99, 2811 (1977).

(25) D.P. Schwendiman and C. Kutal, J. Am. Chem. Soc.,
      99, 5677 (1977).

(26) P. Grutsch and C. Kutal, J. Am. Chem. Soc., 99,
      6460 (1977).

(27) R.B. King and E.M. Sweet, J. Org. Chem., in press.

(28) R.J. Card and D.C. Neckers, J. Org. Chem., 43,
      2958 (1978).

(29) G. Jones, II and B.R. Ramachandran, J. Photochem.,
      5, 341 (1976).

(30) W.R. Bergmark, G. Jones, II, T.E. Reinhardt, and
      A.M. Halpern, J. Am. Chem. Soc., 100, 6665 (1978).

(31) A.W. Adamson, A. Vogler, H. Kunklely, and R. Wachter,
      J. Am. Chem. Soc., 100, 1298 (1978).

(32) M.S. Gordon and W.R. Ware, Ed., "The Exciplex,"
      Academic Press, New York, 1975.

(33) H. Beens and A. Weller, in "Organic Molecular Photo-
      physics," vol 2, J.B. Birks, Ed., John Wiley and
      Sons, New York, 1975.

(34) B.S. Solomon, C. Steel, and A. Weller, Chem. Commun.,
      927 (1969).

(35) S. Murov and G.S. Hammond, J. Phys. Chem., 72, 3797
      (1968).

(36) S.L. Murov, L.-S. Yu, and L.P. Giering, J. Am. Chem.
      Soc., 95, 4329 (1973).

(37) G.N. Taylor, Zeitschrift fur Physikalischer Chemie
      Neue Folge, 101, 237 (1976).

(38) S.L. Murov, R.S. Cole, and G.S. Hammond, J. Am.
      Chem. Soc., 90, 2957 (1968).

(39) R.S. Cooke and G.S. Hammond, J. Am. Chem. Soc., 90,
      2958 (1968).

(40) R. Hoffman, Accounts Chem. Res., 4, 1 (1971).

(41) D. Rehm and A. Weller, Israel J. Chem., 8, 259
      (1970).

(42) T. Majima, C. Pac, A. Nakasone, and H. Sakurai,
      J.C.S. Chem. Comm., 490 (1978).

(43) G.S. Hammond, P. Wyatt, C.D. DeBoer, and N.J. Turro,
      J. Am. Chem. Soc., 86, 2532 (1964).

(44) K. Sandros, Acta Chem. Scand., 13, 2355 (1964).

(45) W.G. Herstroeter, J. Am. Chem. Soc., 97, 4161 (1975).

(47) M. Berger, E. McAlpine, and C. Steel, J. Am. Chem.
      Soc., 100, 5147 (1978).

(48) N.A. Borisevich, V.V. Gruzinski, and A.A. Kotov,
      Izv. Akad. Nauk SSSR, Ser. Fiz., 34, 490 (1970).

(49) W.G. Herkstroeter and G.S. Hammond, J. Am. Chem. Soc.,
      88, 4769 (1966).

(50) F.A. Brooks and W. Miller, in "Introduction to the
      Utilization of Solar Energy," Ed., A.M. Zarem,
      McGraw-Hill Book Company, Inc., New York, 1963,
      chap 3.

NOVEL PHOTOCYCLIZATION REACTIONS OF 1-ALKENYL-2-PYRIDONES

Patrick S. Mariano, Timothy L. Rose
Andrea A. Leone, and Linda Fisher

Department of Chemistry, Texas A&M University
College Station, Texas 77843

INTRODUCTION

The ultimate importance of solar energy utilization in alleviating the world's energy shortage has been emphasized by many authors.[1-6] This renewable, clean and abundant source holds perhaps the greatest potential of any of the emerging energy alternatives to the combustion of fossil fuels. Predictions that up to 25% of the nation's energy demands in the year 2020 could be potentially supplied by solar energy[3] appear reasonable. Of the four major program units designated within the National Solar Energy Program, including direct thermal application, solar electric conversion, fuels from biomass and technology support and utilization, the first has reached the highest level of development and is expected to be the most significant of the solar technologies for the next 40 years. Many demonstration projects applying this technology to solar heating and cooling of buildings are already in operation or planned. In many cases, from 50% to 80% of the necessary energy for hot water and space heating and cooling can be supplied by systems employing solar energy collectors of the flat plate design.[7] A major area still in need of technological advances, however, is storage of the solar energy collected during daylight hours for later use at night or during extended periods of heavy cloud cover. Even more important would

299

be the development of methods for long term storage of
excess heat produced during the seasons of high solar
intensity for regeneration during winter months.

The direct storage of solar energy by photochemical
production of high energy organic compounds has received
relatively sparse attention compared to that given to
other methods of thermal energy storage.[8,9] Importantly,
the unique systems which utilize light-induced valence
isomerization reactions of organic substances producing
highly strained products combine some of the ideal characteristics of other methods for thermal storage such as
high capacity to volume ratios and the use of fluid
systems. Moreover, methods based on this design possess
other advantageous features including a closely controlled
release of energy, the lack of threshold levels associated with collection, insensitivity to ambient temperatures and, importantly, capabilities for extended if not
indefinite storage periods. In photochemical storage
systems, light is used to convert appropriately struc-
tured organic compounds to new, kinetically stable,
molecular systems of higher energy than the precursor.
This endoergonic process, thus, traps solar energy as
potential energy stored in the chemical bonds of the
photoproducts. This energy can be released by trans-
forming products back to reactants using specific catalysts which facilitate retroconversion at low temperatures. Therefore, the method of photochemical entrapment
of solar energy is designed to efficiently use the solar
spectrum to produce high energy, yet stable, compounds
($A+h\nu \rightarrow B$) which can be transformed to initial reactants
($B \rightarrow A+heat$) in exoergonic reactions. The organic components in this system act as working fluids in a cyclic
fashion. Importantly, the unused portion of the solar
spectrum can be harnessed for use in direct thermal
applications. Note, that the strategy here is not to
employ photochemical systems to produce an alternative
fuel for high temperature combustion as in fuels produced
from biomass[10,11] or direct production of hydrogen.[12,13]
Rather, these systems would be designed to serve as an
alternative to or complement of other methods of sensible
and latent heat storage for low temperature applications.

The discussion which follows focuses on two aspects
of solar energy storage using systems based upon organic

photochemical reactions. First, strategies for evaluation of potentially useful systems will be briefly outlined. Following this will be a detailed discussion of
results obtained from initial studies of a potentially
important class of energy storing photoreactions involving
the reversible isomerizations of 1-substituted-2-pyri-
dones.

ENERGY STORING PHOTOISOMERIZATION REACTIONS

Criteria useful in evaluating the efficiency of
solar energy storage as latent heat associated with high
energy organic substances can be easily established by
considering the individual steps of the process involved.
Indeed, the approach employing light-induced valence
isomerization or addition reactions of organic substances
to produce highly strained products can be analyzed in
this way. In essence, the overall efficiency of energy
conversion using the cyclic system A $\rightleftarrows$ B can be factored
into individual efficiencies for electronic excitation
and reaction of A and the efficiency and exothermicity
of reconversion of B back to A. The scientific criteria
for evaluation have been listed in several forms previously.[11,12] More recently, Jones and his co-workers[14]
have reviewed this topic in a reasonably comprehensive
fashion. Thus, answers to one of the major questions
concerning the feasibility of this methodology, i.e.,
solar to thermal energy conversion efficiency, can be
approached through consideration of (1) the absorption
characteristics of the organic substrate undergoing
direct electronic excitation or photosensitizer and the
special characteristics of sun light, (2) the quantum
efficiency of the photoisomerization process, (3) the
transmittance and physical characteristics of the solar
collector and its geographical and local location, (4)
the exothermicity of the reverse reaction, and (5) the
yields of the forward photochemical and reverse thermal
reactions. The last feature is exceptionally significant,
since if yields are less than 100%, the number of effective A $\rightleftarrows$ B cycles possible before replacement of the
organic fuel A is limited. Sasse[15] has pointed out that
a 1% yield of by-product(s) produced in each of the cycles
would after 10-cycles result in a decrease of 18.2% of
the reacting organic components. However, with a per-
cycle by-product yield of only 0.01%, the loss of material

would be as low as 2% after 100-cycles.

Solar Irradiation and Excitation

Solar irradiation reaching the earth's surface differs from that emitted by the sun due to attenuation by several factors.[11] For photochemical reactions of organic materials the region of the solar spectrum of importance lies between 300-700 nm since it is light in this wavelength region which causes production of electronic excited states capable of chemical reaction. Although light of these wavelengths is subject to scattering effects due to gases, dust and water, it is not absorbed by water vapor.[8,16,17]

Flat plate collectors appear ideally suited for employment with this type of solar energy conversion technology. Focusing collectors, while useful when attainment of high temperature is desired, are not needed for the photochemical system and their additional cost is not justified. A flat plate collector receives both direct and scattered radiation and, thus, will permit operation on cloudy days when only diffuse light is available. This feature will be important for energy storage using the photochemical system since a large portion of diffuse light lies in the short wavelength region with a maximum between 440-480 nm.[8,17] Lastly, the flat plate collector system designed for use in this system can be easily incorporated into more conventional technologies, such as hot water and photovoltaics, thus, potentially making the photochemical system compatible with others for a modularized approach to solar energy storage.

Values needed to determine the amount of available solar radiation for operation of the photochemical system, expressed as the average daily (direct and scattered) quantity of sunlight received on a horizontal surface measured in langleys (cal/cm^2), are recorded.[18] It is assumed that with a well-designed collector system these values will equal the amount of radiation penetrating the flat plate cover and impinging upon the surface of the solution containing the organic substance. The effects of cloud cover and daily intensity variations are incorporated in the available solar intensity data given in

Table 1. Also, for more practical calculations it is
convenient to express solar radiation intensities in
energy per unit area as BTU/m^2 (1 langley = 39.68 BTU/m^2).
It should also be noted that the values we have used in
calculations made to derive the results summarized below
are amounts of solar radiation incident upon a horizontal
surface. If a tilted surface is used, the amount of
radiation can be increased by approximately 50% depending
upon the tilt angle and season of the year. Therefore,
in terms of solar radiation the analysis will give a
lower limit for performance of the photochemical system.

Photochemical and Retro Processes

The desirable characteristics for effective storage
systems using photochemical methods have been discussed
by several groups.[8,19,20] The reaction systems considered in the brief introductory analysis given below
consist of substances capable of undergoing photochemical
isomerization or addition reactions following direct
light absorption or sensitized excitation using light
in the available solar spectral region. The reactions
lead to formation of photoproducts which are kinetically
stable and contain high energy chemical bonds. The retroconversion is effected either thermally or through the
use of catalysts.

The absorption characteristics of the photoreactants
or sensitizers used for their activation should as closely
as possible match the solar spectrum impinging on the
earth's surface. The efficiency of light collection, if
a direct absorption process is used to produce electronically excited states, would be related to that fraction
of the solar spectrum which the reactant's absorption
spectrum coincides with. If sensitization is employed,
in which a sensitizer molecule first absorbs light and
transfers electronic excitation to reactant via collision,
the efficiency of light collection will be dependent not
only upon the absorption characteristics of the sensitizer
but also upon the efficiency of energy transfer.

The sensitized or direct photochemical reaction
should have a high quantum efficiency which reflects the
partitioning of excited states produced to product vs.
other pathways leading back to reactant ground state

(radiationless decay and luminescence). This value for
simple isomerization reactions can approach the theoretical limit of unity. Lastly, the thermal or catalyzed
reconversion of photoproduct back to reactant must be
exoergonic. The enthalpy changes, calculated by differences in the heats of formation between product and
reactant and determined experimentally by calorimetry,
represents the stored energy of the system.

Heat Storage Capacities of Selected Organic Systems

A variety of organic systems capable of serving in
solar energy storage cycles have been subjected to
detailed investigation over the past decade. For many
of these, including the norbornadiene-quadricyclene
($\underline{1} \rightleftharpoons \underline{2}$),[21] anthracene -4+4-dimer ($3 \rightleftharpoons \underline{4}$),[22] linked anthra-
cene-cyclization product ($\underline{5} \rightleftharpoons \underline{6}$),[23] naphthalene-photodimer
($\underline{7}$ $\underline{8}$), and azepine-electrocyclization product ($\underline{9} \rightleftharpoons$
$\underline{10}$)[25] (Chart 1), it is possible to calculate the energy
storage capacity using simple procedures based upon the
criteria outlined above. A summary of this data is incorporated in Tables 1-3. A preliminary survey of this
data indicates that the photochemical system involving
the Michler's ketone sensitized conversion of norbornadiene to quadricyclene-Rh(I) catalyzed back reaction is
the most efficient of those explored to date in terms
of its ability to store latent heat. This evaluation
suggests that although the organic photochemical systems
explored to date do not possess proper efficiencies to
meet the daily energy requirements for building heating
and cooling, considerable quantities of energy can be
stored over the summer months in even non-ideal climates.
It can be tentatively concluded that the use of the
photochemical systems could allow for collection and long
term storage of useful quantities of latent heat. Further refinement of several of the system's variables
including collector modification (x1.5), enhanced reaction
efficiencies (x2), improved uv-visible absorption characteristics (x2), and greater reaction enthalpies (x2),
might lead to dramatic improvement in storage capacity and
could have an important impact upon latent heat, solar
energy storage methods.

Chart 1. Organic Photochemical Systems of Potential Use in Solar Energy Storage.

System		$\emptyset_r$	ΔH (kcal/mol)
 1	 2	0.50	-26.2
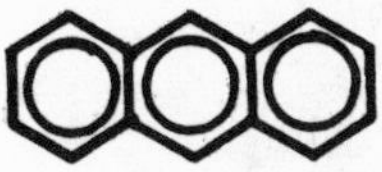 3	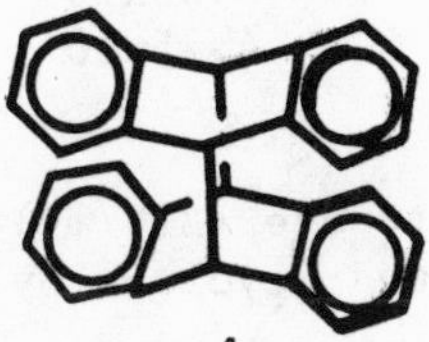 4	0.30	-15.5
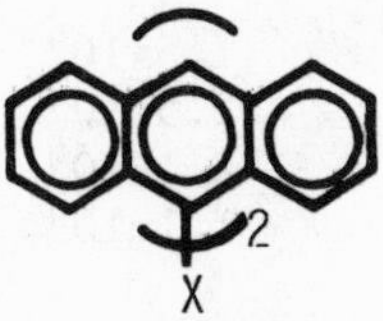 5	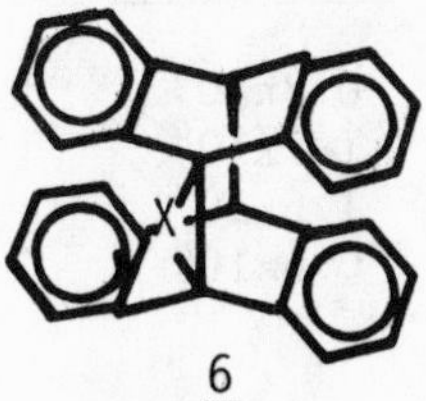 6	0.26	-14.6
 7	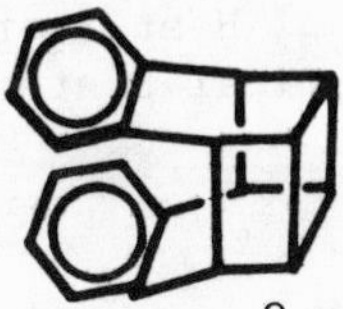 8	(0.30)	-13.6
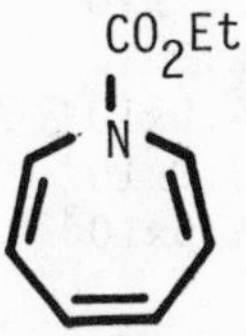 9	 10	0.013	-10.4

Table 1. Norbornadiene-Quadricyclene Storage Capacities.

Location	Annual Solar Irradiation Absorbed (mole photon/ $m^2 \cdot$year)	Annual Quadricyclene Production (mole/$m^2 \cdot$year)	Total Annual Heat Storage (BTU/$m^2 \cdot$year)
El Paso, TX	1.37×10^3	6.88×10^2	7.15×10^4
Geneva, NY	0.87×10^3	4.35×10^2	4.52×10^4
Bismark, ND	1.11×10^3	5.55×10^2	5.77×10^4
Seattle, WA	0.95×10^3	4.75×10^2	4.94×10^4

Table 2. Seasonal Heat Storage (BTU/m^2) for the Norbornadiene-Quadricyclene System.

Months	Geneva, NY	El Paso, TX
January–March	0.8×10^4	1.5×10^4
April–June	1.7×10^4	2.5×10^4
July–September	1.6×10^4	2.1×10^4
October–December	0.6×10^4	1.1×10^4

Table 3. Comparative Annual Heat Storage of Selected Organic Photochemical Systems for El Paso, TX.

System	Annual Heat Storage (BTU/m $\cdot$year)
Norbornadiene-Quadricyclene	7.2×10^4
Anthracene-Anthracene 4+4-Dimer	2.1×10^4
Linked Anthracene-Cyclization Product	4.0×10^4
Naphthalene-Naphthalene Dimer	3.0×10^4
Azepine-Electrocyclization Product	1.6×10^3

THE PHOTOCHEMISTRY OF 1-SUBSTITUTED-2-PYRIDONES

The results from evaluation of solar energy storage efficiencies of organic photochemical systems demonstrate both the potential of the method and the criteria which must be met by a practical system. Accordingly, explorations for new systems must be guided by the requirements summarized above for the foward photochemical and reverse thermal processes. Our recent investigations in this area have focused on an exploration of a new class of potentially useful, energy storing systems involving the reversible valence bond isomerizations of 1-substituted-2-pyridones. The characteristics of the electrocyclic closure of butadienes to cyclobutenes appear suitable for use in storage of solar energy as latent heat contained within the strained cyclobutene ring system. The kinetic stability and thermodynamic instability of cyclobutenes can be enhanced simultaneously by inclusion of the grouping within a small, cis-fused bicyclic ring system, exemplified by 11. This structural constraint would retard thermal electrocyclic opening to the butadiene 12, which occur with low activation energy only when orbital symmetry allowed conrotatory pathways are available.[26] Likewise, fused-systems containing the cyclobutene moiety should possess a greater degree of strain energy, especially when the annelated ring is small (11, n=1 or 2). Moreover, the conversion of cyclobutenes to butadienes can be made more exoergonic if concomittant rearo-

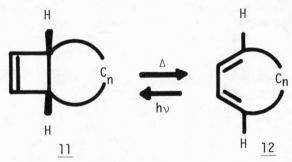

matization occurs, as demonstrated by the comparative enthalpies for the conversion of cyclobutene to butadiene (ΔH = -24 kcal/mol) and hexamethyl-Dewar-benzene to hexamethylbenzene (ΔH = -62 kcal/mol).[27] Lastly, placement of the butadiene moiety into a more extensively conjugated chromophore should enhance longer wavelength, direct light absorption characteristics of these substances.

An accumulation of data suggest that reversible transformations of 1-substituted-2-pyridones (13) and Dewar-pyridones (14) could serve as ideal candidates for initial exploratory studies targeted at the development of practical solar energy storage systems based upon the butadiene-cyclobutene interconversions. Previous work[28] had demonstrated that 2-pyridone and its N-methyl derivative undergo interesting photocyclizations to produce the

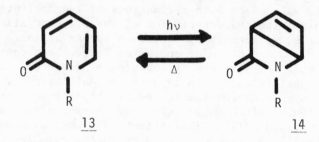

13   14

kinetically stable, Dewar-pyridones. Investigations in our laboratory[29] showed further that incorporation of simple vinyl substituents on nitrogen of these systems leads to shifts of the uv-maxima of these compounds extending further into the solar spectrum (>300 nm) (Table 4). Several additional features of the 1-alkenyl-2-pyridone systems appeared initially attractive. These substances are readily prepared from the available 2-hydroxy-pyridine[29] and most of them possess high water solubility. The latter characteristic would be advantageous for use

Table 4. UV-Spectroscopic Data for a Series of 1-Alkenyl and Related 2-Pyridones (13).[29]

1-Substituent	UV max (ε)
$-CH_3$	301 nm (5,500)
$-CH=CH_2$	322 nm (6,300)
$-CH=CHCH_3$ (trans)	318 nm (6,300)
$-CH=C(CH_3)_2$	308 nm (5,400)
$-CH=CHPh$ (trans)	339 nm (13,000)
$-C_6H_5$	300 nm (7,100)

in flat plate collector systems located on top of or near dwellings. Lastly, the retro processes converting the Dewar-pyridones back to 2-pyridone reactants, although not yet subjected to calorimetric analysis, should be highly exoergonic. Importantly, it is possible that the retroprocess can be both catalyzed and made more exo-ergonic by the addition of acids, as a result of initial production of the aza-Dewar benzene iminium ion 15.

On the basis of this preliminary reasoning, a detailed study of the photochemistry of 1-alkenyl-2-pyridones was initiated. Our interests were to determine the specificity of the Dewar-pyridone forming reactions of these substances and to gather information about the effects of solvent and structure on the efficiencies of the process. During the course of these studies, we have uncovered several interesting aspects of these systems which relate both to the development of solar energy storage utilizations and to the general area of organic photochemistry.[30]

Results of Photochemical Studies

Preparative Photochemistry. Irradiations of solutions of 1-(2-methyl-1-propenyl)-2-pyridone (17)[29] and 1-(1-trans-propenyl)-2-pyridone (18)[29] in a variety of non-polar organic solvents including tetrahydrofuran and benzene led to clean production of the isobutenyl- and trans-propenyl-Dewar-pyridones (19 and 20) and crystalline 4+4-dimers (21 and 22). The Dewar-pyridone to 4+4-dimer product ratio, as expected, was found to be dependent upon the concentration of starting alkenylpyridone; maximal yields of ca. 60-80% for 19 and 20 were obtained using concentrations of 17 and 18 of ca. 4 mM while those for the 4+4-dimers were ca. 30% starting with alkenylpyridone concentrations of <u>ca</u>. 30-50 <u>mM</u>. Low conversion irradia-

17 R₁ = R₂ = CH₃   19   21
18 R₁ = CH₃, R₂ = H   20   22

tion of the trans-propenylpyridone 18 in these solvents
results in cis-trans isomerization.[29] However, none of
the corresponding Dewar-pyridone or 4+4-dimer products
from 18 have the cis configuration about the exocyclic-
propenyl π-bond.

The photochemistry of 17 and 18 is markedly altered
when irradiations are conducted using water or wet methanol as solvent. Accordingly, both 17 and 18 are transformed in a reasonably efficient and high yielding (<u>ca</u>.
40-80%) manner to their corresponding 1-(2-pyridonyl)-
ethanol derivatives 23 and 24 when irradiated in water
using Pyrex-filtered light.

17 or 18 →(hv / H₂O)   23 R₁ = R₂ = CH₃
                        24 R₁ = CH₃, R₂ = H

Interesting changes are noted when the courses of
the aqueous solution photoreactions of 17 and 18 are
monitored using uv spectrophotometric methods. Shifts in
the uv maxima of the respective photolysates derived from
17 and 18 were as follows: 309 and 302 nm before irradiation, 288 and 287 nm immediately after irradiation, and
305 and 303 nm after concentration <u>in vacuo</u>. In addition,
the long wavelength shifts in absorption maxima detected
when the crude photolysates were concentrated could be

mimicked by the addition of excess potassium hydroxide.
These observations suggest that the initial photochemical
step of each reaction leads to generation of a low wavelength absorbing intermediate which is converted to alcohol in a secondary dark reaction occurring during the
room temperature concentration process and, possibly,
involving hydroxide ion. An understanding of these
observations was gained when irradiations of 17 and 18
were performed on aqueous solutions containing 6-fold
molar excesses of perchloric acid. Under these conditions
45-50% yields of two crystalline salts were obtained by
chloroform trituration of the concentrated photolysates.
These substances were shown to have physical and chemical
properties consistent with their assignment as 5,5-
dimethyl- and 5-methyloxazolo[3,2-a]pyridinium perchlorates, 25 (uv_{max} 285 nm) and 26 (uv_{max} 284 nm), respectively. Identical materials could be generated by
irradiation of 17 and 18 in aqueous formic acid followed
by perchlorate ion exchange using Dowex 1-X8.

Further supportive evidence for the structural
assignments of the salts 25 and 26 was gained using
chemical methods. Reactions in methanolic potassium
hydroxide at room temperature led to smooth production
of the respective pyridonylethanol derivatives 23 and 24.
Thermolyses of the pyridinium perchlorate salts at >150°C
led to rapid generation of the known[8] methallyl- and
allylpyridones 27 and 28. Lastly, both 25 and 26 were
efficiently converted to the 1-alkenyl-2-pyridones, 17
and 18 (mixture of cis and trans-isomers), when stirred
with potassium t-butoxide in t-butyl alcohol.

Dark control reactions of the alkenylpyridones 17
and 18 were attempted using conditions which mimic those
present during the photochemical reactions in water or
aqueous acid. In no cases were the pyridonylethanol
derivatives or pyridinium perchlorates detected and
unreacted starting materials were recovered quantitatively.
In addition, irradiations of 1-vinyl-2-pyridone (28),[31]
using conditions found to be ideal for photoconversion of
17 and 18, did not produce isolable quantities of 1-(2-
pyridonyl)-ethanol (32) or the oxazolo[3,2-a]pyridinium
salt 33.

Mechanistic Aspects. Triplet photosensitized reac-

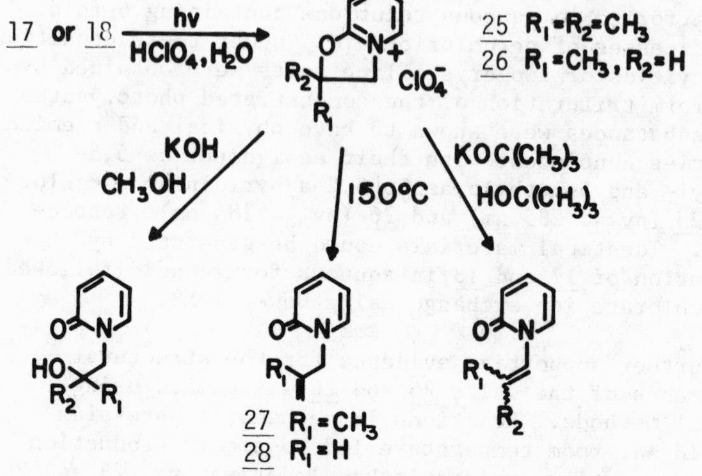
17 or 18
hv
HClO4, H2O
O
N+
R2
R1
ClO4-
25 R1=R2=CH3
26 R1=CH3, R2=H
KOH
CH3OH
150°C
KOC(CH3)3
HOC(CH3)3
O
N
HO
R2
R1
O
N
R1
27 R1=CH3
28 R1=H
O
N
R1
R2

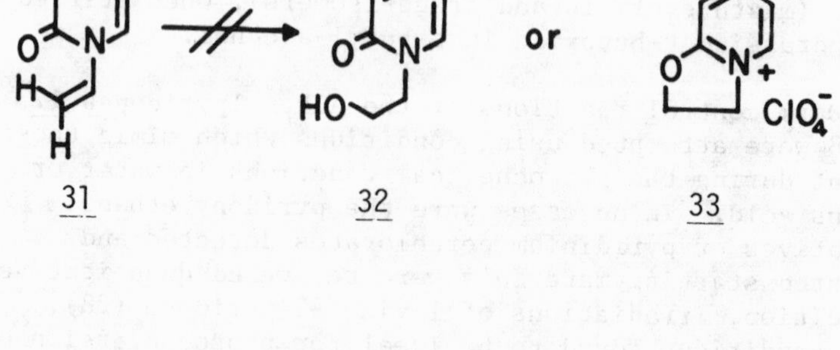
O
N
H
H
31
O
N
HO
32
or
O
N+
ClO4-
33

tions of isobutenylpyridone 17 were attempted using xanthone (E_T = 74.2 kcal/mol) and Michler's ketone (E_T = 62 kcal/mol) as sensitizers in wet methanol as solvent and employing conditions which guaranteed light absorption by the sensitizer and high triplet energy transfer efficiencies. In each case, unreacted starting material was quantitatively recovered. It is certain that energy transfer from triplet xanthone to 17 was occurring, since results from parallel irradiations of methanolic solutions of xanthone and 17 and xanthone alone indicated the ability of 17 to quench the familiar triplet photoreduction reaction.

Additional information necessary to assign the singlet as the reactive 1-alkenyl-2-pyridone excited state undergoing photocyclization and alcohol production was gained from quenching studies using 2,5-dimethyl-2,4-hexadiene, a known singlet quencher. Photoreactions of 17 in methanolic acetonitrile solutions containing this diene in concentrations ranging from 0 to 0.5 $\underline{M}$ were conducted under simultaneous irradiation conditions. A plot of the ratio of unquenched to quenched relative quantum yields for alcohol 19 production (ϕ_{ro}/ϕ_{rq}) vs. diene concentration gave a straight line with slope approximately equal to those obtained from plots of corresponding unquenched to quenched fluorescence quantum yield and singlet lifetime ratios.

33 $R_1 = R_2 = CH_3$
34 $R_1 = CH_3, R_2 = H$

In an attempt to further explore the mechanistic details of the novel photochemical reactions observed, the 1-alkenyl-2-hydroxypyridinium perchlorates, 33 and 34, were independently prepared from the corresponding pyridones by treatment with perchloric acid. Irradiation of aqueous methanolic solutions of 33 and 34, containing 0.54 $\underline{M}$ perchloric acid to insure >99% of the pyridinium salt form, led to complete consumption of these substances. However, none of the corresponding oxazolo-

[3,2-a]pyridinium salts were formed in either case. Likewise, the methallyl- and allylpyridones, 27 and 28,
remained unreacted when irradiated in water for time
periods which would have led to complete reaction of 17
and 18.

On the basis of the results presented thus far, it
is possible to conclude that the singlet transformations
to pyridonyl alcohols and oxazolopyridinium salts require
that the starting pyridone contain an N-vinylamide grouping. An important exception to this is the parent 1-
vinylpyridone which is lacking alkyl-substitution at the
terminal carbon of the exocyclic vinyl grouping. In
order to clarify the apparent critical role played by
alkyl substituents, information about the relative
efficiencies for reaction of the cis- and trans-propenyl-
pyridones was sought. This turned out to be quite difficult owing to the competitive cis-trans isomerization
process which efficiently interconverts both isomers.
However, qualitative evidence was obtained through varying conversion photolyses of the isomeric propenylpyri-
dones. The data obtained demonstrates that transformation of the cis-isomer to alcohol 24 is slightly more
efficient than that of the trans-isomer.

Lastly, the divergent photochemical behavior of the
1-alkenyl-2-pyridones in non-polar organic solvents as
compared to methanol and water required closer inspection.
Dewar-pyridone 19 and alcohol 23 production from fixed-
time irradiations of 17 were monitored as a function of
the amounts of water in the tetrahydrofuran solutions
used. The data obtained from these experiments, plotted
in Figure 1, clearly shows that the efficiency of alcohol
production increases at the expense of Dewar-pyridone as
the percentage of water in the aqueous tetrahydrofuran
solvent is increased. In addition, the quantum efficiency
for production of Dewar-pyridone 19 from isobutenylpyridone 17 was found to decrease dramatically as solvent
polarity increases (Table 5).

Quantitative Photochemical and Photophysical Data.
Quantum efficiencies (ϕ_r) for alcohol 23 and pyridinium
salt 25 production from isobutenylpyridone 17 were measured. The data obtained from irradiations of 17 in
solvents containing varying water and perchloric acid

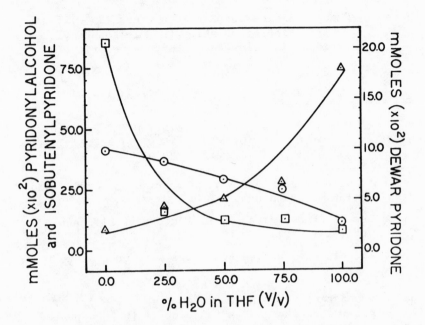

Figure 1. The relative efficiencies for formation of
isobutenyl-Dewar-pyridone (◻), 1-pyridonyl-2-methyl-2-
propanol (Δ), and disappearance of isobutenyl-2-pyridone
(○) as a function of water concentration in aqueous-
tetrahydrofuran solutions.

concentrations (Table 6) show that ϕ_r increases dramatically as either water or acid concentrations increase
(see also Figure 2). In order to examine the source of
these changes, fluorescence quantum yields (ϕ_f) and singlet lifetimes (τ) for 17 were measured under conditions
identical to those employed for the ϕ_r determinations.
ϕ_f and τ were found to be dependent upon $[H_2O]$ but to be
invariant with changes in $[H_3O^+]$ (Table 6). In addition,
through the use of this data it is possible to calculate

Table 5.   Quantum Yields and Rate Constants (25°C) for
           Dewar-Pyridone Formation From 1-Isobutenyl-2-
           pyridone in Solvents of Varying Polarity.

Solvent[a]	Solvent E_T-Value[b]	ϕ_{DP} 1 → 3	τ (nsec)[c]	k_{DP} (25°C)(sec^{-1})
CH$_3$OH	37.4	0.061	7.3±1.0	8.36x10^6
CH$_3$CN	46.0	0.143	16.6	8.61x10^6
THF	55.5	0.260	30.0	8.67x10^6

[a]Solvents were rendered anhydrous by rigorous drying and
distillation prior to use.

[b]E_T-values are those given by K. Dimroth et al., Ann.
Chem., 661, 1 (1962).

[c]Degassed, nitrogen purged solutions were used at 25°C.

the rate constant for fluorescence of 17 ($k_f = \phi_f/\tau$).
The changes noted in ϕ_f as a function of [H$_2$O] appeared
interesting and worthy of further exploration. Accordingly, the fluorescence efficiencies of 17 in a variety
of single-component solvents (Table 7), and ethanol-water
mixtures of differing polarity (Table 8), were recorded.
This data reveals a clear inverse dependence of ϕ_f on
solvent polarity.

DISCUSSION

Mechanistic Features. Several possible mechanisms
can be envisaged to rationalize the photochemical reactivity of 1-alkenyl-2-pyridones in polar protic solvents.
It appears almost certain that the initial photoproducts
from irradiations of 17 and 18 under these conditions
are oxazolo[3,2-a]pyridinium salts. When produced in
acid solution, these salts (25 and 26) possessing the
non-nucleophilic perchlorate counter ion are sufficiently
stable to survive isolation and purification. The primary photoproducts produced from reactions of 17 and 18
in water have uv spectroscopic characteristics indicative
of cyclized pyridinium salt structures 35 and most probably contain hydroxide as the counter ion (Z = OH).
The secondary dark reactions leading to the pyridonylethanol products, occurring upon concentration of or by

Table 6. Experimental and Calculated Photophysical and Photochemical Properties of 1-Isobutenyl-2-pyridone.

Water Concentration[a] (M)	HClO$_4$ Concentration (mM)	Singlet Lifetimes, τ (nsec)	Fluorescence Quantum Yields[b] ϕ_f	Reaction Quantum Yields[c] ϕ_r	Fluorescence Rate Constants $k_f(sec^{-1})$
27.7	-0-	10.3±1.0	0.007±0.001	0.071	7x10^5
50.0	-0-	6.5±1.0	0.004	0.110	6x10^5
55.6	-0-	5.1±1.0	0.003	0.130	6x10^5
55.6	0.50	--	--	0.160	--
55.6	1.00	--	--	0.190	--
55.6	2.00	--	--	0.213	--
55.6	3.00	--	--	0.248	--
55.6	5.00	5.1±1.0	0.003	0.315	6x10^5

[a]Acetonitrile was used as co-solvent for solutions containing less than 55.6 M water.

[b]Fluorescence quantum yields were measured using degassed, nitrogen purged solutions at room temperature. Benzene and naphthalene were used as fluorescence standards.

[c]Reaction quantum yields were measured for product production using degassed, nitrogen purged solutions and an optical bench apparatus. Alcohol product analysis was performed by glc using internal standards. The pyridinium salts obtained from irradiation of acid solutions were first converted to alcohol by treatment with base before analysis. Conversions to product were in the range of 11-17%. Estimated error for ϕ measurements is 5%.

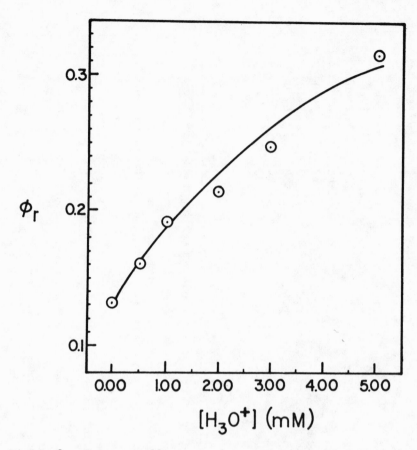

Figure 2. Quantum efficiency for production of 5,5-di-
methyloxazolo[3,2-a]pyridinium perchlorate as a function
of hydronium ion concentration from irradiation of 1-(2-
methyl-1-propenyl)-2-pyridone in aqueous perchloric acid
solutions.

addition of hydroxide to solution containing the pyridinium
salts 35 most probably involve nucleophilic attack at the
pyridinium C-2 position followed by uncoupling of the
hemiketal linkage in the transient dihydropyridine 36.

Table 7. Fluorescence Quantum Yields for 1-Isobutenyl-
2-pyridone in Single Component Solvents of
Varying Polarity.

Fluorescence Quantum Yields, ϕ_f[a]	Solvent	Solvent Z-Value[b]
0.003	H_2O	94.6
0.017	EtOH	79.6
0.024	DMF	68.5
0.031	DMSO	71.1

[a]Solutions used for these measurements were degassed and nitrogen purged. Benzene and naphthalene were used as fluorescence standards. Estimated errors are ±0.001.

[b]Z-Values for solvents are those reported by Kosower, J. Amer. Chem. Soc., 80, 3253 (1958).

Table 8. Fluorescence Quantum Yields for 1-Isobutenyl-
2-pyridone in Aqueous Ethanol Solutions.

Fluorescence Quantum Yields, ϕ_f[a]	Ethanol Content (%, V/V)	Solvent Y-Value[b]
0.017	100	−2.033
0.016	90	−0.747
0.015	80	0.000
0.014	70	0.595
0.013	60	1.124
0.012	50	1.660
0.011	40	2.196
0.009	30	2.720
0.007	20	3.051
0.005	10	3.310
0.003	0	3.493

[a]Solutions used for these measurements were degassed and nitrogen purged. Benzene and naphthalene were used as fluorescence standards. Estimated errors are ±0.001.

[b]Y-Values are those derived by Grunwald and Winstein[32] using measured rate constants for solvolysis on t-butyl chloride in aqueous ethanol solutions.

The more important mechanistic question, however, concerns the nature of the pathway(s) used by singlet 1-alkenyl-2-pyridones for transformation to oxazolopyridinium salts. Simple photohydration[33] of the exocyclic vinyl groupings in 17 and 18 followed by dehydrative cyclization appears an unlikely possibility since the pyridonyl alcohols 23 and 24 show no tendency to cyclize to the pyridinium salts in aqueous or dilute acid solutions used for photoreactions.

Another mechanism, worthy of more serious consideration, finds analogy in the results of studies by Schmid[34] and Horspool[35] of the photocyclization reactions of o-allylphenols to benzodihydrofurans. These transformations are most probably due to the enhanced facilities for intramolecular proton transfer in the excited state manifold of the allylphenols. Accordingly, oxazolopyridinium salts may arise from analogous pathways involving intramolecular proton transfer in initially formed singlet excited hydroxypyridinium salts followed by rapid cyclization of the intermediate pyridonyl ethyl cations 37. Although this mechanism appears attractive since it nicely serves to rationalize the large differences in reactivity between the isobutenyl and vinylpyridone systems, several critical observations appear to limit its serious consideration. The concentrations of 33 and 34 in water or dilute acid solution should be exceptionally low due to the large pK_b of N-substituted pyridones (pK_a's of 1-methyl-2-hydroxypyridinium salts are 0.3).[36] Ultraviolet measurements on aqueous solutions of 17 and 18, and 33 and 34 strongly support this expectation and further demonstrate that the amount of light competitively absorbed by the lower wavelength absorbing pyridinium salts under

33 or 34　　　　　37

the reaction conditions would be exceedingly low. Likewise, proton transfer from water to singlet excited 1-
alkenyl-2-pyridones as a mechanism for formation of 33
and 34 can be deemed improbable on the basis of the
results of Briggs and coworkers,[37] which show that the
singlet state pK_a's of O-protonated-2-pyridones are in
the range of -4, and our observations which demonstrate
that fluorescence of 17 is not quenched by acid. Lastly,
the hydroxypyridinium perchlorates 33 and 34 do not serve
as photochemical precursors of the oxazolopyridinium salts.
Likewise, the methallyl- and allyl-2-pyridones, 27 and 28,
which are closer analogs of the allylphenol systems
studied earlier[34,35] are also unreactive when irradiated
in water or aqueous acid solutions.

In considering other likely candidates for the mechanism operating in converting singlet 1-alkenyl-2-pyri-
dones to oxazolopyridinium salts, it is instructive to
consider the structural and electronic make-up of the
center in these substances at which reaction is occurring.
The N-vinylamide moiety 38, comprising the exocyclic
alkenyl and endocyclic amide groupings, is isoelectronic
with chromophores found in di-vinyl and di-aryl amines,
ethers and sulfides all of which possess patterned excited
state reactivity.[38-40] Thus, the related N-vinylamide
function when present in appropriate environments established by structural features and solvent, could undergo
electrocyclic closure to produce oxygen substituted,
cyclic azomethine ylids 39.[41]

Based upon these thoughts, we suggest that the conversion of 1-alkenyl-2-pyridones to oxazolo[3,2-a]pyridin-
ium salts can best be explained by a mechanism involving

initial singlet excited state electrocyclizations to
generate the pyridinium ylids 40, The ylid intermediates
could then be rapidly trapped by proton transfer from
water or hydronium ion, or in competition could revert
to starting pyridone by the retro-process. Inclusion of
the latter pathway for ylid disappearance is dictated by
our observations on the reactions of 25 and 26 with potassium t-butoxide which most probably proceed via 40.
Additional support for the ring opening process comes from
studies by Huisgen and his co-workers[41b] that demonstrate
a similar fate for monocyclic ylids of general structure
39. Although not explored experimentally, ylid formation
and return to ground state pyridone might well be the
major if not exclusive pathway for singlet cis-trans
isomerization of 1-alkenyl-2-pyridones or, more generally,
of N-vinylamides.[42]

Kinetic Analysis. As is suggested by the mechanism
postulated above, the quantum efficiency for conversion
of 1-alkenyl-2-pyridones to oxazolo[3,2-a]pyridinium salts
should be dependent on the nature and concentration of
proton donors in solution. Partitioning of the ylid to
pyridinium salt rather than to ground state pyridone

should be enhanced by inclusion of higher concentrations of acid in the reaction medium since proton transfer rate constants (k_p) from H_3O^+ to 40 are expected to be close to diffusion controlled ($k_{diff}(H_2O) = 1 \times 10^{10}$ $M^{-1}sec^{-1}$).[43] In contrast, the bimolecular rate constant for protonation of 40 by H_2O (k_{po}) should lie in the range of 1×10^6 $M^{-1}sec^{-1}$ at room temperature[43] due to an expected pK_a of 25 and 26 of ca. 10.[44] Indeed, as inspection of Figure 2 indicates, the quantum efficiency for production of 25 from 17 (ϕ_r) rapidly increases as perchloric acid concentration in the range of 0 to 5 mM is increased.

The dependence of quantum efficiency on $[H_3O^+]$ is expressed mathematically in the relationship derived for ϕ_r (eq. 1) using the kinetic sequence shown in Chart 2, steady state assumptions for 17^{S1} and 40, and the definition of $\tau = 1/(k_d+k_{DP}+k_f+k_c)$. It is instructive to point out here that both the singlet lifetime (τ) and fluorescence quantum yield (ϕ_f) of 17 are independent of $[H_3O^+]$,

Chart 2.

$$17^{S0} \xrightarrow{h\nu} 17^{S1}$$

$$17^{S1} \xrightarrow[k_d]{} 17^{S0}$$

$$17^{S1} \xrightarrow[k_f]{} 17^{S0}$$

$$17^{S1} \xrightarrow[k_{DP}]{} 19$$

$$17^{S1} \xrightarrow[k_c]{} 40$$

$$40 \xrightarrow{k_{ret}} 17^{S0}$$

$$40 + H_2O \xrightarrow{k_{po}} 25$$

$$40 + H_3O^+ \xrightarrow{k_p} 25$$

suggesting that alternate explanations for the dependence of ϕ_r on acid concentration involving protonation of 17^{S1}

cannot be invoked. This kinetic treatment provides us
with a rare opportunity to disect out the individual rate

$$\frac{1}{\phi_r} = \frac{1}{\tau k_c} + \frac{k_{ret}}{k_c} \left[\frac{1}{k_{p_o}[H_2O] + k_p[H_3O^+]} \right] \quad (1)$$

constants for $\underline{17}^{s1}$ cyclization (k_c), ylid $\underline{40}$ return to $\underline{17}$
(k_{ret}), and protonation of $\underline{40}$ by H_2O (k_{po}) and thus to
obtain nearly all the rate constants for processes in-
volved in deactivation and reaction of $\underline{17}^{s1}$. Accordingly,
non-linear least squares analysis of the ϕ_r $\underline{vs.}$ $[H_3O^+]$
data provided in Table 6 gives an excellent fit to equation
1 when the values for k_c, k_{ret} and k_{po} are those shown
in Table 9.

Several important aspects of this kinetic analysis
require comment. The nearness of the value found for k_{po}
to that predicted using the Eigen formulation,[43] inter-
relating the rate constant for proton transfer from H_2O
to $\underline{40}$ to the bimolecular diffusion rate constant, K_w and
K_a of the conjugate acid, suggests that the statistical

Table 9. Summary of the Rate Constants (25°C, H_2O) of
Processes Used in Deactivation and Reaction of
Isobutenylpyridone ($\underline{17}$).

$$k_f = 6 \pm 1 \times 10^5 \text{ sec}^{-1}$$

$$k_c = 3.5 \pm 2 \times 10^8 \text{ sec}^{-1}$$

$$k_{DP} = 8.5 \times 10^6 \text{ sec}^{-1}$$

$$k_d^a = 3.2 \times 10^7 \text{ sec}^{-1}$$

$$k_{ret} = 3.9 \pm 3 \times 10^8 \text{ sec}^{-1}$$

$$k_{p_o} = 6.3 \pm 3 \times 10^5 \text{ M}^{-1} \cdot \text{sec}^{-1}$$

$$k_p^b = k_{diff} = 1 \times 10^{10} \text{ M}^{-1} \cdot \text{sec}^{-1}$$

---

[a] Derived from $k_d + k_{DP}$ value found using solvent associated linear free energy treatment of ϕ_f data and
value of k_{DP}.

[b] Assume for proton transfer from H_3O^+ to $\underline{40}$.

method employed to obtain the individual rate constants
is reliable. The comparison between the value derived
for k_c, although having associated with it a reasonably
large error, and $1/\tau$ ($k_d + k_{DP} + k_f + k_c$) demonstrates
that the major pathway for 17^{S1} decay in aqueous solution
involves cyclization to the dipolar intermediate, 40.
Thus, the postulate made above that a significant contribution to the mechanism for cis-trans isomerization of 18
is through the ylid 40 is convincing. One of the more
surprising features of these results is the exceptionally
large rate constant found for return of the photochemically generated oxazolopyridinium ylid to starting alkenylpyridone.

Solvent Polarity Effects. One of the more significant consequences of observations made during the course
of our investigations relates to the dramatic role played
by solvent in controlling the photochemical reaction pathways followed by 1-alkenyl-2-pyridones. We have shown
that direct irradiation of 17 and 18 in solvent systems
of low polarity, such as CH_3CN and C_6H_6, leads exclusively
to formation of Dewar-pyridone and 4+4-dimer products, in
a concentration dependent ratio, and that pyridonylethanol derivatives result from reactions conducted in wet
methanol and water. Moreover, the efficiency of iso-
butenyl-Dewar pyridone 19 production is reduced in a continuous fashion as $[H_2O]$ in H_2O-THF solutions is increased.
Simultaneously, the efficiencies of disappearance of 17
and production of 23 are increased as $[H_2O]$ is increased.
Moreover, the quantum yield for Dewar-pyridone production
from 17 decreases dramatically as solvent is varied from
the less polar THF to the more polar CH_3CN and CH_3OH.
It is quite likely that these observations are reflective
of an interesting effect of solvent polarity on partitioning of 17^{S1} to ylid 40 and Dewar-pyridone 19, i.e., on
the k_{DP}/k_c ratio. This effect is almost certainly due
to changes in the rate constant (k_c) for formation of
the dipolar ylid caused by variation in medium polarity.
Hammond and Sharp[28d] have shown that the quantum yield for
Dewar-pyridone production from 1-methyl-2-pyridone, and
thus the rate constant, is solvent polarity independent.
Likewise, we have shown that k_{DP} for the isobutenyl-2-
pyridone 17 is solvent polarity independent and, thus,
not responsible for the changes noted.

The solvent polarity dependence of k_c should be
reflected in solvent effects on the singlet lifetimes
($\tau = 1/(k_d + k_{DP} + k_f + k_c)$) and fluorescence efficiencies
($\phi_f = k_f\tau$) of 17 as shown in Tables 6-8. A reasonably
simple, quantitative treatment of the solvent polarity
effects has turned out to be quite informative. The
solvent associated linear free energy of activation
relationship, suggested by Grunwald and Winstein,[32]
applied to the ylid forming process is shown in equation
2, where k_c and k_{c_0} are the cyclization rate constants
in aqueous ethanol and 20% H_2O-EtOH (V/V), m is the reaction parameter reflecting the sensitivity of k_c to
changes in solvent polarity or the degree of charge
separation in the transition state for $17^{S1} \rightarrow 40$, and
Y the solvent polarity parameters. Substitution for
k_c and k_{c_0} by functions containing the easily measured
ϕ_f gives equation 3 in which $c = 1 + ((k_d + k_{DP})/k_f)$.
Analysis of the data for ϕ_f vs. Y shown in Table 8 using
non-linear least squares methods, in which the best m
and c values are found, gives an exceptionally close fit
to equation 3 when the reaction parameter m is 0.49 ± 0.07
and $(k_d + k_{DP})/k_f$ is 63 ± 1.

$$\log k_c = mY + \log k_{c_0} \qquad (2)$$

$$\log \left(\frac{1}{\phi_f} - c\right) = mY + \log \left(\frac{1}{\phi_{f_0}} - c\right) \qquad (3)$$

The large value of m obtained for cyclization of
17^{S1} when compared to those observed for ground state
solvolysis reactions (ca. 1) strongly suggests that a
significant polarity change is occurring in proceeding
from the singlet pyridone to the transition state for
ylid formation. It seems reasonable to postulate that
this is reflective of a late transition state resulting
from conversion of 17^{S1} to the singlet excited ylid 40.
The possibility that the large m-value stems not necessarily from a transition state with a large degree of
charge separation but, rather from a reversed polarity
in 17^{S1} appears unlikely since the fluorescence maximum
and band shape of 17 is unchanged over a large range of
solvent polarities.

In summary, the solvent polarity treatment described

above offers additional support for the proposed mechanism
for conversion of 1-alkenyl-2-pyridones to oxazolopyridinium salts. It is clear from the fact that k_c varies
ca. four orders of magnitude in changing solvent from
$\overline{THF}$[45] to H_2O and ca. two orders of magnitude from EtOH to
H_2O that this process would be an insignificant competi-
tor of Dewar-pyridone formation in photoreactions of
alkenylpyridones conducted in non-polar organic solvents.
Lastly, from the value of $(k_d + k_{DP})/k_f$ obtained from
this treatment it is possible to calculate the rate con-
stant for radiationless decay of $\underline{17}^{S1}$ (k_d) (see Table 9).

   Structural Effects. Another interesting feature
of the results presented above concerns the apparent
effects on the reactivity of singlet 1-alkenyl-2-pyridones.
Our preliminary results indicate that the efficiency for
oxazolopyridinium salt formation is critically dependent
on both the number and geometrical location of substituents on the exocyclic vinyl moiety, i.e., a qualitatively
judged order of reactivity of isobutenyl > cis-propenyl >
trans-propenyl >> vinyl. Importantly, the source of the
reactivity differences appears to lie in substituent
effects on the rates of ylid formation from the singlet
pyridones, as judged by the correlation between reactivity
and solvent polarity effects on singlet lifetimes and
fluorescence efficiencies of the 1-alkenyl-2-pyridones.
Although a rationalization for the substituent effects on
k_c is not readily apparent, information which might be
suggestive of the source of the control is found in
earlier studies[29] on the preferred C-N conformations of
1-alkenyl-2-pyridones. Spectroscopic data presented at
the time showed that the presence of cis-alkyl substitution of the terminal vinyl-carbon of 1-vinyl-2-pyridones
causes bisected conformers, 41, to be preferred. In
conformations of this type, the terminal vinyl carbon is
preparatorily oriented in the direction of motion required
for cyclization. Moreover, the preferred C-N conformations in cases lacking large cis-alkyl substituents, such
as the parent vinylpyridone 31, are known[29] to be planar
and to most probably have the vinyl moiety oriented anti
with respect to the amide carbonyl. Conformations of
this type, 42, for singlet excited systems would of course
be incapable of cyclization. Thus, the substituent
effects may be stereoelectronic or conformational in
nature and may be manifested in terms of least motion

or orbital overlap control. Further studies should
clarify these questions.

                41   42

Summary. The results of studies conducted thus far
in the area of 1-alkenyl-2-pyridone photochemistry have
led to the unmasking of a novel photocyclization reaction
involving the N-vinylamide chromophore. Although only
tentative, the results indicate that this excited state
process may well be quite general and operable in a number of systems containing this and related chromophores.[46-48] In addition, the study has demonstrated the
wealth of information that can be made available from
detailed analyses of photophysical and photochemical data
obtained by employing the modern tools of photochemistry.
Moreover, the 1-alkenyl-2-pyridones should serve as useful models in future studies aimed at probing the detailed
electronic and structural features of excited state
processes.[49]

REFERENCES

(1) A. F. Hildebrandt, "Potential of Solar Energy for
     Texas," report to The State of Texas Governor's
     Energy Advisory Council, Houston, Texas, November
     1974.
(2) S. P. Ellison, Jr., "Summary of Future Energy
     Resources," 79th Annual Meeting of the Texas Academy
     of Sciences, Texas A&M University, College Station,
     Texas, March 1976, #109.
(3) National Solar Energy Research, Development and Demon-
     stration Program, Definition Report (Energy Research
     and Development Administration, Washington, D.C.,
     June 1975).

(4)   S. V. Szoklay, Solar Energy and Building (Halsted
      Press, New York, 1975).

(5)   B. Sorensen, Science, 189, 255 (1975).

(6)   M. Wolf, Science, 184, 382 (1974).

(7)   J. A. Duffie and W. A. Beckman, Science, 191, 143
      (1976).

(8)   For earlier references, mostly to inorganic systems,
      see (a) F. Daniels, Direct Use of the Sun's Energy
      (Yale University Press, New Haven, Conn., 1964)
      Chapter 17; and (b) J. G. Calvert in Introduction
      to the Utilization of Solar Energy, A. M. Zarem
      and D. D. Erway, eds. (McGraw-Hill, New York, 1963)
      Chapter 9.

(9)   G. Jones, II, and B. R. Ramachandran, J. Org. Chem.,
      41, 789 (1976).

(10) M. Calvin, Science, 184, 375 (1974).

(11) P. H. Abelson, Science, 191, 1221 (1976).

(12) J. O'M. Bockris, Energy, the Solar-Hydrogen Alterna-
      tive (Halsted Press, New York, 1975).

(13) V. Balzani, et al., Science, 189, 852 (1975).

(14) G. Jones, T. Reinhardt and W. Bergmarks, Solar
      Energy, 20, 241 (1978).

(15) W. H. F. Sasse, Solar Power and Fuels, ed. J. A.
      Bolton, Acad. Press., N.Y., Chap 8, 1977.

(16) J. C. Denton, Energy Conversion, 16, 181 (1977).

(17) J. C. McVeigh, Sun's Power, Pergamon Press, Oxford
      Eng., 1977.

(18) "Climatological Data: National Survey", National
      Climate Center, Asheville, N.C., Annual Survey, 27,
      125 (1976).

(19) S. Levine, H. Halter, and K. Mannis, Solar Energy,
      2, 11 (1958).

(20) J. A. Bolton, Solar Energy, 20, 181 (1978).

(21) (a) S. L. Murov and G. S. Hammond, J. Phys. Chem.,
      72, 3793 (1968); (b) K. B. Wiberg and H. A. Connon,
      J. Amer. Chem. Soc., 98, 5411 (1976); (c) G. S.
      Hammond, N. J. Turro and A. Fischer, ibid., 83,
      4674 (1961); (d) R. R. Hautala, J. Little and E.
      Sweet, Solar Energy, 19, 503 (1977).

(22) (a) F. Weigert, Elder's Jahrb., 111 (1909); Chem.
      Abst., 4 3170 (1910); (b) F. Weigert, Naturwissenschaften,
      15, 124 (1927); (c) N. C. Yang, D. M. Ghold and B.
      Kim, J. Amer. Chem. Soc., 98, 6587 (1976); (d) P.
      Bender and J. Farber, ibid., 74, 1450 (1952); (e)
      F. D. Greene, Bull. Soc. Chim. Fr., 1356 (1960);

(f) D. E. Applequist, R. L. Little, E. C. Friedrich, and R. E. Wald, J. Amer. Chem. Soc., 81, 452 (1959); (g) G. Guarini and P. Sarti-Fantoni, Mol. Cryst. and Liquid Cryst., 6, 423 (1970); (h) F. D. Greene, S. L. Misrock and J. R. Wolfe, J. Amer. Chem. Soc., 77, 3852 (1955).

(23) (a) W. R. Bergmark and G. Jones, Nouveau J. de Chimie, 1, 271 (1978); (b) R. Livingston and K. S. Wei, J. Amer. Chem. Soc., 89, 3098 (1967); (c) H. Shizuka, Y. Ishii, M. Hoshino and T. Morita, J. Phys. Chem., 80, 30 (1976); (d) G. Kaupp, Angew. Chem. Intl. Ed. Eng., 11, 313 (1972); (e) G. Jones, T. E. Reinhard and W. R. Bergmark, Solar Energy, 20, 241 (1978).

(24) (a) J. S. Bradshaw and G. S. Hammond, J. Amer. Chem. Soc., 85, 3953 (1963); (b) P. J. Collins, D. B. Roberts, G. Sugoudz, D. Wells and W. H. F. Sasse, Tetrahedron Lett., 321 (1972); (c) T. Teitei, D. Wells and W. H. F. Sasse, Tetrahedron Lett., 367 (1974); (d) T. W. Mattingly, J. E. Lancaster and A. Zweig, Chem. Comm., 595 (1971); (e) W. H. F. Sasse, Aust. J. Chem., 24, 2325 (1971).

(25) (a) L. A. Paquette and J. H. Barrett, J. Amer. Chem. Soc., 88, 1718 (1966); (b) G. Jones and L. J. Turbini, J. Org. Chem., 41, 2362 (1976).

(26) R. B. Woodward and R. Hoffman, Conservation of Orbital Symmetry, Springer Verlag, N.Y., 1969.

(27) Private communication from R. Criegee reported in W. Adam and J. Chang, Int. J. Chem. Kin., 1, 487 (1969).

(28) (a) E. J. Corey and J. Streith, ibid., 86, 459 (1964); (b) R. C. Deselms and W. R. Scheigh, Tetrahedron Lett., 3563 (1972); (c) H. Furrer, Chem. Ber., 105, 2780 (1972); (d) L. J. Sharp and G. S. Hammond, Mol. Photochem., 2, 225 (1970).

(29) P. S. Mariano, E. Krochmal, P. L. Huesmann, R. L. Beamer and D. Dunaway-Mariano, Tetrahedron, 34, 2609 (1978).

(30) P. S. Mariano, A. Leone and E. Krochmal, J. Org. Chem., 42, 1122 (1977); P. S. Mariano, A. Leone, and E. Krochmal, Tetrahedron Lett., 2227 (1977); P. S. Mariano and A. A. Leone, J. Amer. Chem. Soc., 100, 3947 (1978).

(31) H. Kaye and S. H. Chang, Tetrahedron, 26, 1369 (1970).

(32) E. Grunwald, S. Winstein, J. Amer. Chem. Soc., 78, 2770 (1956); S. W. Winstein, E. Grunwald and H. W. Jones, ibid., 73, 2700 (1951).

(33) P. J. Kropp, Z. L. F. Gaibel, E. J. Reardon, K. E. Willard and J. H. Hattaway, J. Amer. Chem. Soc., 95, 7058 (1973) and references cited therein.

(34) G. Frater and H. Schmid, Helv. Chem. Acta., 50, 255 (1967).

(35) W. M. Horspool and P. L. Pauson, Chem. Communications, 195 (1967).

(36) A. Albert, Heterocyclic Chemistry, Athlong Press, London, 1968, p. 80.

(37) J. W. Bridges, D. S. Davies and R. T. Williams, Biochem. J., 98, 451 (1966).

(38) (a) Chapman-type photocyclizations[38b] are prevented in 1-alkenyl-2-pyridone systems by structural constraints; (b) O. L. Chapman, G. L. Eian, A. Bloom and J. Clardy, J. Amer. Chem. Soc., 93, 2918 (1971).

(39) (a) A. G. Schultz and M. B. DeTar, J. Amer. Chem. Soc., 96, 296 (1974); (b) S. H. Groen, R. M. Kellog, J. Butler and H. Wynberg, J. Org. Chem., 33, 2218 (1968).

(40) (a) W. A. Henderson and A. Zweig, Tetrahedron Lett., 625 (1969); (b) J. A. Elix, D. P. H. Murphy and M. V. Sargent, Synth. Communications, 2, 427 (1972); A. G. Schultz and R. D. Lucci, J. Org. Chem., 40, 1371 (1975).

(41) (a) Ylids of this nature have been prepared previously by reactions of Munchnones with reactive aldehydes and ketones;[41b] (b) E. Funke, R. Huisgen, and F. C. Schaefer, Chem. Ber., 104, 352 (1971); R. Huisgen, E. Funke, H. Gotthardt and H. L. Panke, ibid., 104, 1532 (1971).

(42) (a) N. C. Yang and G. R. Lenz, Tetrahedron Lett., 4897 (1967); (b) R. W. Hoffman and K. R. Eicken, ibid., 1759 (1968).

(43) M. Eigen, Angew. Chem., I. E., 3, 1 (1964).

(44) I. Zugravescu and M. Petrovand, Nitrogen Ylid Chemistry, McGraw-Hill, New York, 1976, p. 204, and references cited therein.

(45) Based upon a predicted Y-value for THF in the range of ca. -6 using data recorded in C. Reichardt, Angew. Chem., I. E., 4, 29 (1965).

(46) Preliminary studies have shown that water serves to effectively quench the familiar 1,3-acyl shift

reactions of N-phenyl-N-isobutenylacetamide.

(47) This process might be responsible for the conver-
      sion of α-acylamino-β-thioalkylacrylamides to
      amidoyloxazoles; C. J. Veal and D. W. Young, _Tetra-
      hedron Lett_., 2985 (1976).

(48) A heteroatom analog of this cyclization process
      appears to be operating in the cyclization reac-
      tions of 10-phenyl-isoalloxazines occurring upon
      irradiation in acid solution; W. R. Knappe, _Chem_.
      Ber., _107_, 1614 (1974).

(49) Financial support for this research by the Robert
      A. Welch Foundation, the Center for Energy and
      Mineral Resources and, briefly, by the Department
      of Energy is gratefully acknowledged.

THE NORBORNADIENE-QUADRICYCLENE ENERGY STORAGE SYSTEM

Richard R. Hautala, R. Bruce King and
Charles Kutal

Department of Chemistry

University of Georgia, Athens, Georgia 30602

INTRODUCTION

The potential applicability of endoergic photochemical transformations of organic molecules to the storage of solar energy is currently receiving considerable attention. Indeed the topic is the subject of three presentations in this book. Jones and his coworkers have taken the broadest perspective by considering general design principles and by investigating numerous types of systems. Mariano and coworkers have focused attention on a novel photocyclization reaction of pyridone derivatives and have explored the photochemistry of these compounds in depth. We, on the other hand, have selected one of the most promising systems currently available (the norbornadiene-quadricyclene interconversion) and have devoted considerable effort toward aspects of its development. A number of practical problems have been addressed in our work. The present account provides an assessement of the problems, an overview of our objectives, and a summary of our most recent efforts.

CRITERIA

The concept of a fuel source based on photochemical reactions which generate products of high energy content which, in turn, can be reconverted at will to the original material is exceedingly attractive in that an inexhaustible source of

energy (sunlight) is converted into a usable form with no
necessary consumption of resources.   However the possibility

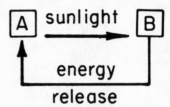

that such a conceptual scheme could be reduced to practice
must be considered within stringent pragmatic constraints.

     Three essential criteria form our basis for the assess-
ment of this possibility.

(1)   Energy Capacity-Do endoergic photochemical transform-
      ations have the potential to produce sufficient quan-
      tities of energy for typical applications?
(2)   Cost-Could endoergic photochemical transformations be
      economically competitive with alternative fuel
      sources?
(3)   Current Feasibility-Can the actual coupling of an en-
      ergy storage reaction and the complementary energy re-
      leasing reaction be demonstrated?

     Each of these factors will be addressed within the con-
text of the norbornadiene (NBD)-quadricyclene (Q) system.
The attractive features of this system, with respect to its

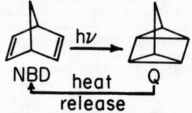

suitability for solar energy storage, have been discussed in
detail previously, and the salient points are enumerated be-
low(1).

  1. norbornadiene is readily available and compara-
      tively inexpensive
  2. both norbornadiene and quadricyclene are liquids
      with moderate boiling points (90 and 115°C re-
      spectively)

3. while the electronic absorption spectrum of
    norbornadiene does not overlap at all with the
    solar insolation spectrum, numerous photosensi-
    tizers are known which sensitize the conversion
    with quantum efficiencies approaching 100% and
    which exhibit absorption of at least portions of
    the solar spectrum.
4. the volumetric energy storage capacity of
    quadricyclene is very large (230 cal per ml) in
    spite of the fact that it is kinetically stable
    (the half-life for reversion to norbornadiene is
    14 hours at 140°C; under ambient conditions it
    appears to be stable indefinitely).
5. numerous catalysts are available which quantita-
    tively and "spontaneously" convert quadricyclene
    into norbornadiene with the concomitant evolution
    of heat.

ENERGY CAPACITY

The concept of energy capacity must be discussed within the context of appropriate applications for this type of
"fuel." Because the stored energy is released in the form
of heat, typically generating temperatures just above the
boiling point of water, this energy source falls in the
category of low-grade heat. To appreciate the magnitude of
energy consumption which could be met by a low grade heat
source, it is instructive to look at the breakdown of U.S.
energy consumption by end-use. Data for the year 1973 have
been assembled by Hippel and Williams(2) and show that, of the
total energy consumed in the U.S., 4% was used for water
heating and 19% for space conditioning of buildings. In
addition another 14% was used for low-temperature industrial
heat. Thus at least 23% of the total energy consumed in the
United States could, in principle, be furnished by a low-
grade heat source. Consequently the heating of domestic
buildings (and water) is the most obvious, and probably the
only, potential application for such a system.

In order to consider the energy capacity, it is first
necessary to determine the energy requirements for a typical
application. Talbert et al. have analyzed the energy requirements for a standard 1800 sq. ft. house in Albuquerque,
New Mexico(3). The daily energy consumption for water heating, space heating and air conditioning varies from 120,000

BTU (October) to 650,000 BTU (January, July), with an annual
average of 430,000 BTU per day. Three factors could lower
this value appreciably:

1. omission of air-conditioning, an unrealistic application
    considering the low efficiencies obtainable utilizing
    a low-grade heat energy source
2. improvements in energy-efficient building designs
3. consideration of locations more moderate in climate
    than Albuquerque, New Mexico.

Under more favorable circumstances an annual daily average
of 200,000 BTU might be a reasonable goal.

     The energy storage capacity of quadricyclene (23 Kcal-
mole^{-1})(4) translates into approximately 3400 BTU-gallon^{-1}.
Thus the daily consumption of quadricyclene would amount to
approximately 60 to 120 gallons. On this basis the energy
capacity of the NBD-Q system looks attractive, for the necessary storage tanks would not be much larger than typical
hot-water heaters (and would not require the insulation).
The NBD-Q system is particularly attractive in comparison
to other chemical systems. It is a fluid system and therefore can be used neat. Solid systems requiring solvents
would reduce the energy density considerably. The molecular
weight of the system is low (MW=92) and the storage per
mole is high (23 Kcal-mole^{-1}). It is highly unlikely that
these specifications will be exceeded by any reasonable organic photochemical system.

     A second aspect of energy storage capacity, however,
is far more formidable and challenging. Given that at least
200,000 BTU of energy are consumed daily, an equivalent
amount must be stored on an average daily basis. In order
to determine whether this is feasible, several factors must
be considered. Because the quantity of energy stored is
constant regardless of the energy of the photon absorbed
(provided the photon is effective in producing quadricyclene), the intensity of available sunlight must be considered in terms of photons (or Einsteins) per unit area. The
solar collection area is limited by the application and we
have used a 900 sq. foot size (for the 1800 sq. ft. dwelling)
in calculating data for the model. Any larger size would
be cumbersome, expensive, unaesthetic and otherwise unrealistic.

The solar insolation data utilized in the model are typical mean daily values for 40° north latitude (e.g. Washington, D.C.) under average atmospheric conditions in the summer. Atmospheric, geographical and seasonal variations will alter these values considerably. It should be noted that these data have been used to model environmental photochemical transformations, and the calculated photolysis "rates" derived from these data have been shown to simulate actual "in field" experiments with excellent agreement[5]. Thus the insolation data are realistic.

The total "collectable" or "storable" energy can then be calculated considering the energy stored per photon absorbed (91.3 BTU-Einstein^{-1}). Assuming that the light absorbing chromophore absorbs all incident light from the solar onset to a designated wavelength cutoff, the maximum storable energy has been calculated for processes operating at 100% quantum efficiency and at 50% quantum efficiency. These data are presented in Table 1. It is immediately clear that the magnitude of energy involved falls short of the 200,000 BTU per day target figure in all but the most extreme (and rather unrealistic) cases. Currently available sensitizers for the NBD-Q systems do not absorb beyond 500 nm, and the prospects for uncovering any with high quantum efficiencies will probably require a major breakthrough.

Aside from the storage value of 23 Kcal-mole^{-1}, the above analysis is not unique to the NBD-Q system but applies to any endoergic system with thermal energy as its product. The likelihood that the energy capacity could be doubled by discovering a system storing 46 Kcal-mole^{-1} is remote if not impossible. Even so, the maximum storable energy would still be marginal with respect to the needs. Does this dampen the prospects for potential use of such systems in solar energy storage? An alternative usage might be quite attractive. Conventional solar collectors used in domestic heating provide only a portion of required heat[6]. During heavy demands (extreme temperatures and/or several successive overcast days) existing systems must be switched into conventional (non-solar) utility sources as a backup. The problem in so doing is that such occasions coincide with peak demands on utilities. It has been pointed out that unfavorable pricing structures and expanded peak capacities for utilities place an inordinately high value on this type of backup energy[7]. Thus a photochemical storage system,

TABLE 1

MAXIMUM COLLECTABLE ENERGY FOR THE NBD-Q ENERGY STORAGE
SYSTEM USING A 900 SQUARE FOOT COLLECTOR WITH TYPICAL SOLAR
INSOLATION

Absorption Cut-off (λ, nm)	Energy[a] Equivalent (Kcal-mole^{-1})	Solar[b] Intensity (Ein-m^{-2}-day^{-1})	Total Storable[c] Energy (BTU - day^{-1}) $\Phi = 1$	$\Phi = 0.5$
335	85.4	0.50	3800	1900
355	80.6	1.21	9400	4700
365	78.4	1.61	12000	6000
385	74.3	2.53	16000	8000
395	72.4	3.11	24000	12000
405	70.6	3.93	30000	15000
425	67.3	6.13	47000	23000
435	65.8	7.19	55000	28000
455	62.9	9.87	75000	38000
485	59.0	14.3	110000	55000
500	57.2	16.4	125000	63000
550	51.9	24.1	180000	90000
600	47.6	32.1	240000	120000
650	44.0	40.1	310000	155000
700	40.8	48.4	370000	185000

a) minimum energy equivalent of absorption cutoff wavelength (for reference only).
b) total intensity from solar onset (300 nm) to absorption
cutoff; based on average daily intensity; 40° north
latitude; adapted from data in reference 5.
c) based on the storage of 23 Kcal-mol^{-1} (equivalent to
91.3 BTU-Ein^{-1}) calculated for quantum yields of 1.0
and 0.5 assuming the system is totally absorbing over
the wavelength range.

operating simultaneously and in tandem with a conventional collector and providing backup for periods of heavy demand and/or low input, would be quite attractive. The energy capacity burden on the photochemical system would be relieved considerably, making it viable even with presently available sensitizers. Such a system would operate, as indicated in Figure 1, by absorbing and storing as much

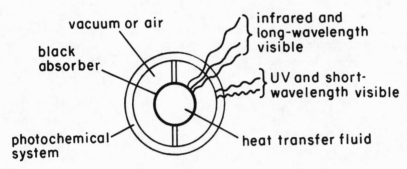

Figure 1.   A concentric-pipe solar collector system

energy as possible, allowing the remainder to operate a conventional system. A schematic illustration of a solar collector, a working model of which is currently under investigation in our laboratories, utilizing this design principle is illustrated in Figure 2. Because of the dual function we have dubbed this the DUPLEX collector. Numerous practical and scientific problems require further investigation before a meaningful assessment can be made of the concept. Presently it offers an attractive, <u>albeit</u> hypothetical, solution to a serious drawback in conventional solar collector design.

In summary, the factor of energy capacity has a number of facets. The high energy density of endoergic photochemical systems, particularly the norbornadiene-quadricyclene system, is a very attractive attribute. The prospect of such systems providing sufficient energy for reasonable applications is marginal unless coupled with conventional

solar heating units. However it should be noted that the
marginal nature of such systems is not unique, but rather
represents one of the extremely challenging aspects of research efforts to harness and convert <u>any</u> of the diffuse
energy sources such as sunlight and wind.

COST

To analyze the projected cost of a photochemical energy
storage system it is useful to relate the cost of the individual components to a common denominator, cost per energy
unit produced (e.g. dollars per million BTU). This facilitates comparison with alternative energy sources for producing low-grade heat. The necessity for doing so derives
from the fact that no system, regardless of how attractive
it might be conceptually, will ever be employed unless it

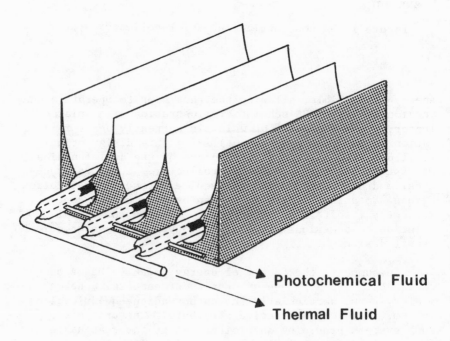

Photochemical Fluid

Thermal Fluid

Figure 2. The DUPLEX solar collector

is economically advantageous to do so.

The overall cost will be treated in terms of two components, (a) the collector and system hardware and (b) the
photochemical fluid system. It has been pointed out by
Talbert and coworkers(3) that the cost of the hardware for a
photochemical based solar collector system would be significantly less than that of a conventional solar system for
several reasons. A reasonable estimate of the system cost
for a 900 ft^2 collector is approximately $2700. Based on
a replacement lifetime of ten years and assuming an energy
production of 200,000 to 400,000 BTU per day, the average
cost over ten years would be $1.85 to $3.70 per million BTU
(MBTU).

The second and most volatile aspect of the cost relates
to the photochemical fluid. The norbornadiene-quadricyclene
system stores 3400 BTU per gallon of quadricyclene. At the
rate of $5.70 per gallon(8) the cost for producing one mil-
lion BTU is $1680. However, since the system is recyclable,
the cost reduces to $168 per MBTU if 10 cycles are completed
before replacement, $16.80 per MBTU for 100 cycles, $1.68
per MBTU for 1000 cycles and so on. The factor which of
course dictates the number of cycles attainable is the inevitable loss to side reactions per cycle. If the fluid
were replaced when its composition reached 50% loss, each
cycle must lose no more than 6.7% for a 10-cycle lifetime,
0.7% for a 100 cycle lifetime, 0.07% for a 1000 cycle lifetime and so on. In computing the real cost per energy unit
it is necessary to take into account the fact that, over
the lifetime of the fluid (as it degrades from 100% to 50%),
only 75%, on the average, is effective. Thus the cost must
be mulitplied by 1.33 to account for this factor. For NBD
this raises the value to $2230 per cycle. Each of these
factors is illustrated graphically in Figure 3. The solid
line (labeled N) represents the analysis for the NBD-Q system. The dashed lines (labeled N/2, 2N and 10N) represent
hypothetical systems with cost factors twice as favorable,
half as favorable and ten times less favorable than NBD-Q.
The cost factor, as indicated above, relates to the cost
per unit volume divided by the energy storage per unit volume (eq 1). In comparison with other photochemical systems,
norbornadiene is extremely favorable in both respects.
Again it is highly unlikely that any forseeable organic
photochemical system will exceed or even meet the favorable

"cost"specifications of the NBD-Q system, and as such it

$$\text{cost per energy unit} = 1.33\frac{\text{cost}}{\text{volume}}(\$/\text{gal}) \div \frac{\text{energy}}{\text{volume}}(\text{MBTU}/\text{gal}) \quad [1]$$

represents a useful standard for comparison.

It is abundantly clear that the degree of loss per cycle is a major, if not the major, factor controlling the economics of systems of this type. A third cost factor arises from sensitizers and catalysts used in the system. At this point it is difficult to accurately assess the cost of these materials. Efforts are in progress to discover and develop examples (vide infra) which are sufficiently inexpensive to be negligible relative to the fluid cost(9).

In view of the importance of long-term repetitive cycling it is interesting to speculate on the feasibility of being economically competitive with alternative conventional fuel sources. The current cost of such fuels (natural gas, fuel oil, electrical heating, coal, etc.) is approximately $3 per MBTU on the average. Dramatic increases in this value are expected over the next two decades, and estimates in the $10-20 per MBTU, range are common. These levels for conventional fuels are included (dotted lines) in Figure 3. Thus the NBD-Q system would require cycling in the 100-1000 range to become attractive economically. This stringent demand for any reversible chemical system is indeed formidable, but nonetheless possible. Presently, conditions are known where the individual steps in the cycle appear to be quantitative, i.e. the mass balance is good and side products have not been detected. No long-term cycling studies have been carried out, however.

An intriguing means of extending the lifetime, or number of attainable cycles, might be based on the concept of a repair mechanism, commonly encountered in biological chemical systems. Such a scheme utilizing highly specific catalysts is shown in Figure 4. As side products begin to emerge in a cycling system they would be catalytically converted back into one or the other of the two principal components (NBD or Q). In addition to specific catalysts, the total system would make available several different "settings" for effecting conversions of this type. For example, a

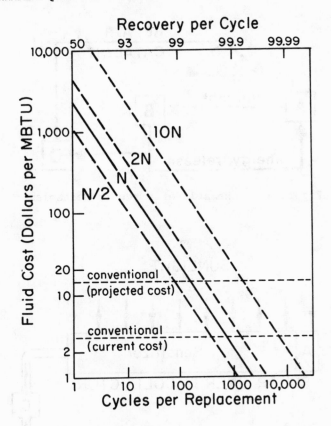

Figure 3.   Cost analysis of a photochemical energy
            storage system (see text)

small portion of the solar collector could be devoted to
repair and would contain specific photoinduced catalytically
active materials. Alternatively, if high temperatures were
necessary to drive a specific catalytic repair reaction, a
portion of the system near the heat exchanger would be appropriate. For catalytic systems operating best under ambient conditions, anywhere else in the system would be appropriate. Each of these hypothetical "repair" centers has
been included in Figure 5 where an overall system scheme is
illustrated. The idea of specific "repair" centers built
into the system represents an interesting potential solution
for easing the awesome task of repetitive cycling. If so,

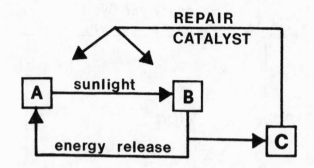

Figure 4.   Schematic of a repair mechanism

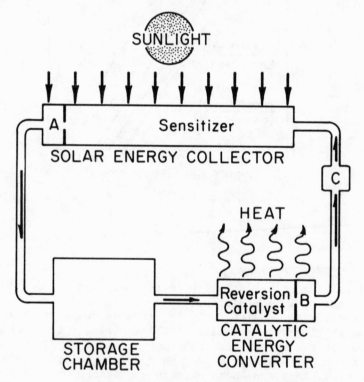

Figure 5.   Schematic diagram of a total photochemical
            solar energy storage system. System repair
            centers are designated by A: photoactivated
            catalytic repair center; B: thermally ac-
            tivated catalytic repair center; C: ambient
            catalytic repair center

the criteria necessary for economic viability might be attainable.

DEVELOPMENT OF THE NORBORNADIENE-QUADRICYCLENE ENERGY STORAGE SYSTEM

We have explored a variety of fundamental and practical areas related to the development of the NBD-Q energy storage system. For example, we have been concerned with photosensitization mechanisms involving both organic and inorganic sensitizers in order to design new sensitizers with improved properties. Efforts to systematically study catalysts for the reverse reaction have been directed toward uncovering inexpensive examples which exhibit high activity, high product specificity and long-term durability. Finally, considerable attention has been devoted to immobilizing both sensitizers and catalysts by chemical grafting onto inert solid supports. By so doing, these components can be confined to the region of function in an overall system. Such confinement of the reversion catalyst to that portion of the system where heat is to be delivered to a load is an absolute requirement. Restricting the sensitizer to a photochemical chamber located on the roof of a building is likewise desirable, since it greatly reduces the requisite amount of this component and also prevents any deleterious interactions with the catalyst. Recent results in each of these areas will be presented below.

ORGANIC PHOTOSENSITIZERS

For organic compounds which effectively sensitize the transformation of norbornadiene to quadricyclene, no compelling evidence exists for the operation of any mechanism other than triplet-triplet energy transfer (eq 2). Consequently the triplet energy of norbornadiene (ca. 70 Kcal-

$$\text{sens} \xrightarrow{h\nu} \text{sens}^{1*} \longrightarrow \text{sens}^{3*} \xrightarrow{\text{NBD}} \text{sens} + \text{NBD}^{3*} \longrightarrow Q \quad [2]$$

mole^{-1}) sets a limit, albeit a somewhat flexible one, to the longest wavelength of light which can be used in effecting the photochemical reaction. This wavelength limit is determined by the onset of the singlet absorption band of the senitizer. For those sensitizers with triplet energies equal to or exceeding that of norbornadiene, the wavelength limit would be near 380 nm (equivalent to ca.75 Kcal-mole^{-1}

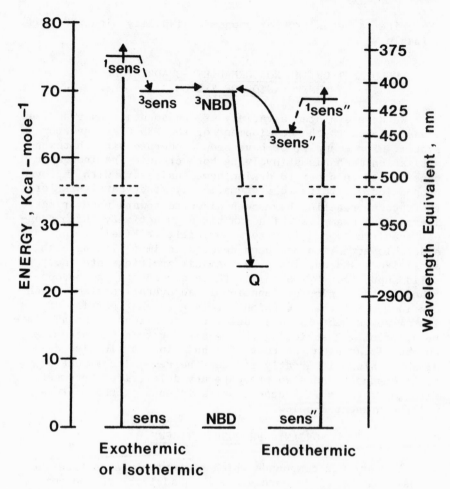

Figure 6.   Schematic illustration of the relationship
             between energy absorbed and energy stored
             in photosensitized processes.

with the excess energy of approximately 5 Kcal-mole^{-1} necessary for the inevitable minimal loss encountered in relaxation of the sensitizer from the singlet to the triplet
state). Fortunately this limit is not a rigorous one, and
several very efficient sensitizers with triplet energies
well below that of norbornadiene are known. The energy de-

ficiency is ostensibly made up through thermal activation or "thermal upconversion"(see Figure 6). A more detailed discussion of endothermic energy transfer is available in the chapter by Jones.

Several derivatives of benzophenone (E_T<68 Kcal-mole^{-1}) represent examples of such "energy deficient" sensitizers which have been studied in some detail. We have focussed considerable attention on 4-(N,N-dimethylamino)benzophenone and have grafted this chromophore onto several solid supports. In a recent publication we have reported detailed comparisons of both a polystyrene and a silica functionalized derivative with the comparable homogeneous sensitizer counterparts(10). The structures, along with convenient acronyms, are given in Figure 7. A study of the quantum efficiency for sensitization of quadricyclene formation by the polystyrene derivative, (P)-AK, reveals that this is a remarkably efficient system. Not only is the heterogeneous sensitizer comparable to the homogeneous counterpart, MeAK, but the limiting quantum efficiency is near unity (see Figure 8).

In contrast, the heterogeneous silica sensitizer, (Si)-AK, was found to be significantly less efficient than the homogeneous counterpart, Si-AK. The limiting quantum yield for (Si)-AK was 0.24 whereas that of the homogeneous silane was found to be unity (Figure 9). The poorer performance of the silica derivative was disappointing cause, in many respects, it is an ideal solid sensitizer. The photostability of the silica support is excellent whereas that of the polystyrene support is not. Furthermore the optical properties of the silica supported sensitizer are remarkably good, particularly with neat norbornadiene, making it an ideal solid sensitizer for use in the DUPLEX collector.

The reason for the diminished quantum efficiency for (Si)-AK is related to the highly polar local environment of the silica surface. The dimethylaminobenzophenone chromophore is quite sensitive to medium polarity, the effect of which is to lower the triplet yield (for example the triplet yield in hexane is 100% whereas in acetonitrile it is 6%)(1). A second effect of the medium polarity is to reduce the energy of the lowest triplet state. Triplet energies can be obtained from low temperature phosphorescence spectra. Such spectra for (P)-AK and MeAK are presented in Figure 10 and the corresponding spectra for (Si)-AK and Si-AK in Figure 11.

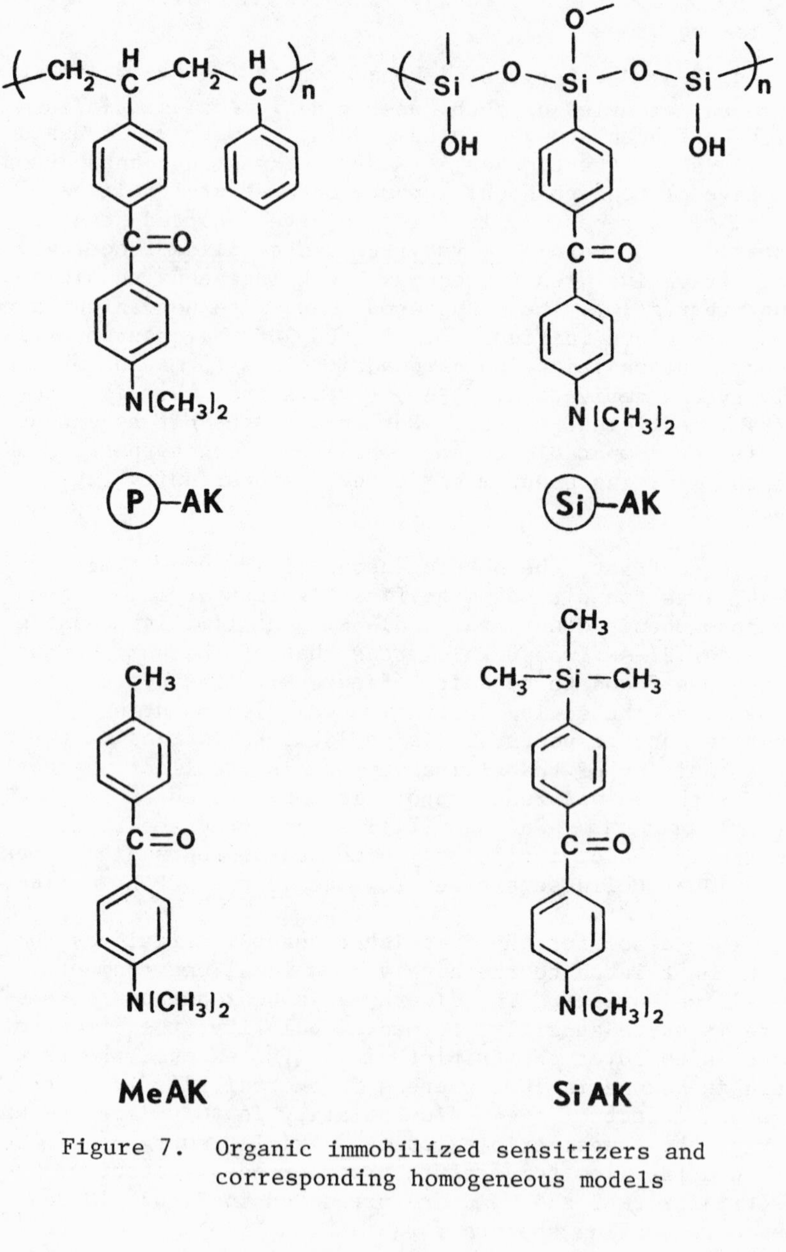

Figure 7.   Organic immobilized sensitizers and
corresponding homogeneous models

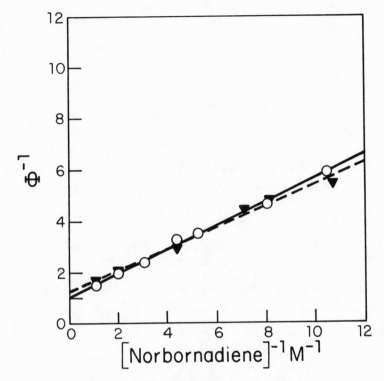

Figure 8. Reciprocal plot of the quantum yield for quadricyclene formation as a function of norbornadiene concentration for the polysty-rene immobilized sensitizer and homogeneous model. Solvent: benzene. Irradiation wavelength: 366 nm. Conversions: approximately 5%. Solutions were not deoxygenated.

————o————: MeAK ---∇---: (P)-AK (from ref. 10)

The pronounced red shift for (Si)-AK, relative to the other sensitizers, is immediately clear. The actual triplet energies are tabulated in Table 2. The value of 57.0 Kcal-mole^{-1} for (Si)-AK is substantially lower than the rest, and would require that energy transfer to norbornadiene be endothermic by some 13 Kcal-mole^{-1}. Thus it is remark-

TABLE 2

TRIPLET ENERGIES AND SENSITIZATION PERFORMANCE INDICES OF
HETEROGENEOUS PHOTOSENSITIZERS AND MODELS
(ref.10)

Sensitizer	Emission[a] λ, nm	Triplet[b] Energy Kcal-mole^{-1}	Stilbene[c] PSS [cis]:[trans]	NBD[d] limiting Φ
(P)-AK	452	63.1	1.50	0.7ᴱ
MeAK	442	64.7	1.78	1.0
(Si)-AK	500	57.0	2.33	0.24
Si AK	446	64.1	1.86	1.0

a)   0-0 phosphorescence band; methylcyclohexane glass
     at 77°K
b)   from emission spectra
c)   stilbene isomerization - photostationary state ratio;
     in benzene
d)   limiting quantum yield for the norbornadiene to quad-
     ricyclene isomerization extrapolated to infinite nor-
     bornadiene concentration; in benzene

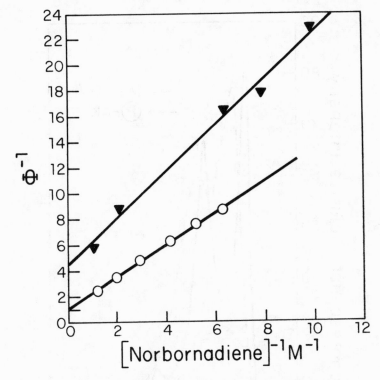

Figure 9. Reciprocal plot of the quantum yield for quadricyclene formation as a function of norbornadiene concentration for the silica functionalized sensitizer and homogeneous model. Solvent: benzene. Irradiation wavelength: 366 nm. Conversions: approximately 5%. Solutions were not deoxygenated.

———o———: SiAK ———▽———: (Si)-AK (from ref. 10)

able that the sensitization proceeds with any reasonable efficiency. An alternative method for checking the triplet energy values was employed. The photostationary state ratio of cis to trans stilbene is related to the triplet energy of the sensitizer used to photoisomerize stilbene. Carefully measured values for each of the sensitizers above were determined and compared with values from the literature. As is apparent from the Saltiel plot in Figure 12, the photostationary state ratio of 2.33 (cis:trans) is con-

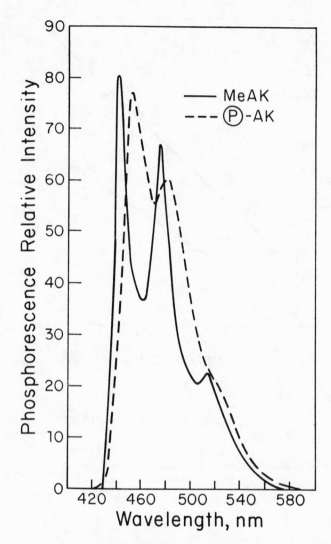

Figure 10.   Phosphorescence spectra of the polystyrene
grafted sensitizer and homogeneous model.
Spectra were taken at 77°K in a methylcyclohexane glass.   The emission slit width
was 1.0 nm for the polymer and 1.5 nm for the
homogeneous sensitizer. (from ref. 10)

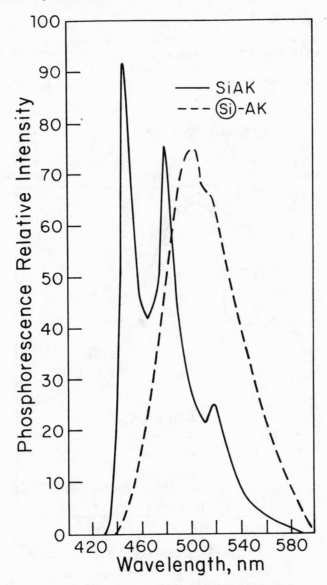

Figure 11. Phosphorescence spectra of the silica
functionalized sensitizer and homogeneous
model. Spectra were taken at 77°K in a
methylcyclohexane glass. The emission slit
was 1.7 nm for the silica and 1.6 nm for the
homogeneous sensitizer. (from ref. 10)

sistent with the measured triplet energy (data are listed
in Table 2).   Treatment of the silica surface with silylating
agents following sensitizer grafting could result in a less
polar microenvironment for the sensitizer while retaining
the desirable properties of the silica support.   Such experiments are in progress.

INORGANIC PHOTOSENSITIZERS

Prior to 1975, virtually nothing was known about the
ability of transition metal compounds to sensitize the

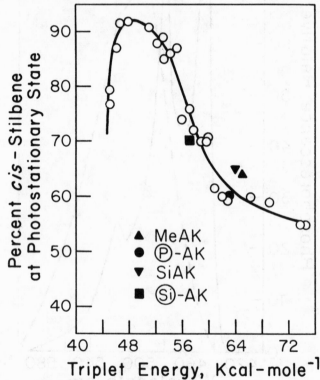

Figure 12.   Saltiel plot of the photostationary state
of stilbene isomerization sensitized by the
immobilized sensitizers and homogeneous models.
Open circles represent other data taken from
Hammond(15).   Photostationary state ratios
were obtained using both cis and trans stilbene with agreement better than ±1% .   Benzene was used as the solvent.(from ref. 10)

photochemical conversion of NBD to Q. Since then, we have
shown that a diverse assortment of copper(I) compounds and
an intriguing class of iridium(III)ions can function as
effective photosensitizers for this process(11,12). In
addition, evidence has been presented for the operation of
two fundamentally different sensitization mechanisms: (I)
photoexcitation of a ground state complex between the sensitizer and NBD, (II) interaction of the electronically
excited sensitizer with ground state NBD. The salient
characteristics of each pathway as they pertain to transition
metal sensitizers are summarized below.

  Mechanism I. A substantial number of transition metals
form stable complexes with norbornadiene(13). The Dewar-
Chatt-Duncanson model(14) of the bonding in such complexes
identifies two interrelated components: (a) donation of π
electron density from the olefin to a σ-acceptor metal orbital, and (b) back donation from a filled metal d orbital
to an empty π^* orbital on the olefin. An important consequence of these orbital interactions is the creation of new
states of the metal-NBD charge transfer variety. Irradiation
into a charge transfer absorption band of the transition
metal-norbornadiene complex may, depending upon the direction
of electron flow, populate the lowest unoccupied (charge
transfer to olefin) or depopulate the highest occupied
(charge transfer to metal) π-type molecular orbital pre-
dominantly localized on NBD. In either case the resulting
redistribution of electron density weakens the bonding between carbon atom pairs C_2-C_3 and C_5-C_6 while enhancing
that between C_2-C_6 and C_3-C_5 (Scheme 1)(15-17). These
photoinduced changes generate a reactive species which can
relax back to NBD or complete the structural reorganization
necessary to form Q.

  While photoexcitation of a transition metal-norbornadiene complex would thus appear to provide an attractive
pathway for the production of quadricyclene, two additional

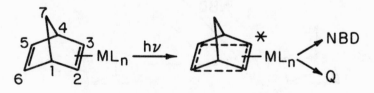

Scheme 1

factors tend to limit the range of metals which can function
in this capacity. Quite clearly, compounds which can catalyze the energy release step are of little practical value
as sensitizers. This restriction excludes transition metals
which undergo facile oxidative addition (common among d^8
and d^{10} metals) or function as good Lewis acids (for example, Ag^+), since they have been shown to be effective
catalysts for the reversion of Q to NBD(18).

A second complication arises from the coordination of
two or more NBD molecules to a single metal center. The
template effect provided by the metal facilitates undesirable thermal or photochemical olefin oligomerization process-
es(19,20) which compete with the energy storage reaction.
To avoid such multiple cooordination, transition metal compounds containing more than two ligands which can be readily
displaced by NBD should be avoided.

Based upon the considerations noted above, it is perhaps not surprising that a rather limited selection of
transition metal compounds have been found which possess the
requisite properties to function as effective sensitizers
for the conversion of NBD to Q via mechanism I. To date,
the only well-authenticated examples involve compounds of
copper(I). Catalytic amounts of simple CuX (X is Cl, Br,
I, $C_2H_3O_2$) salts, for example, efficiently sensitize the
production of Q in a variety of solvents(11). Detailed
spectral and quantum yield studies of the CuCl system
support a mechanism (Scheme 2) involving a photoactive

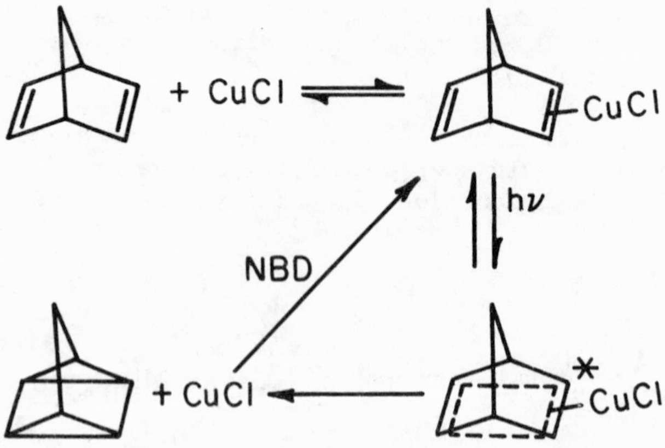

Scheme 2

ClCu–NBD complex. Since the Q formed has little affinity
for copper(I)(21), the photoactive complex is regenerated
and the cycle repeated. This type of mechanism is also
probable for sensitization by $Cu[HB(pz)_3]CO$ ($HB(pz)_3$ is hy-
drotris(1-pyrazolyl)borate), since displacement of the sub-
stitutionally labile CO group by NBD would generate a complex similar to the postulated photoactive species in Scheme
2.

Presently the main shortcoming of Cu(I) sensitizers
which function via mechanism I is their generally poor absorption characteristics in the wavelength region of available solar radiation. Attempts to overcome this deficiency
by incorporating a strongly absorbing chromophore (such as
1,10-phenanthroline or 2,2'-bipyridine) into the sensitizer
structure while retaining the ability of the metal to co-
ordinate a molecule of NBD have thus far met with little
success.

Mechanism II. The enormous interest in the ability of
photoexcited transition metal compounds to undergo bimolecular energy transfer and/or charge transfer process-
es(22,23) in solution prompted us to examine whether such
interactions could be of practical value in sensitizing the
conversion of norbornadiene to quadricyclene. Thus far we
have found two intriguing classes of inorganic sensitizers
which appear to function via this type of mechanism:
(i) copper(I) phosphines such as $Cu(PPh_3)_2BH_4$ and $Cu(PPh_2-$
$Me)_3BH_4$, and (ii) the iridium(III) ions, $Ir(bipy)_2(bipy')-$
OH^{+2} and $Ir(bipy)_2(bipy')OH_2^{+3}$ (Structure I), where the prime

Structure I

denotes monodentate coordination by the bipyridine ligand.
The Cu(I) compounds sensitize the production of quadricyclene
with high quantum efficiency but possess marginal absorption
characteristics(24). The $Ir(bipy)_2(bipy')OH_2^{+3}$ ion is considerably more attractive in that it couples high sensitization efficiency ($\sim$0.7 for 366 nm irradiation) with appreciable absorption(25) in the visible wavelength region;
furthermore, it is easily immobilized onto a transparent
heterogeneous support (vide infra).

Our current efforts in this area are directed toward
defining the nature of the interaction between norbornadiene
and the excited states of these transition metal sensitizers.
Some interesting possibilities which arise in this context
merit brief consideration. Thus in the case of $Ir(bipy)_2$-
$(bipy')OH_2^{+3}$, electronic energy transfer from the lowest
state of the sensitizer to norbornadiene would be endothermic by some 6-10 Kcal. While an energy barrier of this
magnitude should drastically decrease the rate constant for
energy transfer, the overall sensitization efficiency via
this pathway can still be appreciable provided that (i) the
donor excited state possesses a sufficiently long lifetime
(on the order of microseconds or longer), and (ii) the acceptor state (presumably the lowest triplet in NBD) reacts
rapidly enough to limit the extent of back energy transfer
(26). Since the characteristics of the $Ir(bipy)_2(bipy')$-
OH_2^{+3}-NBD system are qualitatively compatible with these
requirements, endothermic energy transfer (see Figure 6)
represents a potentially viable mechanism for the high sensitization efficiency observed. The alternative pathway
involving formation of an exciplex or charge transfer complex between the photoexcited Ir(III) ion and norbornadiene
also merits consideration. Thus the recent finding by
Watts(27) that the lowest excited state in $Ir(bipy)_2(bipy')$-
OH^{+2} is a good oxidant suggests that charge transfer interactions could play an important role in the sensitization
process. This possibility has the intriguing and potentially significant consequence that sensitization is not necessarily constrained to proceed via the lowest triplet state
in the norbornadiene molecule. Detailed mechanistic studies
will be required to distinguish between these alternative
sensitization pathways.

The desirability of anchoring a sensitizer onto an insoluble support has been mentioned earlier. In this con-
nection we have recently devised a procedure (Scheme 3) for

preparing carboxylated silica gel for use in immobilizing charged sensitizers. Beside providing a chemically inert support, silica gel has the added advantage of being relatively transparent to visible light. Initial studies have shown that $Ir(bipy)_2(bipy')OH_2^{+3}$, when immobilized onto this functionalized silica gel, retains a high sensitization efficiency ($\sim$0.6 for 366 nm irradiation) for the conversion of norbornadiene to quadricyclene. This finding is particularly significant in that it represents the first evidence that a charged transition metal compound immobilized onto a cation exchange type support can function as an effective sensitizer for this process.

CATALYST STUDIES

Reversion of quadricyclene to norbornadiene with the release of the stored energy in the form of heat requires a catalyst with the following two properties:

(1) Rapid and specific conversion of Q to NBD. The occur-

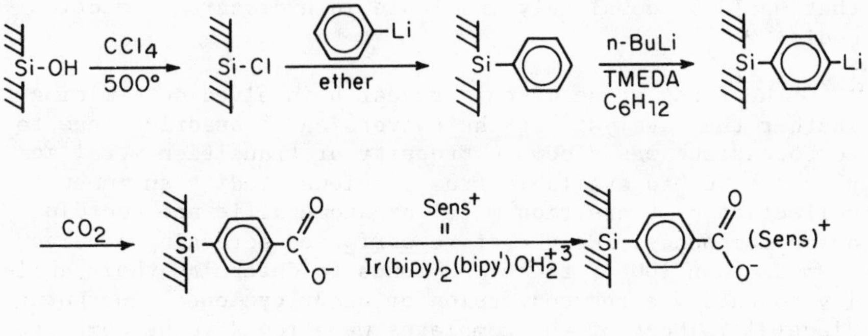

Scheme 3

rence of side reactions would result in the accumulation of unwanted and eventually destructive impurities
in the system after repeated recycling unless an appropriate "repair process" could be devised.

(2)   Insolubility in the reaction medium. A soluble catalyst would, in effect, short-circuit the energy storage reaction since its presence in the photochemical
chamber would prevent any build-up of quadricyclene.

For these reasons we have been seeking catalysts for the
clean and rapid conversion of quadricyclene to norbornadiene
which could be chemically bonded to an insoluble polymer
matrix.

When this research program was initiated in 1975 the
known efficient catalysts for the conversion of quadricyclene to norbornadiene were limited to square planar derivatives of iron(II)(28), cobalt(II)(28), nickel(II)(28),
rhodium(I)(29), and palladium(II)(29). Catalysis by sil-
ver(I) had also been reported(30), but side reactions were
too significant for silver(I) to receive serious consideration in a system where repeated recycling would be necessary. All of the reported catalysts for the conversion
of quadricyclene to norbornadiene were at least somewhat
hydrocarbon soluble, thereby rendering them unsuitable for
use in the proposed energy storage system. It was therefore minimally necessary to find a way to anchor a known
active catalyst onto a polymer or to find a new catalyst
that would be completely insoluble in hydrocarbon reaction
media.

The first stage of this research involved determining
whether the catalysis of the conversion of quadricyclene to
norbornadiene was a common property of transition metal complexes. We had available from previous studies an immense
collection of transition metal organometallic and coordination compounds. We therefore performed screening studies
on more than 100 of these complexes to determine their ability to catalyze the conversion of quadricyclene to norborna-
diene(31). Most of the complexes were found to be completely inactive with two interesting exceptions:

(1)   The trigonal prismatic molybdenum dithiolene complex
$[(CF_3)_2C_2S_2]_3Mo$ (Structure II).

(2)   Several metal carbonyl iodides on prolonged standing
      (e.g. $C_3F_7Fe(CO)_4I$, $C_7H_9Fe(CO)_2I$, $C_5H_5Co(CO)I_2$, and
      $C_5H_5Co(CO)(C_2F_5)I$). Apparently the activity of these
      complexes resulted from the liberation of free iodine,
      which in a separate experiment was shown to be an ac-
      tive catalyst for the conversion of quadricyclene to
      norbornadiene.

Structure II   Structure III

      The observation of catalytic activity for $[(CF_3)_2C_2$-
$S_2]_3Mo$ was an intriguing one for several reasons: (1) This
was the first example of a trigonal prismatic metal complex
functioning as a catalyst; (2) The observed catalytic activity for this complex combined with the activities noted
above for square planar complexes of iron(II), cobalt(II),
nickel(II), rhodium(I), and palladium(II) suggested that
square faces in the coordination polyhedron of a metal complex might be conducive to catalytic behavior by providing
quadricyclene a means of access to the metal atom; (3) Discovery of similar catalytic activity for anionic metal
dithiolenes(32) $[(R_2C_2S_2)_3M]^-$ (M = Mo and W), would provide
a basis for an insoluble catalyst by use of an ion exchange
column. These considerations led to a more detailed study
of the metal dithiolenes as catalysts for quadricyclene
reactions(33). However, this study was disappointing for
the following reasons: (1) The conversion of Q to NBD using
$[(CF_3)_2C_2S_2]_3Mo$ led to undesirable side reactions resulting
in the formation of a polymer; (2) Reduction of $[(CF_3)_2C_2$-
$S_2]_3Mo$ or the related nickel complex $[(CF_3)_2C_2S_2]_2Ni$ to the
corresponding anions led to complete loss of activity. The
latter observation indicates that the catalytic activity of
$[(CF_3)_2C_2S_2]_3Mo$ cannot depend on oxidative addition to the
metal, since reducing the metal oxidation state through
formation of the corresponding anion leads to a loss rather
than an increase in activity. Mechanisms based on oxidative
addition of quadricyclene to sulfur atoms rather than the

metal in $[(CF_3)_2C_2S_2]_3Mo$ are possible(33) but not necessary.

In view of the known(28) catalytic activity of various square planar rhodium(I) complexes such as $[Rh(CO)_2Cl]_2$ and $[(diene)RhCl]_2$, a variety of related rhodium derivatives were screened in order to find systems containing relatively tightly bonded ligands that could be used to anchor an active rhodium(I) system onto an insoluble polymer(31). However, introduction of triphenylphosphine, cyclopentadienyl, or acetylacetone ligands into rhodium(I) derivatives (which would suggest simple means of chemically anchoring rhodium(I) onto a polystyrene matrix or similar polymers)led to a significant decrease in catalytic activity. Thus the complexes $[Rh(CO)_2Cl]_2$, $[1,5-C_8H_{12}RhCl]_2$, $[1,5-C_8H_{12}RhOCOCH_3]_2$, $1,5-C_8H_{12}RhCl(NH_2C_6H_4CH_3-p)$, and $1,5-C_8H_{12}RhCl(HNC_5H_{10})$ all rapidly and completely converted quadricyclene to norbornadiene. The complexes $[(C_6H_5)_3P]_3RhCl$, $1,5-C_8H_{12}RhCl[P(C_6-H_5)_3]$, $(acac)Rh(CO)_2$ (acac = acetylacetonate), $(acac)Rh-(1,5-C_8H_{12})$, $1,5-C_8H_{12}RhC_3H_5$, and $[(C_6H_5)_3P]_4RhH$ were less effective, although the conversion was complete after several days or more. In the cases of the still less active complexes $[(C_2H_4)_2RhCl]_2$, $\{[(C_6H_5)_3P]_2RhCl\}_2$, $[(C_6H_5)_3P]_2Rh-(CO)Cl$, $(acac)Rh(CO)P(C_6H_5)_3$, $C_5H_5Rh(CO)_2$, $1,5-C_8H_{12}RhC_5H_5$, $C_5H_5Rh(CO)I_2$, $[(C_6H_5)_3P]_3RhNO$, and $[Rh(OCOCH_3)_2]_2$ the conversion of quadricyclene to norbornadiene was incomplete even after two weeks at room temperature indicating that the presence of rhodium in a complex, even in the +1 oxidation state, was not a guarantee of catalytic activity. These screening studies on rhodium complexes revealed that β-diketonate, tertiary phosphine, and cyclopentadienyl linkages could not be used for anchoring a rhodium structure onto an insoluble polymer. The high catalytic activities of $1,5-C_8H_{12}RhCl(NH_2C_6H_4CH_3-p)$ and $[1,5-C_8H_{12}RhOCOCH_3]_2$ suggest the possibilities of anchoring catalytically active rhodium units onto amino- or carboxy-polystyrene, respectively, but these possibilities have not yet been explored.

Shortly after the start of this project Noyori and co-workers(34) reported that the nickel(0) complexes $(1,5-C_8H_{12})_2Ni$ and $(CH_2=CHCN)_2Ni$ catalyze the conversion of quadricyclene to norbornadiene. Unfortunately, these nickel complexes are far too air-sensitive to survive in a practical energy storage system. In searching for a more air-stable nickel(0) catalyst we discovered that the triphenylcyclopropenylnickel derivatives(35) $(C_6H_5)_3C_3Ni(CO)X$

(Structure III: X = Cl and Br) effectively catalyze the conversion of quadricyclene to norbornadiene(36). These complexes are essentially insoluble in hydrocarbon solvents, suggesting the possibility of using them directly in the energy storage system without supporting them onto insoluble polymers. However, more detailed study revealed that the $(C_6H_5)_3C_3Ni(CO)X$ complexes are still too air-sensitive to be attractive catalyst candidates for a practical solar energy storage system, although they appear to be considerably more air stable than the instantly oxidized nickel(O) complexes used by Noyori and co-workers(34). In an attempt to obtain triphenylcyclopropenylnickel derivatives that would retain the catalytic activity of $(C_6H_5)_3C_3Ni(CO)X$ but would be more air-stable and more readily chemically bonded to insoluble polymers, the reactions of $(C_6H_5)_3C_3Ni(CO)X$ with various Lewis base ligands were investigated(36). A variety of complexes of the type $(C_6H_5)_3C_3NiL_2Cl$ were prepared (L = tetrahydrofuran, $(C_6H_5)_2PCH_3$, $(C_6H_5)_2PCl$, C_6H_5-PCl_2, and $P(OR)_3$ (R = phenyl, isopropyl, and methyl and $L_2 =$ 2,2'-bipyridyl). However, all of these complexes were either catalytically inactive or at best exhibited activities considerably less than those of the original $(C_6H_5)_3C_3Ni$-(CO)X derivatives. For all of these reasons triphenylcyclopropenylnickel complexes appear to be unattractive candidates for catalysts in the NBD-Q energy storage system.

Among the types of metal complexes known at the start of this project to be active catalysts for the conversion of quadricyclene to norbornadiene were cobalt(II) porphyrins and related square planar cobalt(II) complexes(28). Furthermore, a report by Rollmann(37) described methods for anchoring cobalt(II) porphyrins onto polystyrene. Accordingly, an obvious source of heterogeneous catalysts were these polymer-anchored cobalt(II) tetraarylporphyrins. We therefore studied the preparation of cobalt(II) tetraarylporphyrins anchored onto macroreticular polystyrene using methods similar to those reported by Rollmann(37). We found(38,39) that by using the carboxamide and sulfonamide-linked systems(Structure IV: Z = -CO-, R = CO_2CH_3 and Z = -SO$_2$-, R = SO_3CH_3, respectively) it was possible to prepare polystyrene-anchored cobalt(II) tetraarylporphyrins with significantly higher metal contents (0.28 to 0.44% cobalt) than the amine- and ketone-linked systems ($\leq$ 0.2% cobalt) reported previously(37). Furthermore, these new polystyrene-anchored cobalt(II) tetraarylporphyrins were found to be highly active catalysts for the conversion of quadricyclene

to norbornadiene with turnover numbers in excess of 10^4.
Recycling studies reveal the presence of a deactivation
pathway which apparently involves the slow oxidation of
cobalt(II) to cobalt(III), since the activity of used catalyst can be enhanced by treatment with a strong reducing
agent such as titanium(III). In a closed system this mode
of deactivation would presumably be limited to the first
few cycles. For all the above reasons, these carboxamide-
and sulfonamide-linked cobalt(II) tetraarylporphyrins are

Structure IV

very promising candidates for the catalyst in an energy
storage system based on the NBD-Q interconversion.

Square planar palladium(II) complexes are generally air
stable and are known to be active catalysts for the conversion of quadricyclene to norbornadiene(29,40). We therefore prepared some polystyrene-anchored beads containing
palladium(II) by stirring macroreticular polystyrene containing diphenylphosphino substituents with palladium(II)
chloride in acetonitrile solution(38,41). The resulting
lemon-yellow beads were found to be active catalysts for
the conversion of quadricyclene to norbornadiene. However,

despite the fact that their palladium content (0.70 to 1.36%) was considerably higher than the cobalt content (0.28 to 0.44%) of the polystyrene-anchored cobalt(II) tetraarylporphyrins (Structure IV), the palladium catalysts were considerably less active than the cobalt catalysts, even after the latter had suffered some deactivation from several uses as noted above. Furthermore, the palladium catalysts were not immune to loss of their catalytic activity upon repeated use, and, in fact, were significantly worse than the cobalt catalysts in this regard. Therefore, the polystyrene-supported palladium catalysts are considerably inferior to the polystyrene-supported cobalt catalysts in all respects and thus do not appear to merit serious consideration for use in the NBD-Q energy storage system.

Our experience with a variety of catalysts for the conversion of quadricyclene to norbornadiene suggests that such catalysts can be divided into two categories:

(1) Square planar complexes of relatively late transition metals lacking at least two electrons of the favored rare gas configuration. These 15 or 16 electron species include the cobalt, rhodium, nickel, and palladium complexes discussed in this paper. Such complexes can activate quadricyclene through some variation of an oxidative addition mechanism. Details of the mechanism may vary depending on whether the charge of the metal oxidation state is one unit (e.g. cobalt(II) to cobalt(III)) or two units (e.g. rhodium-(III)).

(2) One electron oxidizing agents. These include silver(I), iodine (possibly functioning through dissociation into iodine atoms), and the dithiolene $[(CF_3)_2C_2S_2]_3Mo$ (Structure II). These catalysts can oxidize quadricyclene to an unstable cation which, following rearrangement, can generate norbornadiene upon reduction with the reduced form of the catalyst.

Furthermore, our observations suggest that the square planar metal complex catalysts (type 1) are relatively specific in effecting the Q to NBD conversion, whereas the one electron oxidizing agent catalysts (type 2) lead to side reactions, including polymer formation, apparently owing to poor selectivity in the mode of decay of the quadricyclene cation.

The most promising of the above catalyst systems appears

to be the cobalt(II) tetraarylporphyrins. Thus these compounds exhibit high catalytic activity, contain an inexpensive transition metal, and are reasonably stable to
repeated cycling. Furthermore, methods have been developed
for immobilizing the catalytically active species onto an
insoluble polystyrene matrix. The resulting polystyrene-
anchored cobalt(II) tetraarylporphyrins are highly active
catalysts for the conversion of quadricyclene to norbornadiene. The initial activity of the catalyst is so high
that the small amount of deactivation through oxidation
which would occur in a closed system can be easily tolerated.
While further improvement in the performance of these polystyrene-anchored cobalt(II) tetraarylporphyrin catalysts is
likely, they appear to be adequate for a first generation
photochemical energy storage device.

CONCLUSIONS

The norbornadiene-quadricyclene interconversion is an
excellent system for evaluating the feasibility of using
reversible photoisomerization reactions for the conversion
and storage of solar energy. When the features of the system are tested against the stringent practical criteria imposed by the nature of the energy source (sunlight) and the
applications, the likelihood for potential use appears to
rest upon surmounting several obstacles. In order to provide sufficient capacity, the light-gathering ability of
such a system must be increased significantly. A possible
means of easing this burden somewhat, through employment of
the DUPLEX design, has been proposed. In order for any such
system to be feasible from an economic point of view, all
aspects of the chemistry must be rigorously quantitative.
A rather hypothetical repair scheme to "assist" an "unwil-
ling" system in this regard has been proposed. Highly effective photosensitization and catalysis under heterogeneous
conditions are essential.

It is abundantly clear that meeting these obstacles is
not simply a matter of technological refinement, but rather
an imposing challenge to fundamental scientific research.
A successful pursuit may well be worth the effort.

ACKNOWLEDGMENTS

We gratefully acknowledge the financial support of the

National Science Foundation and the Department of Energy. Furthermore we are particularly grateful for significant contributions to this work by S. C. Chang, Dr. Paul Grutsch, Dr. Ronnie Hanes, Dr. Shigeru Ikai, James L. Little, Dr. Mahalingam Santhanam, Dr. Dwight Schwendiman, Audrey Shields, and Dr. Edward M. Sweet.

REFERENCES AND NOTES

1. R. R. Hautala, J. L. Little and E. Sweet, Solar Energy, 19, 503 (1977); T. Laird, Chem. and Ind., 186 (1978).

2. F. von Hippel and R. H. Williams, Bulletin of the Atomic Scientists, October, 1977, p. 12.

3. S. G. Talbert, D. H. Frieling, J. A. Eibling and R. A. Nathan, Solar Energy, 17, 367 (1975).

4. Values of 22 Kcal-mole^{-1} (a); 24 Kcal-mole^{-1} (b); and 26 Kcal-mole^{-1} (c); have been measured. We have used a somewhat conservative "average" value of 23 Kcal-mole^{-1}.
   (a) D. S. Kabakoff, J.-C. G. Bunzli, J. F. M. Oth, W. B. Hammond, and J. A. Berson, J. Am. Chem. Soc., 97, 1510 (1975).
   (b) R. B. Turner, P. Goebel, B. J. Mallon, W. V. E. Doering, J. F. Coburn, and M. Pomerantz, J. Am. Chem. Soc., 90, 4315 (1968).
   (c) K. B. Wiberg and H. A. Connon, J. Am. Chem. Soc., 98, 5411 (1976).

5. R. G. Zepp and D. M. Cline, Env. Sci. and Technol., 11, 359 (1977).

6. S. A. Klein, W. A. Beckman and J. A. Duffie, Solar Energy, 18, 113 (1976).

7. S. L. Feldman and B. Anderson, Solar Energy, 17, 339 (1975).

8. This figure represents the projected cost of NBD quoted to us by a company interested in producing the compound on a large scale.

9. This analysis has neglected the value of the "spent" fuel (containing 50% NBD) which could be purified by a

simple distillation. Perhaps the "trade-in" value of
such would compensate for the catalyst and sensitizer
costs.

10. R. R. Hautala and J. L. Little in "Interfacial Photo-
    processes: Applications to Energy Conversion and Syn-
    thesis", ACS Advances in Chemistry Series, ed. by M. S.
    Wrighton, American Chemical Society, Washington, D.C.,
    in press.

11. C. Kutal, Adv. Chem. Ser., 168, 158 (1978).

12. C. Kutal, D. P. Schwendiman, and P. Grutsch, Solar En-
    ergy, 19, 651 (1977).

13. H. W. Quinn and J. H. Tsai, Adv. Inorg. Chem. Radiochem.,
    12, 217 (1968).

14. M. J. S. Dewar, Bull. Soc. Chim. Fr., 18, C79 (1951);
    J. Chatt and L. A. Duncanson, J. Chem. Soc., 2939 (1953).

15. R. Hoffmann, Acc. Chem. Res., 4, 1 (1971).

16. G. Jones, II, S.-H. Chiang, and P. T. Xuan, J. Photo-
    chem., in press.

17. D. P. Schwendiman and C. Kutal, Inorg. Chem., 16, 719
    (1977).

18. K. C. Bishop, Chem. Rev., 76, 461 (1976).

19. G. N. Schrauzer, Adv. Catal., 18, 373 (1968).

20. B. Hill, D. Math, D. Pillsbury, G. Voecks, and W. Jen-
    nings, Mol. Photochem., 5, 195 (1973).

21. D. P. Schwendiman and C. Kutal, J. Am. Chem. Soc., 99,
    5677 (1977).

22. N. Sutin and C. Creutz, Adv. Chem. Ser., 168, 1 (1978).

23. P. J. Delaive, J. T. Lee, H. Abruna, H. W. Sprintschnik,
    T. J. Meyer, and D. G. Whitten, Adv. Chem. Ser., 168,
    28 (1978).

24. P. Grutsch and C. Kutal, J. Amer. Chem. Soc., 99, 6460 (1977).

25. R. J. Watts, J. S. Harrington, and J. Van Houten, J. Am. Chem. Soc., 99, 2179 (1977).

26. P. S. Engel and B. M. Monroe, Adv. Photochem., 8, 245 (1971).

27. R. J. Watts, personal communication.

28. J. Manassen, J. Catal., 18, 38 (1970).

29. H. Hogeveen and H. C. Volger, J. Am. Chem. Soc., 89, 2486 (1967).

30. B. C. Menon and R. E. Pincock, Can. J. Chem., 47, 3327 (1969).

31. R. B. King and S. Ikai, unpublished results.

32. J. A. McCleverty, Prog. Inorg. Chem., 10, 49 (1968).

33. R. B. King and S. Ikai, J. Mol. Catal., 4, 361 (1978).

34. R. Noyori, I. Umeda, H. Kawauchi, and H. Takaya, J. Am. Chem. Soc., 97, 812 (1975).

35. S. F. Kettle and E. W. Gowling, Inorg. Chem., 3, 604 (1964).

36. R. B. King and S. Ikai, Inorg. Chem., in press.

37. L. D. Rollmann, J. Am. Chem. Soc., 97, 2132 (1975).

38. R. B. King, E. M. Sweet, and R. M. Hanes, Preprints Div. Petr. Chem. Am. Chem. Soc., 22, 1201 (1977).

39. R. B. King and E. M. Sweet, J. Org. Chem., in press.

40. P. G. Gassman and D. S. Patton, J. Am. Chem. Soc., 90, 7276 (1968).

41. R. B. King and R. M. Hanes, J. Org. Chem., in press.

42. C. A. Tolman, Chem. Soc. Rev., 1, 337 (1972).

NEW AB$_5$ HYDRIDES AND THEIR APPLICATION IN CHEMICAL HEAT PUMP SYSTEMS*

Dieter M. Gruen, Marshall H. Mendelsohn, and Irving Sheft

Chemistry Division, Argonne National Laboratory Argonne, Illinois 60439

Austin E. Dwight

Department of Physics, Northern Illinois Univ. DeKalb, Illinois 60115

INTRODUCTION

The discovery of the ability of AB$_5$ compounds (A = rare earth, B = transition metal) to form hydrides with quite unique properties has stimulated research and development in several areas (1-4). Although measurements of the physico-chemical properties of a number of AB5 hydrides have been made, a detailed understanding of the factors that determine the hydrogen dissociation pressures of this class of materials is just now beginning to emerge.

The rapid kinetics of the reactions,

$$AB_5 + nH_2 = AB_5H_{2n} \qquad [1]$$

the high volumetric hydrogen storage densities and the wide range of hydrogen decomposition pressures available with the AB5 hydrides have led to proposals for their use as chemical compressors, cryogenic devices and as media for thermal energy storage, space heating and cooling and the

*Work performed under the auspices of the Division of Basic Energy Sciences of the U.S. Department of Energy.

conversion of low grade heat to work (5). The use of AB_5
hydrides as chemical or thermally driven heat pumps is
particularly intriguing (6). Since two different AB_5
alloys are involved, the relationship between their respective hydrogen decomposition pressures as a function of
temperature is the key parameter determining the thermodynamics of heat pump action.

For the above mentioned as well as possible future
applications of these materials it is crucial to discover
the relationship between the chemical composition of the
AB_5 compounds and their hydrogen decomposition pressures.

Intermetallic compounds of lanthanide and transition
metals form a very interesting class of structures. The
AB_5 series normally crystallize in the hexagonal $CaCu_5$ (P6/mmm)
structure (1). Generally, radius ratios (r_A/r_B) greater
than 1.30 form the $CaCu_5$ type configuration, whereas compounds with r_A/r_B less than 1.30 prefer the cubic UNi_5
structure. Compounds formed by rare earths to the right of
lanthanum in the periodic table display the lanthanide
contraction resulting in a decrease in AB_5 unit cell volume.
The AB_5 phase is generally stable over the composition
range $(AB_{4.8-5.2})$ (5,7).

The hexagonal AB_5 compounds form orthorhombic hydrides.
The changes in structure appear to be a result of expansion
of the basal plane, caused by hydrogen occupation of interstitial sites (4). Both ordered and disordered structures
can occur. For example, Kuijpers and Loopstra (4) found by
neutron diffraction that the deuterium atoms in $PrCo_5D_4$
were ordered on certain interstitial octahedral and tetrahedral sites. A full understanding of site occupation by
hydrogen in the various AB_5 hydrides is lacking due to
insufficient neutron diffraction data, however, the available information is considered in the section on configurational entropies.

$$RT\ell nP = \Delta G = \Delta H - T\Delta S \qquad\qquad [2]$$

expresses the relationship between the dissociation pressure,
P; the free energy of formation, ΔG; the enthalpy change,
ΔH and the entropy change, ΔS of reaction (1).

Phase diagrams, consisting of pressure-composition
curves, are available for a number of AB_5 intermetallic

Table I.

Plateau Pressure and Alloy Cell Volume Data

Compound	P_{eq} (at 20°) (atm)	Cell Volume ($\overset{\circ}{A}^3$)
$LaCo_5$	0.04	89.74
$CeCo_5$	1.55	84.30
$PrCo_5$	0.51	87.13
$NdCo_5$	0.68	86.79
$SmCo_5$	3.30	85.67
$GdCo_5$	24	85.19
YCo_5	17.1	83.96
$LaNi_5$	1.5	86.54
$PrNi_5$	8	84.73
$SmNi_5$	30	83.44
$GdNi_5$	120	82.58
$LaCo_4Ni$	0.055	89.01
$LaCo_3Ni_2$	0.12	88.44
$LaCo_2Ni_3$	0.22	87.80
$NdNi_5$	12.7	84.32

hydrides (5). In some cases ($CeCo_5$, $SmCo_5$ and YCo_5) single plateaus are observed approaching the composition AB_5H_3. In other cases ($PrCo_5$ and $NdCo_5$) two plateaus occur with a maximum composition approaching AB_5H_4. Finally certain AB_5 intermetallics possessing one ($LaNi_5$), two ($NdNi_5$) or even three ($LaCo_5$) plateaus absorb more than four atoms of hydrogen up to a maximum of nine atoms at pressures approaching 1,250 atm for the compound $LaCo_5H_9$ (8).

From plots of $\ln P_{plateau}$ vs. $1/T$ several workers have determined the experimental heats and entropies of reaction (1). The enthalpies thus determined are generally taken to be the heats of formation for the β-phase hydride from the α-phase hydrogen saturated alloy. Data from a variety of sources collected in Table I show that replacing La by another lanthanide or Ni by a different transition metal can bring about order of magnitude changes in hydrogen decomposition pressures at some fixed temperature.

However, it is crucial for the successful use of metal

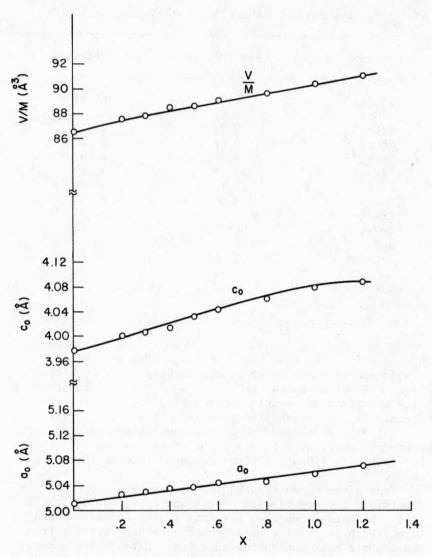

Figure 1. Unit cell parameters and cell volumes in the
LaNi$_{5-x}$Al$_x$ Ternary Alloy System.

hydrides as chemical heat pumps and in other applications
that alloy systems be developed which allow one to have
excellent control over the free energies of formation so as
to be able to change dissociation pressures in a continuous
fashion over a wide range. Recent work in our laboratory
has shown that Group III A and IV A substitutions for Ni in
LaNi$_5$ are particularly effective in lowering the plateau
hydrogen pressures. In this paper, the results of Al, Ga,
In, Si, Ge and Sn substitutions will be presented and discussed in the light of several important theoretical considerations.

The use of these materials in chemical heat pump applications will also be elucidated.

THE LaNi$_{5-x}$Al$_x$ TERNARY ALLOYS

It had been known for some time that aluminum can partially replace the transition element component in AB$_5$ alloys
without altering their crystal structure. Thus, the hexagonal CaCu$_5$ structure of ThNi$_5$ is maintained up to the
composition ThNi$_2$Al$_3$ without undergoing a phase transition
(9). Cubic UCu$_5$ transforms to the hexagonal MgZn$_2$ structure
at the composition UCu$_{4.5}$Al$_{0.5}$, and to the hexagonal CaCu$_5$
structure at the composition UCu$_{3.5}$Al$_{1.5}$ (10). Cubic ZrNi$_5$
and cubic UNi$_5$ both form single phase solid solutions up to
the compositions ZrNi$_4$Al and UNi$_4$Al (11,12). Of these systems, only the hydrogen absorption properties of Th(Ni,Al)$_5$
ternaries had been studied (13).

A careful x-ray diffraction study of Al substitutions
in LaNi$_5$ had shown that the a$_o$ and c$_o$ lattice parameters
as well as the crystallographic unit cell volumes are
strongly increasing functions of the Al content up to the
composition LaNi$_{4.0}$Al$_{1.0}$ as illustrated in Fig. 1 (14,15).
Intensity calculations show that Al tends to enter the 3g
sites in preference to the 2c sites (15). This finding is
consistent with an earlier investigation of the YCo$_5$-YNi$_5$
system (16). Further substitution of Al causes c$_o$ to level
off but a$_o$ increases at a faster rate. The initial increase
in c$_o$ may be due to a lengthening of the bond between layers,
while the subsequent leveling off in c$_o$ may result from the
beginning of a trend for the large La atom to collapse into
the hexagonal ring of Ni(Al) atoms above and below it.

On the basis of a cell volume-decomposition pressure

Table II.

Thermodynamic Values of Al
Substituted AB_5 Hydrides

Alloy	ΔH	ΔS	Cell Volume
$LaNi_5$	-7.2	-26.1	86.8
$LaNi_{4.8}Al_{0.2}$	-8.3	-27.3	87.28
$LaNi_{4.6}Al_{0.4}$	-8.7	-26.1	87.89
$LaNi_{4.5}Al_{0.5}$	-9.2	-26.6	88.77
$LaNi_4Al$	-12.3	-28.7	90.51
$LaN_{3.5}Al_{1.5}$	-14.5	-29.6	92.81
$CFMNi_{4.8}Al_{0.2}$[*]	-7.5	-26.5	86.44

[*] CFM = cerium-free mischmetal

correlation to be discussed more fully below one would predict that aluminum substitutions in $LaNi_5$ should lower the
hydrogen decomposition pressures by one order of magnitude
for every $2\overset{\circ}{A}^3$ increase in cell volume.

Hydriding properties of Al substituted $LaNi_5$ have been
measured in several laboratories (15,17,18,19,20,21). The
decrease in hydrogen decomposition pressures observed in
these measurements is in reasonable agreement with predictions based on cell volume correlations. Furthermore, the
hydriding kinetics are unimpaired, the hydrogen to metal
ratios for the saturated β-phases are not seriously decreased, hysteresis effects are small and the resistance to
air oxidation is better for the Al substituted alloys than for
pure $LaNi_5$. The important new result is that Al substitutions allow a wide range of hydrogen decomposition pressure
to be spanned in a continuous fashion. In the range 0-20%
Al, the plateau pressures of the $LaNi_5$-$LaNi_4Al$ hydride
system are reduced by a factor of ∿300 (15). The enthalpies

and entropies for the hydriding reactions (1) are listed in
Table II for a series of Al substitutions. Table II also
lists the cell volumes for the various alloy compositions.
The linear relation between the enthalpies and the Al content displayed in Fig. 2 has been rationalized on the
basis of an "average" interstitial site bonding energy
model (20).

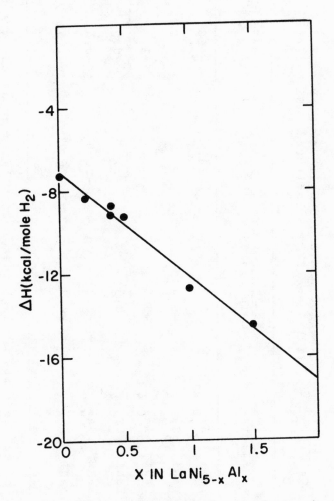

Figure 2. Enthalphy of the α-β Transformation in the
$LaNi_{5-x}Al_xH_y$ System.

Table III

Thermodynamic and Cell Volume Data for III A and IV A Substituted LaNi$_5$ Alloys

Alloy	$\Delta H_{\alpha \to \beta} \left(\dfrac{\text{Kcal}}{\text{mole } H_2} \right)$	$\Delta S_{\alpha \to \beta} \left(\dfrac{\text{cal}}{\text{deg-mole } H_2} \right)$	$\Delta G \left(\dfrac{\text{calc. for } 20°}{} \right)$	Cell Vol. (Å^3)
LaNi$_{4.6}$Al$_{0.4}$	-8.7	-26.1	-1.05	87.9
LaNi$_{4.6}$Ga$_{0.4}$	-8.4*	--	-0.7	88.1
LaNi$_{4.6}$In$_{0.4}$	-9.5	-26.5	-1.7	91.0
LaNi$_{4.6}$Si$_{0.4}$	-8.5	-27.3	-0.5	86.9
LaNi$_{4.6}$Ge$_{0.4}$	-8.2	-26.5	-0.4	87.8
LaNi$_{4.6}$Sn$_{0.4}$	-9.2	-26.2	-1.5	91.0

*calculated from a single desorption isotherm assuming a value of -26.3 for $\Delta S_{\alpha \to \beta}$.

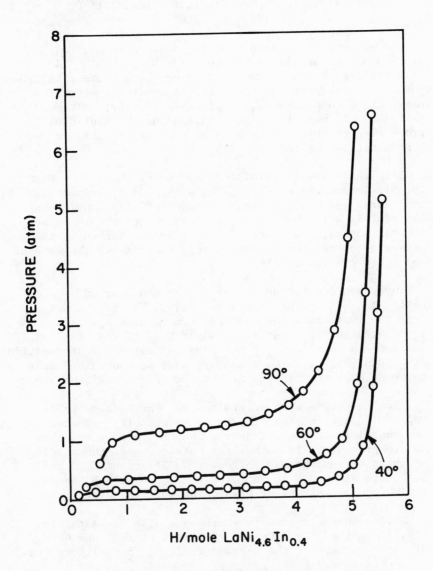

Figure 3. Pressure-Temperature Isotherms in the
LaNi$_{4.6}$In$_{0.4}$H$_y$ Systems.

THE EFFECT OF OTHER GROUP III A AND IV A
SUBSTITUTIONS ON THE HYDRIDING PROPERTIES OF
$LaNi_{5-x}M_x$ ALLOYS

Recent work has shown that in addition to Al, Ga, In,
Si, Ge and Sn can also substitute for Ni in $LaNi_5$ (14).
The detailed crystallographic data on these systems will be
published elsewhere but the cell volumes for the various
$LaNi_{4.6}M_{0.4}$ compositions are given in the last column of
Table III. It was of interest to examine the hydriding
properties of this series of alloys so as to obtain a
systematic set of results (22).

Alloys of similar stoichiometry $LaNi_{4.6}M_{0.4}$ (M = Ga,
In, Si, Ge and Sn) were prepared by standard arc-melting
procedures and characterized as substantially single phase
material by x-ray powder diffraction patterns. The experimental apparatus and procedure for determination of the
pressure-composition isotherms were essentially the same as
described previously (15).

Hydrogen desorption isotherms for the hydride
$LaNi_{4.6}In_{0.4}H_y$ are presented in Fig. 3. As can be noted
from the isotherms, the hydrogen composition approaches
y = 6 at higher pressures. A van't Hoff plot of $\ln P$ vs $1/T$
was used to calculate the enthalpy and entropy of transition
($\Delta H_{\alpha \to \beta}$ and $\Delta S_{\alpha \to \beta}$). The pressures used were obtained near
the middle of the plateau at the hydride composition y = 2.75.

Two isotherms were determined for the hydride $LaNi_{4.6}$
$Sn_{0.4}H_y$. In this case, the hydrogen capacity appeared to be
somewhat less than 6 and approached the value y = 5.8 near
20 atm pressure. The pressures obtained near the middle
of the plateaus were as follows: 0.071 atm at 20°C and
0.195 atm at 40°C. Again, from these two values, an enthalpy and an entropy of transition were calculated.

Hydrogen desorption isotherms are also presented in
Fig. 4 for the Si and Ge substitutions at 30°C and 60°C.
It can be seen that the β-phase limiting compositions are
at approximately $LaNi_{4.6}Si_{0.4}H_4$ and $LaNi_{4.6}Ge_{0.4}H_5$
respectively.

The derived thermodynamic quantities for the hydriding
reactions involving Group III A and IV A substitutions of
$LaNi_5$ are brought together in Table III. It is to be noted,

that on an equal mole fraction basis, both indium and tin
are more effective than either aluminum or gallium in reducing the hydrogen dissociation pressure of their corresponding hydrides (more negative ΔG). Figure 5 illustrates
the linear relationship between $\ln P$ and composition found
for ternary Al substituted alloys, and the assumed linear

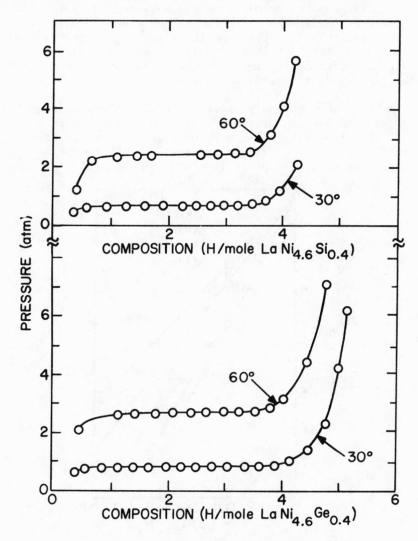

Figure 4.   Pressure-Temperature Isotherms in the
LaNi$_{4.6}$Si$_{0.4}$H$_y$ and LaNi$_{4.6}$Ge$_{0.4}$H$_y$ Systems.

relationships for Ga, Sn, and In substituted alloys for comparison.

It is also interesting to note the approximate equality of the measured entropies for the hydriding reactions involving the six alloys listed in Table III. According to a previously proposed model of configurational entropies (23) this result implies that the number and type of occupied interstitial sites is the same for all of these hydrides. A neutron diffraction investigation of aluminum substituted LaNi₅ is currently in progress in our laboratory to determine the validity of this conclusion.

A phenomenon commonly observed in all alloys of general composition AB₅ studied to date is hysteresis in the sense that the hydrogen absorption pressures are greater than the hydrogen desorption pressures in the two-phase plateau region of the pressure-composition diagram. The

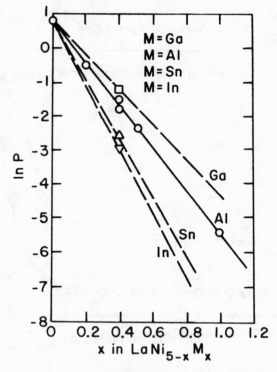

Figure 5.   lnP vs. Composition in Group III A and
            Group IV A Substituted $LaNi_{5-x}M_xH_y$ Systems.

magnitude of the hysteresis has been shown to be reproducible for a given sample (2). Single absorption point measurements for both $LaNi_{4.6}In_{0.4}$ and $LaNi_{4.6}Sn_{0.4}$ indicate
little or no hysteresis for these materials. The reason
for the dramatic reduction in hysteresis compared to $LaNi_5$
is not clear, but is an important benefit in some of their
proposed practical applications.

A comparison of the cell volumes of these alloys with
the free energies of formation of the corresponding hydrides
confirms, in general, the previously mentioned approximately
linear correlation of decreasing free energy of formation
of AB_5 hydrides with increasing cell volume. However, a
detailed examination of the data reveals a reversal of the
trend for the two hydride pairs $LaNi_{4.6}Al_{0.4}H_y$ – $LaNi_{4.6}Ga_{0.4}H_y$ and $LaNi_{4.6}Si_{0.4}H_y$ – $LaNi_{4.6}Ge_{0.4}H_y$. For both pairs,
there occurs a change in the electronic structure of the
substituted element from empty 3d orbitals to filled 3d
orbitals. Thus, the observed reversal in hydride free
energies may indicate the involvement of specific metal-
hydrogen bonding properties in addition to the previously
noted importance of metal-hydrogen and hydrogen-hydrogen
distances on the dissociation pressure of the alloy
hydrides.

HYDROGEN ABSORPTION BY AB_5 ALLOYS
WITH THE CUBIC UNi_5 STRUCTURE

All alloys of general composition AB_5 which have been
found reversibly to absorb and desorb large quantities of
hydrogen have had the hexagonal $CaCu_5$ type structure.
YNi_4Mn has been shown to crystallize with the cubic UNi_5
type structure (24). Recently, it has been possible to
activate the alloy YNi_4Mn and to observe for the first time
rapid hydrogen absorption and desorption in an alloy of this
structure (25).

The YNi_4Mn alloy, prepared by standard arc melting procedures, was found by x-ray powder diffraction, to be cubic
with a = 6.972 Å.

The alloy was activated by cooling it to 0°C and then
exposing the sample to about 1500 PSI hydrogen. After
standing overnight, the absorbed hydrogen was removed by
pumping on the sample at room temperature.' At pressures of
1200 PSI, the composition of the hydride approached

$YNi_4MnH_{4.5}$. There is apparently a fairly wide solid solution region for hydrogen in the β-phase hydride as the
plateau extended only to the composition $YNi_4MnH_{\sim 1.8+0.2}$.
The constant pressure plateau region is considerably
shorter than, for example, in the case of $LaNi_5H_6$. The
wide β-phase solid solution range may be characteristic of
hydrides derived from the cubic UNi_5 structure. The hydrogen dissociation pressure near the middle of the plateau
(H=1.0) was about 0.4 atm at 21°C. After four hydriding-
dehydriding cycles, the sample was observed to have broken
down to a powder, whose x-ray diffraction pattern however
was identical with that of the original material. In another
experiment, hydrided alloy was transferred without special
precaution to an x-ray capillary. The powder x-ray diffraction pattern indicated that most of the hydrogen had de-
sorbed from the sample since many of the observed lines
were the same as for the alloy. However, two new weak
lines corresponding to the two strongest alloy lines, but
shifted to larger d-spacings were also observed. From
these two lines, the hydride structure was deduced to be
cubic with lattice parameter estimated to be a = 7.3 Å.
This conclusion is tentative since it is based on very
limited data.

An objective of these studies has been to correlate
the structural and thermodynamic properties of the AB_5
hydrides. The hydrides based on cubic AB_5 alloys for example, present a new class of materials whose thermodynamic
properties such as their entropies of transition need to be
examined in the light of configurational entropy considerations for the hydrides based on the hexagonal AB_5 alloys
(23). Again, the YNi_4Mn hydride does not follow the cell
volume-dissociation pressure correlation established for
the hexagonal AB_5 hydrides. It remains to be determined
whether the anamolous behavior is associated with the change
in structure, or is due to other factors involving the way
in which hydrogen is incorporated into the YNi_4Mn lattice.

SOME FACTORS DETERMINING THE THERMODYNAMIC
STABILITIES OF ALLOY HYDRIDES

In the absence of a unifying theory capable of predicting the thermodynamic stabilities of alloy hydrides, several attempts have been made at semi-empirical correlations.
Four of these: an "average" interstitial site bonding
energy model, the rule of reversed stability, configurational

entropy considerations and a free energy-cell volume correlation will be briefly reviewed here.

There have been several discussions in the literature
of the relationships of structure to hydride stability
(23,26). It is of particular interest here to examine the
influence of Al on the structure and the plateau pressure of
LaNi$_5$ hydrides in the hope that these considerations can be
generalized to include the other Group III A and IV A substituted AB$_5$ alloys.

As already mentioned, it had been shown on the basis of
x-ray diffraction intensities that Al atoms in LaNi$_{5-x}$Al$_x$
ternaries preferentially occupy 3g sites which consist of
planes of Ni atoms free of La atoms (15). Making the assumption (without experimental proof) that the Al atoms are
ordered in the 3g sites, results in the nearest neighbor
distributions given in Table IV for six different types of
interstitial sites which hydrogen could occupy in the
hydrides (20).

Several results on neutron diffraction investigations
of LaNi$_5$D$_6$ (27,28,29) have been reported. In each of these
papers, it is concluded that the symmetry of the alloy is
lowered to space group P31m on deuteration and that the
deuterium atoms occupy 3c and 6d tetrahedral sites. These
sites in the deuteride are closely related spatially to
the 12n and 12o tetrahedral sites of the alloy having P6/
mmm symmetry. Concentrating on the n and o sites only, one
notes that the fraction of those sites containing one Al
nearest neighbor is 1/2 for LaNi$_4$Al, and 1/4 for LaNi$_{4.5}$Al$_{0.5}$.
Thus, if Al atoms in fact replace Ni atoms in the 3g
sites in an ordered manner, then the fraction of n and o
sites with one Al nearest neighbor is a linear function of
the Al composition. If one further assumes that hydrogens
in n and o sites are bound with energies averaged over all
available sites, then the "average" interstitial site bonding
energy would also be expected to be proportional to the
Al composition, thus providing a rationale for the observed
linear relationship between ΔH and Al composition shown in
Figure 2.

Shinar, et al. [30] have proposed that specific interstitial site occupations may be determined by associating
different binding energies with hydrogen atoms in different
interstitial sites. The binding energies would be propor-

Table IV.

Site desig.	LaNi$_5$ neighbors La	LaNi$_5$ neighbors Ni	LaNi$_5$ # sites	LaNi$_4$Al neighbors La	LaNi$_4$Al neighbors Ni	LaNi$_4$Al neighbors Al	LaNi$_4$Al # sites
b	2	6	1	2	4	2	1
f	2	4	3	2	2	2	1
				2	4	0	2
h	0	4	4	0	3	1	4
m	2	2	6	2	2	0	2
				2	1	1	4
n	1	3	12	1	2	1	4
				1	3	0	8
o	1	3	12	1	2	1	8
				1	3	0	4

tional to $\Delta H'$, the sum of the heats of formation of (imaginary) binary hydrides formed with the A and B atoms surrounding a particular site. On the basis of this scheme, one would predict that the H atoms prefer sites having an Al nearest neighbor since the heat of formation of aluminum hydride is considered to be more negative than that of nickel hydride [31]. However, the most recent measured values of ΔS for the aluminum containing alloys are almost identical to those of $LaNi_5$, (See Table II) and thus the configurational entropies should be closely similar [23]. The distribution of sites occupied by hydrogen should therefore be much the same for $LaNi_5$ as for the $LaNi_{5-x}Al_x$ alloys. It seems therefore more reasonable to conclude that the hydrogens occupy tetrahedral sites in $LaNi_5H_6$, and that the neutron diffraction results support an "average" interstitial site bonding energy model over models which assume discrete bonding energies associated with every crystallographically distinct site occupied by hydrogen. In any event the question of site occupation of hydrogen in the AB_5 hydrides remains an intriguing one and cannot be definitively answered until more precise structural and thermodynamic measurements on which to base more accurate theoretical calculations become available.

A scheme for correlating hydride stabilities, the so-called "rule of reversed stability" [see e.g., 32,33], states that for a series of homologous alloys, the more stable the alloy, the less stable (i.e., higher dissociation pressure) the corresponding hydride. Using Miedema's formula [34] the calculated heat of formation for $LaNi_5$ is -11.2 kJ/mole and for $LaAl_5$ is -42.1 kJ/mole. Since $LaAl_5$ is more stable (more negative ΔH) than $LaNi_5$, the rule of reversed stability predicts the $LaNi_{5-x}Al_x$ hydrides to be less stable than $LaNi_5H_6$ contrary to observation. Similarly, Shinar, et al. [30] have found disagreement with the rule of reversed stability for the $LaNi_{5-x}Cu_x$ hydrides.

It appears that in the case of the Al and Cu ternaries, as in many other AB_5 hydride systems, the factors determining the hydrogen dissociation pressures are closely correlated with cell volumes. Free energies derived from data given in Tables I and II and several other sources have been plotted versus the alloy cell volumes in Figure 6. A similar plot of ΔH versus cell volume shows a greater deviation from a straight line relationship because of the varying values of ΔS. Lundin, et al., [26] have recently discussed

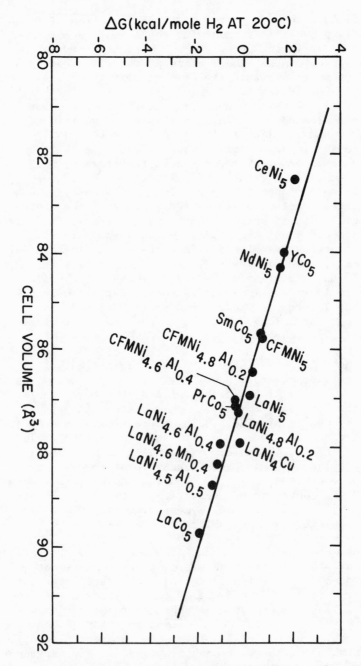

Figure 6.   Free Energy-Cell Volume Correlation for AB₅
             Binary and Ternary Alloy Systems.

a linear correlation of ΔG with tetrahedral hole size and it was of interest to see how closely the two correlations agree with one another, even though cell volume and hole size are not simply related. We have intercompared the correlations between cell volume, hole size and ΔG for 7 binary AB_5 alloys where appropriate data are available. The correlation coefficient for a least squares fit of the 7 data points was -0.99 using cell volume data and -0.97 using hole size data. Because the calculation of hole sizes is somewhat tedious and difficult, cell volume is a more convenient parameter to use particularly for ternary or quarternary alloys. The use of cell volume as a correlation parameter is currently being incorporated into theoretical treatments of H-H repulsion energies (35) and M-H bonding energies in metal hydride lattices (36).

METAL HYDRIDES AS CHEMICAL HEAT PUMPS

Consider the following general chemical reaction:

$$(\text{Condensed phase})_1 + \text{gas} = (\text{condensed phase})_2 + \text{heat}$$

Any reversible chemical reaction of this type with good kinetics is a candidate for chemical heat pump applications. However the remarkable properties of the AB_5 hydrides make these materials especially useful for such applications. In a chemical (thermally driven) heat pump, which could be used for cooling or heating purposes, thermal energy need not first be converted to mechanical energy to drive, for example, a compression refrigerator thus avoiding moving parts and vibration. More importantly, low grade heat sources such as provided by solar collectors, automobile exhausts or reject heat from power plants could be used to drive hydride chemical heat pumps.

Such systems based on the HYCSOS concept, have been under development at ANL since 1975. They involve the use of two different metal hydrides, having different free energies of formation. The first hydride (M_1H) has a lower hydrogen dissociation pressure than the second hydride (M_2H) at the same temperature. Hydrogen can therefore be made to flow from M_1H to M_2; from M_2H to M_1; or from M_1H to M_1 under the influence of appropriate thermal gradients (6).

It is clear that to optimize the choice of metal hydride
pairs for a particular mode of heat pump operation, the relationship between the thermodynamic quantities governing
reactions (1) and the high (T_h), intermediate (T_m) and low
(T_ℓ) temperatures needs to be established.

It was shown that the lowest temperature, T_ℓ, is reached
for a given M_1, M_2 pair when $\Delta S_{M2} = \Delta S_{M1}$, thus making this
special case of particular interest for chemical heat pump
applications (37).

For the special case $\Delta S_{M_2} = \Delta S_{M_1}$ then

$$\frac{\Delta H_{M_2}}{\Delta H_{M_1}} = \frac{T_m}{T_h} = \frac{T_\ell}{T_m} \qquad [3]$$

from which it follows that

$$T_m^2 = T_h T_\ell \qquad [4]$$

It is of interest to determine the solar collector temperatures required to operate a solar energy driven metal
hydride air conditioner. A reasonable value for T_ℓ is 8°C
(46°F) and for T_m is 40°C (104°F), making T_h = 76°C (169°F).
Another set of temperatures might be T_ℓ = 8°C (46°F) and
T_m = 45°C (113°F), making T_h = 86°C (189°F). The input
temperatures are, therefore, in the *flat plate* collector
range.

The desirable state of affairs outlined above can, of
course, only be realized if one has available pairs of metal
hydrides with $\Delta S_{M2} = \Delta S_{M1}$ and having the appropriate ratio
$\Delta H_{M_2}/\Delta H_{M_1}$ dictated by equation 3.

For the two sets of temperatures given above for oper-
ating an air conditioner, the $\Delta H_{M2}/\Delta H_{M1}$ ratios would have to
be 0.895 and 0.884, respectively. If one of the pair of
metal hydrides were chosen to be LaNi$_5$ having ΔH = 7.2
kcal/mole H$_2$, the second hydride would have to have a
ΔH = 8.04 kcal for the first set and a ΔH = 8.15 kcal to satisfy the second set of temperature conditions. The linear
relationship between x in the LaNi$_{5-x}$Al$_x$system and ΔH
specifies the compositions LaNi$_{4.8}$Al$_{.16}$ and LaNi$_{4.80}$Al$_{0.20}$
for the second hydride to be used in the two cases. The

Al/Ni ratios therefore sensitively determine the temperature regime for optimum efficiency of heat pump action.

A further option opened up by the "fine tuning" capabilities available with the $LaNi_{5-x}Al$ system is that for a specified set of T_h, T_m and T_ℓ a whole range of pairs of metal hydrides is available, each pair operating within a characteristic temperature regime. This is an important added flexibility from the point of view of optimizing the engineering design and performance characteristics of a hydride chemical heat pump system.

In the laboratory, it has been demonstrated that the hydrides can release and subsequently reabsorb hydrogen in a few minutes. As a result of the favorable kinetics, the bed size and cost can be reduced by rapid cycling. The metal hydride chemical heat pump and energy conversion system therefore looks attractive for several applications including solar energy use (38).

The experiment to determine the scientific feasibility of the HYCSOS concept and to generate data for an engineering analysis consists of four vessels each containing $\sim$5 kg of two different AB_5 hydrides. Three separate heat transfer fluid flow loops maintain appropriate tank temperatures ($T_h \cong 100°C$; $T_m \cong 40°C$; $T_\ell \cong 8°C$). Temperatures, pressures, liquid flow and hydrogen flow rates ($\sim$33 moles H_2 total) are measured and accessed through a data display and handling system (39).

It is of interest to note that the four tanks of the HYCSOS chemical heat pump are in fact equivalent to the four elements, evaporator, absorber, generator and condenser of an absorption refrigerator.

Starting in the evaporator, refrigerant absorbs heat from the space to be cooled and vaporizes. The refrigerant vapor combines with the absorbent in the absorber, and the heat of reaction is removed. The absorbent-refrigerant mixture is heated in the generator and separated. Pure absorbent is returned to the absorber, while refrigerant vapor continues to the condenser. Heat is removed from the refrigerant in the condenser. Finally, the refrigerant is returned to the evaporator where the cycle is repeated (37).

LITERATURE CITED

1. van Vucht, J. H. N., Kuijpers, F. A., and
    Bruning, H. C. A. M., Philips Res. Repts., (1970)
    $\underline{25}$, 133.
2. Kuijpers, F. A., and van Mal, H. H., J. Less-Common
    Metals, (1971), $\underline{23}$, 395.
3. Zijlstra, H., Chem. Technol. (1972), $\underline{2}$, 280.
    (The discovery of the remarkable hydrogen absorption
    properties of AB_5 compounds was made at the Philips
    Laboratories in the Netherlands in the course of de-
    termining the magnetic properties of these materials.
    Acid etching of $SmCo_5$ crystals led to coercivity
    changes traceable to hydrogen. Ultimately, magnetic
    measurements on AB_5 compound were carried out in
    hydrogen atmosphere conditions which surprisingly, re-
    sulted in uptake of large quantities of gas.)
4. Kuijpers, F. A., and Loopstra, B. O., J. Phys. Chem.
    Solids, (1974), $\underline{35}$, 301.
5. Newkirk, H. W., "Hydrogen Storage by Binary and
    Ternary Intermetallics for Energy Applications",
    UCRL-52110 (August 2, 1976).
6. Gruen, D. M. McBeth, R. L., Mendelsohn, M., Nixon, J. M.,
    Schreiner, F., and Sheft, I., Proc. 11th IECEC, (1976),
    681.
7. Buschow, K. H. J., and van Mal, H. H., J. Less-Common
    Metals, (1972), $\underline{29}$, 203.
8. Steward, S. A., Lakner, J. F., and Uribe, F., (1976),
    UCRL-52039 and 77455.
9. Ban, Z., Sikirica, M., and Raseta, R., J. Less-Common
    Metals, (1967), $\underline{12}$, 478.
10. Blazina, Z., and Ban, Z., Z. Naturforsch, (1973), $\underline{28b}$,
    561.
11. Blazina, Z., and Ban, Z., J. Less-Common Metals, (1973),
    $\underline{33}$, 321.
12. Blazina, Z., and Ban, Z., Croat. Chem. Acta, (1971),
    $\underline{43}$, 59.
13. Takeshita, T., and Wallace, W. E., J. Less-Common Metals,
    (1977), $\underline{55}$, 61.
14. Dwight, A. E., Unpublished Results.
15. Mendelsohn, M. H., Gruen, D. M., and Dwight, A. E.,
    Nature, (1977), $\underline{269}$, 45.
16. Dwight, A. E., J. Less-Common Metals, (1975), $\underline{43}$, 117.
17. Achard, J. C., Percheron-Guegan, A., Diaz, H.,
    Briancourt, F., and Denany, F., 2nd Int'l Cong. on
    Hydrogen in Metals, (1977), Paris, France.

18. Takeshita, T., Malik, S. K., and Wallace, W. E., J. Solid
    State Chem., (1978), 23, 271.
19. Gruen, D. M., Mendelsohn, M. H., and Dwight, A. E.,
    Advances in Chemistry Series No. 167; American Chemical
    Society, Washington, D.C., 1978.
20. Mendelsohn, M. H., Gruen, D. M., and Dwight, A. E.,
    J. Less-Common Metals, (in press).
21. Sandrock, G. D., Proc. 2nd World Hydrogen Energy
    Conference, (1978), Zurich, Switzerland.
22. Mendelsohn, M. H., Gruen, D. M., and Dwight, A. E.,
    Materials Res. Bull. (In press).
23. Gruen, D. M., and Mendelsohn, M. H., J. Less-Common
    Metals, (1977), 55, 149.
24. Dwight, A. E., The Rare Earths in Modern Science and
    Technology; Plenum Press: New York, N.Y., 1978.
25. Mendelsohn, M. H. and Gruen, D. M., Chem. Comm. (1978),581.
26. Lundin, C. E., Lynch, F. E., and Magee, C. B., J. Less-
    Common Metals, (1977), 56, 19.
27. Andresen, A. F., Proc. International Symp. on Hydrides
    for Energy Storage, August, 1977, Geilo, Norway.
28. Fischer, P., Furrer, A., Busch, G., and Schlapbach,
    ibid.
29. Bowman, A. L., Anderson, J. L., and Nereson, N. G.,
    Proc. 10th Rare Earth Res. Conf., (1973), Carefree,
    Arizona.
30. Shinar, J., Jacob, I., Davidov, D., and Shaltiel, D.,
    International Symp. on Hydirdes for Energy Storage,
    (1977), Geilo, Norway.
31. Mueller, W. M., Blackledge, J. P., and Libowitz, G. G.,
    "Metal Hydrides", Academic Press, (1968), New York.
32. van Mal, H. H., Buschow, K. H. J., and Miedema, A. R.,
    J. Less-Common Metals, (1974), 35, 65.
33. Buschow, K. H. J., van Mal, H. H. and Miedema, A. R.,
    J. Less-Common Metals, (1975) 42, 163.
34. Miedema, A. R., J. Less-Common Metals, (1976), 46, 67.
35. Switendick, A. C., Presentation at Hydride Meeting,
    Sandia Labs., March 8, 1978, Albuquerque, New Mexico.
36. Machlin, E. S. (Private Communication).
37. Gruen, D. M., Mendelsohn, M. H., and Sheft, I., Solar
    Energy, (1978), 21, 153.
38. Gorman, R., and Moritz, P. S., Proc. AIAA Solar Energy
    Meeting, November, 1978, Phoenix, Arizona.
39. Gruen, D. M., Sheft, I., Lamich, G., and Mendelsohn, M.,
    (1977), ANL-77-39.

THE ROLE OF THE SOLAR ENERGY RESEARCH INSTITUTE

J. S. Connolly and J. C. Grosskreutz, Assistant
Director

Solar Energy Research Institute*
Golden, Colorado   80401

INTRODUCTION

Heightened concern for environmental quality and
limited domestic capacity to meet energy demands with
fossil fuels have brought the United States to the realization that renewable sources of energy must be given a
new priority. In 1973 the urgency of this problem was
dramatically impressed upon the nation with the Middle
East oil embargo. It became clear that the nation could
not remain complacent in its dependence on fossil fuels.

Positive first steps were taken through the passage
of several important pieces of energy-related legislation. Among these was the Solar Energy Research,
Development, and Demonstration Act of 1974 (PL 93-473).
The need for a comprehensive program aimed at developing
solar energy as a contributor to the nation's energy
supply prompted the creation of the Solar Energy Research
Institute (SERI).

After evaluation of 19 proposals representing sites
in 16 states, the Energy Research and Development
Administration announced in March 1977 that SERI would be

---

*A Division of Midwest Research Institute. Operated for
the Department of Energy on Contract EG-77-C-4042.

managed and operated at a site in Golden, Colorado, by
the Midwest Research Institute. MRI is an independent,
not-for-profit organization headquartered in Kansas City,
Missouri. SERI formally opened on July 5, 1977. Three
months later ERDA's functions were assumed by the newly
created Department of Energy (DOE).

BUILDING A MULTIDISCIPLINARY STAFF

SERI's organization reflects the realization that
energy issues are not solely research problems but
involve a combination of economic, political, environmental, and social considerations as well. Thus, about
40% of the professional staff are scientists and engineers; the rest are comprised of sociologists, econo-
mists, market analysts, lawyers, architects, and
others. This diversity of staff creates the proper environment to generate innovative ideas, provide objective
analyses of solar energy options, and identify realistic
goals.

Much of the first year's effort went toward building
a long-term capability to serve the needs of the solar
energy program. Beginning operations in July 1977 with a
staff of 40, SERI has attracted qualified men and women
from universities, government, research organizations,
and private industry.

By October 1978 the staff numbered 370. This is the
largest group ever assembled for work devoted exclusively
to the development of solar energy. The current staffing
plan calls for an organization of over 700 professional
and support personnel by the end of the second fiscal
year (September 1979).

THE SERI ORGANIZATION

The SERI staff is organized into six divisions.

The Research Division is responsible for carrying
out in-house research programs, both basic and applied.
Basic research is needed to provide advanced options for
collection, conversion, and storage of solar energy. The
results of applied research are used to identify and
reduce technical and cost barriers to solar

technologies. In addition, the Division supplies technical information to other groups in SERI, DOE, and elsewhere to assist them in making appropriate choices among various solar options.

The Analysis and Assessment Division represents the economic and social science arm of SERI. Activities in this division complement the technical orientation of the Research Division by investigating critical issues which underlie public and private decisions influencing the use of solar energy. The A&A Division uses the results of these investigations to support DOE in planning the national solar energy effort.

The Academic and International Programs Division coordinates U.S. participation in international solar research and development programs and focuses on utilizing the unique resources of the university community in solar energy research.

The Information Systems Division provides information and data to individuals and organizations who must make decisions concerning solar options. As part of this effort, the Solar Energy Information Data Bank incorporates a comprehensive data base, an information center, and an active dissemination program, all of which utilize modern computer and communications techniques.

The Technology Commercialization Division assists in the national development of widespread and stable markets for solar technologies by providing information and other assistance to decision makers in the consumer sector, the commercial and industrial sectors, and at various levels of government.

The Administrative and Technical Services Division provides management and support services for the operating divisions. This Division includes accounting, contracts, purchasing, publishing services, and facilities planning and management.

FACILITIES

SERI moved into interim facilities in leased office space in Golden, Colorado (20 minutes west of downtown

Denver), in July 1977. Additional interim facilities
consisting of laboratories, a computer center, and administrative offices were under construction during the
first year. This will be SERI's home until the early
1980s when the permanent Institute will be completed atop
nearby South Table Mountain. Both the interim and future
facilities provide areas for outdoor field experiments.
The interim laboratory building houses 60,000 square feet
of research space. The laboratories will be fully
operational by June 1979.

An architectural and engineering firm has been
selected to prepare the conceptual design and master site
plan for the permanent facilities on South Table
Mountain. The permanent Institute will be a national
showpiece of innovative solar energy design and energy
conservation techniques and will complement the environmental and geographic features of the area. The head-
quarters will include, in addition to offices, research
laboratories, a conference center, field experimentation
facilities, an information data bank, a library, and a
public visitor's center. Construction will begin in late
1979 or early 1980 and is expected to span about two
years.

THE SERI MISSION

SERI's mandate, as stated by Congress, is to perform
research, development, and related functions in support
of the national solar energy program. ERDA had further
defined the mandate by calling for SERI to contribute to
the establishment of a solar energy industrial base that
will foster the widespread use of solar technology.
Thus, the SERI mission is a comprehensive one. It
involves all solar energy technologies and all aspects of
moving a technology from initial research through demonstration to utilization in the marketplace.

To accomplish this broad mission, SERI performs
these major roles:

o Conducts and manages national and international
  solar research, development, and demonstration
  projects.

o Assists DOE in developing national solar energy
   program plans and strategies.

o Serves as an objective evaluator and analyst of
   the content, progress, and direction of the
   national solar energy effort.

o Conducts programs in solar information dissemination, education and training, and technology
   utilization and applications.

PRIORITIES

An operating plan to address the unattended needs of
the national solar program was developed for implementa-
tion during FY79. SERI's programs will complement and
supplement the research and development activities now in
progress at many institutions, public and private, around
the nation. These programs are targeted to the objectives of the nation's solar energy effort, not only in
the near term (to 1990) but also in the long term (after
1990).

Research priorities were established within the
areas of photovoltaics, biological and chemical conversion, wind, passive techiques, storage, industrial process heat, and centralized as well as decentralized
systems. To accomplish the goals in these areas, programs were established in basic and applied research,
economic and social science research, planning and analysis, the national Solar Energy Information Data Bank,
international programs, education and training, program
management, and quality assurance and standards.

Specific projects in these areas have been initiated
in existing nearby laboratory facilities, and subcon-
tracts have been let to major institutions to support the
early in-house efforts. Initial research focuses on
three solar energy conversion areas: photovoltaics
(direct conversion of sunlight into electricity); biological and chemical conversion (including photoconversion,
and thermochemical conversion of biomass into fuels,
chemicals, heat, or electricity); and thermal conversion
(solar heating and cooling and industrial process

heat). Research to remove materials limitations related to solar energy conversion has also begun.

These areas were selected for initial emphasis because of their high potential for significant results and the need for additional research in areas where a void seemed to exist. At the same time, social and economic research and program management efforts involving all the solar technologies were initiated.

NATIONAL SOLAR PROGRAMS

At the request of DOE, SERI has prepared national R&D program plans for absorber coatings, reflector coatings, polymer applications, open-cycle OTEC (ocean thermal energy conversion), and dessicant cooling. These plans were created after interaction with other experts in the field and in the user community.

Informed decisions about new program plans and strategies must be supported by a comprehensive and ongoing effort of analysis and evaluation. SERI assesses current effectiveness and potential impacts of existing and alternative programs to identify problems and to influence program decisions in the future. This is a role that had been specified by ERDA in the request for proposals. During the first year, SERI performed many analyses on all facets of solar energy activity--from research programs to market readiness of existing and emerging technologies.

MANAGING NATIONAL SOLAR RESEARCH

SERI is responsible for two types of solar research projects: those initiated in-house, as previously discussed, and management of research programs for DOE. As the SERI organization is completed, a greater role in program management will be assumed on behalf of DOE to provide coordination, technical direction and support, and managerial expertise.

SERI assumed management of several major DOE-contracted programs, including assessment studies of solar electric options for the North Central, South

Central, and Southwest regions; research in absorber surface coatings; hydrogen production by photoelectrolysis and directed photosynthesis; and an innovative wind energy systems program. SERI was also assigned the responsibility for managing the Solar Thermal Test Facility User Association subcontract with the University of Houston.

In addition, SERI assumed management of the nation's lead effort in photovoltaic materials research and development. The Photovoltaics Program Office of the Research Division has the responsibility for managing all DOE contracts in the photovoltaics R&D area.

Contracts actively managed for DOE by SERI during the first year totaled approximately $2.5 million. In 1979 SERI will be responsible for management of research contracts totaling over $60 million.

In order for a new consumer industry to evolve and flourish, it must be supported by a vigorous basic research program from which new ideas will emerge. Accordingly, SERI has established a University Research Program to ensure that resources of the academic community are involved in the development of solar energy. A University Advisory Council, consisting of faculty and research administrators from throughout the country, was convened to help SERI develop interactions with the academic community. Emphasis in this program will be on encouraging innovative solar research programs. In 1979 SERI will initiate about 125 new university research contracts, totaling approximately $7.5 million.

In addition to its national contract-management function, SERI has a role in guiding U.S. participation in international solar programs. The major activity in this area to date is the U.S.-Saudi Arabian agreement. In October 1977 the two governments signed a $100 million five-year agreement calling for a jointly-funded research and development program in solar technologies of mutual interest. SERI participated in the original negotiations and was assigned the lead role in developing the management plan.

Although the agreement with Saudi Arabia is the largest cooperative international solar program involving the United States, there are many others in which SERI plays an important role. Exchanges of solar scientists between the U.S. and Israel, Japan, and the Soviet Union have been carried out or are planned under SERI auspices. Some currently active projects also involve the International Energy Agency, Spain, Iran, and India.

INFORMATION DISSEMINATION

To be effective, SERI's efforts in research, development, and demonstration must provide useful information to those who need it. One of the principal methods for fulfilling this need is the Solar Energy Information Data Bank (SEIDB), which was mandated by Congress in the legislation creating SERI. Major demands for information and data about solar energy are being made not only by the technical community, but also by federal, state, and local offices and legislative bodies, architects, manufacturers, the financial community, builders, and homeowners.

The SEIDB will provide a centralized and comprehensive resource of information and data easily accessible in usable formats to meet the varied requirement of its users. This data bank will be incorporated into a network involving the Regional Solar Energy Centers (which will have the prime responsibility for meeting state and local information requirements) and the existing National Solar Heating and Cooling Information Center.

SERI will serve in a coordination role to ensure the orderly development of the manpower base needed for commercialization of each solar technology. For example, the Institute will conduct workshops for university faculty and vocational school teachers to develop new curricula and training programs. In 1978 SERI participated in a national effort to establish a computerized data base of all educational institutions offering solar-related courses. After review of this information, SERI will provide colleges and universities with direction and assistance in establishing needed new courses to train the solar experts of the future.

Programs have been initiated for sabbatical visits and summer internships to provide both faculty and students with the opportunity to interact with SERI staff on a variety of solar projects. Fifteen students interned at SERI during the summer of 1978 in the areas of law, research, and social science. This program is expected to expand as SERI grows.

In addition to the academic community, SERI's conference and workshop program reached other groups interested in the development of solar energy. During 1978, 50 conferences were organized under this program. This year the conference program will be expanded to approximately 80 workshops and conferences. In 1980 SERI will host the Third International Conference on Photochemical Conversion and Storage of Solar Energy.

An event reflecting the enthusiasm for and the potential of solar energy took place during SERI's first year on May 3, 1978--SUN DAY. SERI played a major role in disseminating information in support of the national and international Sun Day effort. A SERI-produced presentation was distributed throughout the nation and the Institute provided technical support to the Sun Day committee in its efforts to provide the public with objective information on solar energy.

The highlight of these activities was the visit of President Carter to SERI's permanent site. His visit focused the attention of the nation on SERI and placed new emphasis on the importance of solar technologies in the country's emerging energy picture.

INDEX

A

Acetate ion, 94, 101, 161
Acetic acid, 83, 84, 94,
    108, 111
Acetylacetone, 362
Acetylene, 263
Acid-base reactions, 174
Acridine orange, 288, 292
Actinometers, 189, 199
Action spectrum, 53, 64
Air contamination effect, 56
Allylphenols, 320
Aluminum, 375, 381, 387
  hydriding properties, 376
Aluminum arsenide, 238, 240
    245
Aluminum arsenide/gallium
    arsenide cell, 257
  series resistance, 257
Aluminum trimethyl, 250
Amines
  oxidation, 127
Amino acids, 44
Aminoperylene, 143
Aminopyrene, 143
Ammonia synthesis, 261
Amorphous carbon, 216
Anaerobic digestion, 111
Anatase, 107, 265
Anthracene, 146, 304, 306
Argon lasers, 200
Arsine, 245, 249
Ascorbate, 214
Asymmetric bilayer mem-
    brane, 161
Azepine-electrocyclization,
    304, 306
Azomethine ylids, 321

B

Back electron transfer
    reactions, 153, 178
    195

Back electron transfer
    reactions, cont.
  retardation of, 120-121
    126-128, 178-180
Back reaction, 32, 36, 72
  activation energy, 36
Backup energy, 337
Bacterial reaction center
    protein, 42, 43
Bacteriochlorophyll, 42
Bacteriopheophytin, 43
Band bending, 93, 162, 208
Band gap, 217
Benzanthrone, 292
Benzo(a)pyrene, 5
Benzodihydrofurans, 320
  photocyclization reactions,
    320
Benzophenone, 288, 347
  4-(N,N-dimethylamino), 347
Benzoquinone, 136
Bilayer lipid membrane, 23,
    212
  pigmented, 212
Bimolecular processes, 197
  disproportionation, 178
  energy transfer, 357
  quenching, 176
  reactions, 129, 178
Biochemical inhibitions, 74
Biomass, 83, 97, 111
Biosynthetic sequence, 19
Biphotonic process, 48
Biradical partitioning, 292
Boundary layer, 230
Butadienes
  electrocyclic closure, 307
Butyric acid, 83, 107, 108

C

Cadmium sulfide, 239

405